Communications
in Computer and Information Science   1628

More information about this series at https://link.springer.com/bookseries/7899

Yang Wang · Guobin Zhu · Qilong Han ·
Hongzhi Wang · Xianhua Song ·
Zeguang Lu (Eds.)

Data Science

8th International Conference
of Pioneering Computer Scientists, Engineers
and Educators, ICPCSEE 2022
Chengdu, China, August 19–22, 2022
Proceedings, Part I

Springer

Editors
Yang Wang
Southwest Petroleum University
Chengdu, China

Qilong Han
Harbin Engineering University
Harbin, China

Xianhua Song
Harbin University of Science and Technology
Harbin, China

Guobin Zhu
University of Electronic Science
and Technology of China
Chengdu, China

Hongzhi Wang
Harbin Institute of Technology
Harbin, China

Zeguang Lu
National Academy of Guo Ding Institute
of Data Sciences
Beijing, China

ISSN 1865-0929   ISSN 1865-0937 (electronic)
Communications in Computer and Information Science
ISBN 978-981-19-5193-0   ISBN 978-981-19-5194-7 (eBook)
https://doi.org/10.1007/978-981-19-5194-7

This Springer imprint is published by the registered company Springer Nature Singapore Pte Ltd.
The registered company address is: 152 Beach Road, #21-01/04 Gateway East, Singapore 189721, Singapore

Preface

As the chairs of the 8th International Conference of Pioneer Computer Scientists, Engineers and Educators 2022 (ICPCSEE 2022, originally ICYCSEE), it is our great pleasure to welcome you to the conference proceedings. ICPCSEE 2022 was held in Chengdu, China, during August 19–22, 2022, and hosted by the Southwest Petroleum University, the University of Electronic Science and Technology of China, and the National Academy of Guo Ding Institute of Data Sciences, China. The goal of this conference series is to provide a forum for computer scientists, engineers, and educators.

This year's conference attracted 263 paper submissions. After the hard work of the Program Committee, 63 papers were accepted to appear in the conference proceedings, with an acceptance rate of 24.7%. The major topic of this conference is data science. The accepted papers cover a wide range of areas related to basic theory and techniques for data science including mathematical issues in data science, computational theory for data science, big data management and applications, data quality and data preparation, evaluation and measurement in data science, data visualization, big data mining and knowledge management, infrastructure for data science, machine learning for data science, data security and privacy, applications of data science, case studies of data science, multimedia data management and analysis, data-driven scientific research, data-driven bioinformatics, data-driven healthcare, data-driven management, data-driven e-government, data-driven smart city/planet, data marketing and economics, social media and recommendation systems, data-driven security, data-driven business model innovation, and social and/or organizational impacts of data science.

We would like to thank all the Program Committee members, a total of 261 people from 142 different institutes or companies, for their hard work in completing the review tasks. Their collective efforts made it possible to attain quality reviews for all the submissions within a few weeks. Their diverse expertise in each research area helped us to create an exciting program for the conference. Their comments and advice helped the authors to improve the quality of their papers and gain deeper insights.

We thank the team at Springer, whose professional assistance was invaluable in the production of the proceedings. A big thanks also to the authors and participants for their tremendous support in making the conference a success.

Besides the technical program, this year ICPCSEE offered different experiences to the participants. We hope you enjoyed the conference.

June 2022

Liehui Zhang
Hongzhi Wang
Yang Wang
Guobin Zhu
Qilong Han

Organization

The 8th International Conference of Pioneering Computer Scientists, Engineers and Educators (http://2022.icpcsee.org) was held in Chengdu, China, during August 19–22, 2022, and hosted by the Southwest Petroleum University, the University of Electronic Science and Technology of China and the National Academy of Guo Ding Institute of Data Sciences, China.

General Chairs

Liehui Zhang	Southwest Petroleum University, China
Hongzhi Wang	Harbin Institute of Technology, China

Program Chairs

Yang Wang	Southwest Petroleum University, China
Guobin Zhu	University of Electronic Science and Technology of China, China
Qilong Han	Harbin Engineering University, China

Program Co-chairs

Xiaohua Xu	University of Science and Technology of China, China
Zhao Kang	University of Electronic Science and Technology of China, China
Yingjie Zhou	Sichuan University, China
Jinshan Tang	George Mason University, USA
Jingfeng Jiang	Michigan Technological University, USA
Xiaohu Yang	Hebei University, China

Organization Chairs

Jie Gong	Southwest Petroleum University, China
Yishu Zhang	Southwest Petroleum University, China
Jiong Mu	Sichuan Agricultural University, China
Xianhua Niu	Xihua University, China
Wei Pan	China West Normal University, China

Organization Co-chairs

Bo Peng	Southwest Petroleum University, China
Jian Zhang	Southwest Petroleum University, China
Fei Teng	Southwest Jiaotong University, China
Xin Yang	Southwestern University of Finance and Economics, China
Yongqing Zhang	Chengdu University of Information Technology, China
Hongyu Han	Sichuan Normal University, China
Chuanlin Liu	Southwest Petroleum University, China

Publication Chairs

Xianhua Song	Harbin University of Science and Technology, China
Zeguang Lu	National Academy of Guo Ding Institute of Data Sciences, China

Publication Co-chairs

Xiaoou Ding	Harbin Institute of Technology, China
Dan Lu	Harbin Engineering University, China

Forum Chairs

Lei Chen	Chengdu Supercomputing Center, China
Xiaoliang Chen	Xihua University, China
Hai Li	iQIYI Inc., China
Pinle Qin	North University of China, China

Oral Session Chairs

Fan Min	Southwest Petroleum University, China
Ping Li	Southwest Petroleum University, China
Xin Wang	Southwest Petroleum University, China

Registration and Financial Chair

Zhongchan Sun	National Academy of Guo Ding Institute of Data Sciences, China

Academic Committee Chair

Hongzhi Wang	Harbin Institute of Technology, China

Academic Committee Vice President

Qilong Han	Harbin Engineering University, China

Academic Committee Secretary General

Zeguang Lu	National Academy of Guo Ding Institute of Data Sciences, China

Academic Committee Under Secretary General

Xiaoou Ding	Harbin Institute of Technology, China

Academic Committee Secretaries

Dan Lu	Harbin Engineering University, China
Zhongchan Sun	National Academy of Guo Ding Institute of Data Sciences, China

Academic Committee Executive Members

Cham Tat Huei	UCSI University, Malaysia
Xiaoju Dong	Shanghai Jiao Tong University, China
Lan Huang	Jilin University, China
Ying Jiang	Kunming University of Science and Technology, China
Weipeng Jiang	Northeast Forestry University, China
Min Li	Central South University, China
Junyu Lin	Institute of Information Engineering, CAS, China
Xia Liu	Hainan Province Computer Federation, China
Rui Mao	Shenzhen University, China
Qiguang Miao	Xidian University, China
Haiwei Pan	Harbin Engineering University, China
Pinle Qin	North University of China, China
Xianhua Song	Harbin University of Science and Technology, China
Guanglu Sun	Harbin University of Science and Technology, China
Jin Tang	Anhui University, China
Ning Wang	Xiamen Huaxia University, China

Xin Wang	Tianjin University, China
Yan Wang	Zhengzhou University of Technology, China
Yang Wang	Southwest Petroleum University, China
Shengke Wang	Ocean University of China, China
Yun Wu	Guizhou University, China
Liang Xiao	Nanjing University of Science and Technology, China
Junchang Xin	Northeastern University, China
Zichen Xu	Nanchang University, China
Xiaohui Yang	Hebei University, China
Chen Ye	Hangzhou Dianzi University, China
Canlong Zhang	Guangxi Normal University, China
Zhichang Zhang	Northwest Normal University, China
Yuanyuan Zhu	Wuhan University, China

Steering Committee

Jiajun Bu	Zhejiang University, China
Wanxiang Che	Harbin Institute of Technology, China
Jian Chen	ParaTera, China
Wenguang Chen	Tsinghua University, China
Xuebin Chen	North China University of Science and Technology, China
Xiaoju Dong	Shanghai Jiao Tong University, China
Qilong Han	Harbin Engineering University, China
Yiliang Han	Engineering University of CAPF, China
Yinhe Han	Institute of Computing Technology, Chinese Academy of Sciences, China
Hai Jin	Huazhong University of Science and Technology, China
Weipeng Jing	Northeast Forestry University, China
Wei Li	Central Queensland University, Australia
Min Li	Central South University, China
Junyu Lin	Institute of Information Engineering, Chinese Academy of Sciences, China
Yunhao Liu	Michigan State University, USA
Zeguang Lu	National Academy of Guo Ding Institute of Data Sciences, China
Rui Mao	Shenzhen University, China
Qiguang Miao	Xidian University, China
Haiwei Pan	Harbin Engineering University, China
Pinle Qin	North University of China, China

Zheng Shan	The PLA Information Engineering University, China
Guanglu Sun	Harbin University of Science and Technology, China
Jie Tang	Tsinghua University, China
Feng Tian	Institute of Software, Chinese Academy of Sciences, China
Tao Wang	Peking University, China
Hongzhi Wang	Harbin Institute of Technology, China
Xiaohui Wei	Jilin University, China
Lifang Wen	Beijing Huazhang Graphics & Information Co., Ltd, China
Liang Xiao	Nanjing University of Science and Technology, China
Yu Yao	Northeastern University, China
Xiaoru Yuan	Peking University, China
Yingtao Zhang	Harbin Institute of Technology, China
Yunquan Zhang	Institute of Computing Technology, Chinese Academy of Sciences, China
Baokang Zhao	National University of Defense Technology, China
Min Zhu	Sichuan University, China
Liehuang Zhu	Beijing Institute of Technology, China

Program Committee

Witold Abramowicz	Poznan University of Economics, Poland
Chunyu Ai	University of South Carolina Upstate, USA
Jiyao An	Hunan University, China
Ran Bi	Dalian University of Technology, China
Zhipeng Cai	Georgia State University, USA
Yi Cai	South China University of Technology, China
Zhao Cao	Beijing Institute of Technology, China
Wanxiang Che	Harbin Institute of Technology, China
Wei Chen	Beijing Jiaotong University, China
Hao Chen	Hunan University, China
Xuebin Chen	North China University of Science and Technology, China
Chunyi Chen	Changchun University of Science and Technology, China
Yueguo Chen	Renmin University, China
Siyao Cheng	Harbin Institute of Technology, China
Byron Choi	Hong Kong Baptist University, China

Vincenzo Deufemia	University of Salerno, Italy
Gong Dianxuan	North China University of Science and Technology, China
Xiaofeng Ding	Huazhong University of Science and Technology, China
Jianrui Ding	Harbin Institute of Technology, China
Hongbin Dong	Harbin Engineering University, China
Lei Duan	Sichuan University, China
Xiping Duan	Harbin Normal University, China
Xiaolin Fang	Southeast University, China
Ming Fang	Changchun University of Science and Technology, China
Jianlin Feng	Sun Yat-sen University, China
Jing Gao	Dalian University of Technology, China
Yu Gu	Northeastern University, China
Qi Han	Harbin Institute of Technology, China
Meng Han	Georgia State University, USA
Qinglai He	Arizona State University, USA
Wei Hu	Nanjing University, China
Lan Huang	Jilin University, China
Hao Huang	Wuhan University, China
Feng Jiang	Harbin Institute of Technology, China
Bin Jiang	Hunan University, China
Cheqing Jin	East China Normal University, China
Hanjiang Lai	Sun Yat-sen University, China
Shiyong Lan	Sichuan University, China
Hui Li	Xidian University, China
Zhixu Li	Soochow University, China
Mingzhao Li	Royal Melbourne Institute of Technology, Australia
Peng Li	Shaanxi Normal University, China
Jianjun Li	Huazhong University of Science and Technology, China
Xiaofeng Li	Sichuan University, China
Zheng Li	Sichuan University, China
Min Li	Central South University, China
Zhixun Li	Nanchang University, China
Hua Li	Changchun University of Science and Technology, China
Rong-Hua Li	Shenzhen University, China
Cuiping Li	Renmin University of China, China
Qiong Li	Harbin Institute of Technology, China

Yanli Liu	Sichuan University, China
Hailong Liu	Northwestern Polytechnical University, China
Guanfeng Liu	Macquarie University, Australia
Yan Liu	Harbin Institute of Technology, China
Zeguang Lu	National Academy of Guo Ding Institute of Data Sciences, China
Binbin Lu	Sichuan University, China
Junling Lu	Shaanxi Normal University, China
Jizhou Luo	Harbin Institute of Technology, China
Li Mohan	Jinan University, China
Tiezheng Nie	Northeastern University, China
Haiwei Pan	Harbin Engineering University, China
Jialiang Peng	Norwegian University of Science and Technology, Norway
Fei Peng	Hunan University, China
Yuwei Peng	Wuhan University, China
Shaojie Qiao	Southwest Jiaotong University, China
Li Qingliang	Changchun University of Science and Technology, China
Zhe Quan	Hunan University, China
Yingxia Shao	Peking University, China
Wei Song	North China University of Technology, China
Yanan Sun	Oklahoma State University, USA
Minghui Sun	Jilin University, China
Guanghua Tan	Hunan University, China
Yongxin Tong	Beihang University, China
Xifeng Tong	Northeast Petroleum University, China
Vicenc Torra	University of Skövde, Sweden
Leong Hou	University of Macau, China
Hongzhi Wang	Harbin Institute of Technology, China
Yingjie Wang	Yantai University, China
Dong Wang	Hunan University, China
Yongheng Wang	Hunan University, China
Chunnan Wang	Harbin Institute of Technology, China
Jinbao Wang	Harbin Institute of Technology, China
Xin Wang	Tianjin University, China
Peng Wang	Fudan University, China
Chaokun Wang	Tsinghua University, China
Xiaoling Wang	East China Normal University, China
Jiapeng Wang	Harbin Huade University, China
Huayu Wu	Institute for Infocomm Research, Singapore

Yan Wu	Changchun University of Science and Technology, China
Sheng Xiao	Hunan University, China
Ying Xu	Hunan University, China
Jing Xu	Changchun University of Science and Technology, China
Jianqiu Xu	Nanjing University of Aeronautics and Astronautics, China
Yaohong Xue	Changchun University of Science and Technology, China
Li Xuwei	Sichuan University, China
Mingyuan Yan	University of North Georgia, USA
Yajun Yang	Tianjin University, China
Gaobo Yang	Hunan University, China
Lei Yang	Heilongjiang University, China
Ning Yang	Sichuan University, China
Xiaochun Yang	Northeastern University, China
Bin Yao	Shanghai Jiao Tong University, China
Yuxin Ye	Jilin University, China
Xiufen Ye	Harbin Engineering University, China
Minghao Yin	Northeast Normal University, China
Dan Yin	Harbin Engineering University, China
Zhou Yong	China University of Mining and Technology, China
Lei Yu	Georgia Institute of Technology, USA
Ye Yuan	Northeastern University, China
Kun Yue	Yunnan University, China
Xiaowang Zhang	Tianjin University, China
Lichen Zhang	Shaanxi Normal University, China
Yingtao Zhang	Harbin Institute of Technology, China
Yu Zhang	Harbin Institute of Technology, China
Wenjie Zhang	University of New South Wales, Australia
Dongxiang Zhang	University of Electronic Science and Technology of China, China
Xiao Zhang	Renmin University of China, China
Kejia Zhang	Harbin Engineering University, China
Yonggang Zhang	Jilin University, China
Huijie Zhang	Northeast Normal University, China
Boyu Zhang	Utah State University, USA
Jian Zhao	Changchun University, China
Qijun Zhao	Sichuan University, China
Bihai Zhao	Changsha University, China

Xiaohui Zhao	University of Canberra, Australia
Jiancheng Zhong	Hunan Normal University, China
Fucai Zhou	Northeastern University, China
Changjian Zhou	Northeast Agricultural University, China
Min Zhu	Sichuan University, China
Yuanyuan Zhu	Wuhan University, China
Wangmeng Zuo	Harbin Institute of Technology, China

Contents – Part I

Machine Learning for Data Science

Multimedia Data Management and Analysis

Contents – Part II

Infrastructure for Data Science

Education Track

Regulatory Technology in Finance

Big Data Mining and Knowledge Management

Self-attention Based Multimodule Fusion Graph Convolution Network for Traffic Flow Prediction

Lijie Li[(✉)], Hongyang Shao, Junhao Chen, and Ye Wang[(✉)]

College of Computer Science and Technology, Harbin Engineering University, Harbin, China
{lilijie,shaohongyang,junhaochen,wangye2020}@hrbeu.edu.cn

Abstract. With rapid economic development, the per capita ownership of automobiles in our country has begun to rise year by year. More researchers have paid attention to using scientific methods to solve traffic flow problems. Traffic flow prediction is not simply affected by the number of vehicles, but also contains various complex factors, such as time, road conditions, and people flow. However, the existing methods ignore the complexity of road conditions and the correlation between individual nodes, which leads to the poor performance. In this study, a deep learning model SAMGCN is proposed to effectively capture the correlation between individual nodes to improve the performance of traffic flow prediction. First, the theory of spatiotemporal decoupling is used to divide each time of each node into finer particles. Second, multimodule fusion is used to mine the potential periodic relationships in the data. Finally, GRU is used to obtain the potential time relationship of the three modules. Extensive experiments were conducted on two traffic flow datasets, PeMS04 and PeMS08 in the Caltrans Performance Measurement System to prove the validity of the proposed model.

Keywords: Flow prediction · Temporal-spatial correlation · Graph convolution network · Self-attention mechanism

1 Introduction

In recent years, with the introduction of the term intelligent transportation, traffic flow prediction has become increasingly important, and spatiotemporal prediction is one of the fundamental problems of traffic flow prediction. The traffic data record the continuous time data of an area, which makes it easy to analyze the dynamic correlation of adjacent nodes. With the in-depth study of traffic flow prediction, many models have emerged. Traffic flow prediction can be practically subdivided into several directions: deterministic methods, transfer learning methods [1], statistical methods [2–5], and deep learning methods [6, 7].

Figure 1 shows the spatial correlation between each point, and the color of the line represents the strength of the influence between the two points. Points with dark edges have high correlation, and vice versa. As shown in Fig. 1, at the same time node, every surrounding node has a different influence on A. If node D is blocked, the state of node A has a high probability of entering the situation of congestion or poor road. If node C is

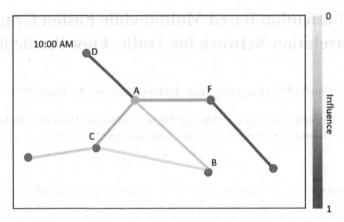

Fig. 1. Spatial correlation of traffic flow

congested, then node A may be unobstructed because there is little correlation between node A and node C.

As shown in Fig. 2, the traffic flow information of multiple historical time periods is correlated for the traffic flow prediction for the coming time. Comparing weekly time data, daily time data, recent time data and ground truth values reveals that their traffic flow has the same trend, which indicates that the traffic flow data of these three time periods have a great impact on the traffic flow in the coming time. By fusing the features of these time periods, more effective feature values can be obtained.

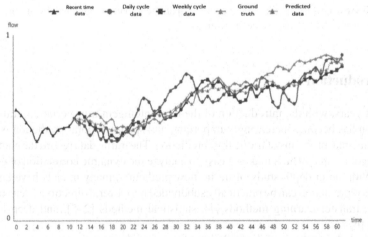

Fig. 2. Temporal correlation of traffic flow

In this study, we present a self-attention based multi-module fusion graph convolution network for the traffic flow prediction (SAMGCN) method to solve the traffic flow prediction problem. The model is based on attention spatiotemporal decoupling weighted fusion of multiple decoupling modules, and finally completes the fusion of multiple temporal granularity information through the time series GRU network to learn the dynamic correlation of traffic data. There are three main contributions:

- Our model obtains more node features through spatiotemporal decoupling. We use self-attention to mine spatial features and temporal features separately and obtain each node and moment feature. In this way we achieve fine-grained division of spatiotemporal features.
- Our model fuses the information of the three modules, recent time data, daily cycle data and weekly cycle data. The model weights and integrates the features by an attention mechanism, so that the recent time module, the daytime module, and the weekly module are fused to obtain a better model training effect.
- The GRU module is built in the output position of the last three modules. The relevant timing information of each module can be effectively captured by GRU.

2 Spatiotemporal Prediction in Deep Learning

2.1 Time Correlation Research

Many researchers use deep learning (DL) for spatiotemporal prediction due to its powerful performance in sequence modeling. Long short-term memory (LSTMs) [8] and gated recurrent unit (GRUs) are variants of Recurrent Neural Network (RNNs) that can effectively learn the temporal dependencies by using the self-loop mechanism. Duan et al. [9] combined convolutional neural networks (CNNs) and LSTM to forecast traffic flow using the trajectory data. However, they ignored the cyclical correlation of temporal data of traffic flow. Zheng et al. [10] used LSTM to achieve a higher prediction effect in predicting passenger flow. However, the performance of LSTM heavily relies on the large amount of training data and parameters, and the computational power of the model is limited by the computer memory. GRU can alleviate the shortcomings of LSTM. R et al. first utilized the GRU to forecast passenger car flow, and in the follow-up, Wang et al. [11] continued to improve the GRU to achieve a better effect. Inspired by the ST-GCN, Guo et al. [6] remodeled the MST-GCN to achieve higher prediction accuracy. They found that time has a certain periodicity, and it was divided into three modules for training, to obtain a more accurate prediction value.

2.2 Time Correlation Research

The traditional CNN can effectively mine local patterns. However, it can only be implemented for standard grid data. Recently, Yu et al. proposed the graph convolutional network (GCN) to solve the flow forecast problem. GCN generalizes convolution operations to graph-structured data, and it also shows excellent performance in node classification and network representation. Spectral GCN is defined in the spectral domain, and these methods are derived from Bruna et al. [12]. Kipf et al. [13] simplified ChebNet into a simpler form of GCN and achieved better performance on various tasks. Spatial

GCN generalizes traditional CNN from Euclidean space to vertex domain. In addition, for the classification of nodes in the graph, Jia et al. [14] used self-attention to process graph-structured data and achieved good results, Then, Velickovic P. et al. [15] found that the cooperation of attention mechanism and GCN can accomplish better results and they proposed the GAT (graph attention network) which used an attention layer to dynamically accommodate the weight of neighbor nodes. Fu et al. [16] focused on the metapath level aggregation method, which aggregated the information between different metapath instances for the first time, and then aggregated the node and edge data in every metapath. Hong et al. [17] designed the type-aware attention method without leveraging metapaths for information fusion. Although these models present great success in HIN embedding, there are still some limitations.

3 Prediction Model of Traffic Flow Based on Multi-module Fusion

The transportation network is defined as an undirected graph $G = (V, E, A)$, where V represents the set of finite nodes, E denotes the edge connecting two nodes, and $A \epsilon R^{N \times N}$ is an adjacency matrix, that denotes the connectivity of the nodes. $f \epsilon (1, 2, \ldots, F)$ denotes the f-th time series of each node on graph G, F represents the eigenvalue of a node at a certain moment, and the eigenvalues contained in F are normalized. Each node in the traffic graph G has f time series data. $x_t^{c,i} \in R$ represents the c eigenvalue of node i at time t. $X_t(x_t^1, x_t^2, \cdots, x_t^N) \in R^{N \times F}$ are all the eigenvalues of all the nodes at time t. $\chi = (X_1, X_2, \cdots, X_t)^T \epsilon R^{N \times T \times \tau}$ denotes all the eigenvalues of all nodes during period τ. $y_t^i = x_t^{f,i} \in R$ is the predicted value of node i at time t. The model predicts future traffic flow from historical measurements at all nodes. $Y = (y^1, y^2, \cdots y^N) \in R^{N \times T_p}$ denotes the sequence of all the nodes over the T_p time period. $y^i = (y_{\tau+1}^i, y_{\tau+2}^i, \cdots, y_{\tau+T_p}^i)$ denotes the future predicted value of node i from τ to $\tau + T_p$. Each block contains GCN, S-attention, Conv and T-attention, and a new feature matrix is obtained every time a block is passed. The output of each round of the three modules is fused through the weight value, input to the next module, and the block is continuously iterated.

3.1 Model Frame Diagram

The SAMGCN model consists of three independent components that have the same structure, modeling recent time data, daily cycle data, and weekly cycle data (Fig. 3).

Supposing that the current time is t_0, the sampling frequency is q times per day, and the prediction window size is T_p, the inputs of the three modules are defined as follows:

Recent time data: This is the earlier time series pieces immediately adjacent to the prediction period:

$$\chi_h = (X_{t_0-T_h+1}, X_{t_0-T_h+2}, \cdots, X_{t_0}) \in R^{N \times F \times T_h} \tag{1}$$

Daily cycle data: This consists of time series pieces in the earlier days of the same time period as the forecast period:

$$\chi_d = (X_{t_0-(Td/T_p)*q+1}, X_{t_0-(Td/T_p)*q+2}, \cdots, X_{t_0-q+T_p}) \in R^{N \times F \times T_d} \tag{2}$$

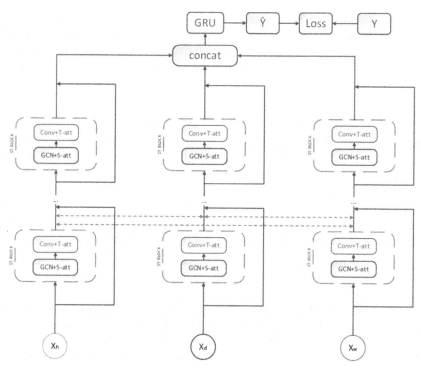

Fig. 3. SAMGCN model framework diagram

Weekly cycle data: This consists of data segments from the earlier weeks that have the same weekly attribute time period as the forecast period:

$$\chi_w = (X_{t_0-7*(T_w/T_p)*q+1}, X_{t_0-7*(T_w/T_p)*q+2}, \cdots, X_{t_0-7*q+T_p}) \in R^{N \times F \times T_w} \qquad (3)$$

In the above three formulas, T_h, T_d, and T_w denote the time series segment lengths of the recent, daily-period and weekly-period components respectively, and they are all integer multiples of T_p.

By referring to the DenseNet dense connection [19], the output data of each block are connected. Although the three modules share the same network structure, they finally obtain three different output results because of different initial data. The data of the three modules are different but there is a correlation in time series logic, and their features are further obtained through GRU training. Finally they are fed to the loss function to obtain the final predicted value.

3.2   Space-Time Decoupling

At present, most of the methods couple space and time together by weighting the attention of time and space. However, this is not actually suitable for such a high-latitude space-time model. For example: when performing a 3×3 convolution, the 3×3 data will not be convolved directly, but they will be convolved by 3×1, 1×3 and 1×1, because the convolution effect is much better than the 3×3 convolution. Moreover, the decoupling method, is no longer limited to the time-space correlation of a time, and achieves a more fine-grained time feature acquisition, to obtain the time characteristics of each moment. The acquisition of these features can predict the results more accurately, so the model achieves a better training effect by means of spatiotemporal decoupling.

3.3 Spatial Convolution

Spectral graph theory-based graph convolution is employed to directly process the signal at each time slice, exploiting the signal correlation on the transportation network in the spatial dimension. Spectral methods convert graphs into algebraic form to analyze graph topological properties, such as connectivity in graph structures. In spectral analysis, Laplace matrix can be used to represent a graph. By analyzing the Laplacian matrix and its eigenvalues, the properties of the graph structure can be obtained. Laplacian Matrix Definition for Graphs $L = D - A$, The normalization method is $L = (I_N - D^{\frac{1}{2}}AD^{\frac{1}{2}})$. In this paper, the traffic network is essentially a graph, and the features of each node can be regarded as signals on the graph. Therefore, to take full advantage of the topological properties of the transportation network, spectral graph theory-based graph convolution is employed to directly process the signal at each time slice, exploiting the signal correlation on the transportation network in the spatial dimension. Spectral methods transform graphs into algebraic form to analyze graph topological properties, such as connectivity in graph structures. Finally, we obtain its convolution formula:

$$g_{\theta*G}x = g_\theta(L)x = g_\theta(U\Lambda U^T)x = Ug_\theta(\Lambda)U^T x \qquad (4)$$

where $*G$ represents convolution operation on the graph, and the result is obtained through a Fourier transform.

$$g_{\theta*G}x = g_\theta(L)x = \sum_{k=0}^{K-1} \theta_k T_k(\tilde{L})x \qquad (5)$$

where θ_k is a learnable parameter, representing the vector of Chebyshev polynomial coefficients. $\tilde{L} = 2L/\lambda_{max} - I_N$ is the scale of the eigenvector matrix, and the size is [-1, 1]. λ_{max} is the maximum feature of L.

The graph convolution module adopts a recursive linear unit (ReLU) as the final activation function, and C_{r-1} is defined as multiple channels. For example, in the nearest component, the input is $\hat{\chi}_h^{r-1} = (\hat{X}_1, \hat{X}_2, \cdots, \hat{X}_{T_{r-1}}) \epsilon R^{N \times C_{r-1} \times T_{r-1}}$, and T_{r-1} is the size of the time dimension at layer r. When $r = 1$, the module is $T_0 = T_h$ at the latest time.

3.4 Spatial Self-attention

This section mainly uses spatial self-attention to obtain the spatial high-order neighbor features and the spatial features between nodes. In the spatial dimension, not only do the adjacent nodes influence the predicted target node, but other nodes also influence the target node. The dynamic correlation between all nodes is obtained through the spatial self-attention mechanism. The specific formula is:

$$K = K(e_s^{ti}) = W_k e_s^{ti} \qquad (6)$$

$$Q = Q(e_s^{ti}) = W_q e_s^{ti} \tag{7}$$

$$V = V(e_s^{ti}) = W_V e_s^{ti} \tag{8}$$

where $e_s^{ti} \epsilon R^{C \times N}$ denotes the feature vector of each node. N is the number of space node, C is the number of channels, and ti denotes the i-th time step. $K(x), Q(x)$ and $V(x)$ are obtained by a one-dimensional convolution of each node. We obtain its spatial attention matrix through Q, K, V:

$$S = Softmax \frac{(Q^T * K)}{\sqrt{d_k}} \tag{9}$$

where Q and V are obtained from formulas (7) and (8), and $S \epsilon R^{N \times N}$ is the spatial attention matrix. $S_{i,j}$ denotes the degree of influence of location i on location j. In this way, we can obtain the feature weight between each node. The weight matrix $\hat{x} = VS(x \epsilon R^{N \times N})$ which contains the information of high-order neighbors can be obtained. Finally, the feature matrix e_{sa} containing information of the high-order neighbor is acquired.

3.5 Temporal Convolution

After using graph convolution to obtain the adjacent information of each node in the spatial dimension, we leverage standard convolutional layers in the temporal dimension to further update the feature of the node by merging the information on adjacent time slices. An example of the operation of the r-th layer in the nearest component is as follows.

$$e_t^r = ReLU(\Phi * (ReLU(g_{\theta*G} e_{sa}))) \epsilon R^{N \times C_r \times T_r} \tag{10}$$

where $*$ is the standard convolution operation. ϕ denotes the time dimension convolution kernel parameters, and $ReLU$ is used as the activation function. e_t^r presents the temporal convolution output.

3.6 Time Self-attention

In this paper, temporal self-attention is used to capture the temporal correlation of a node at different times in the transportation network space. In the time dimension, traffic flow forecasting often focuses on time trend analysis. Each time series has implicit periodicity and correlation. Not only do the adjacent time segments have an impact on the predicted target node, but the data of other time nodes also have a strong correlation with the predicted target node. The temporal dynamic correlation between all nodes is obtained through the temporal self-attention mechanism, and the specific formula is:

$$K = K(e_t^{Ni}) = W_k e_t^{Ni} \tag{11}$$

$$Q = Q(e_t^{Ni}) = W_q e_t^{Ni} \tag{12}$$

$$V = V(e_t^{Ni}) = W_V e_t^{Ni} \tag{13}$$

where $e_t^{Ni} \epsilon R^{C \times T}$ denotes the feature vector of the i node. T denotes the set of each node for an amount of time, and C is the channel count. $Q(x), K(x)$ and $V(x)$ are the feature matrices in the time dimension of a single node. Through $Q(x)$, $K(x)$ and $V(x)$, its time attention matrix can be obtained

$$E = Softmax \frac{(Q^E * K)}{\sqrt{d_k}} \tag{14}$$

where $E \epsilon R^{T \times T}$ is the attention matrix. $E_{i,j}$ is the weight of influence of a node in time slice i on time slice j. The self-attention mechanism over time can obtain $\hat{x} = VE$ ($x \epsilon R^{C \times T}$), and in this way, the feature weight between each moment of each node is obtained. Finally, the result of a space-time block is obtained and represented by e_{ta}^r, where r is the layer count in this module.

3.7 Information Fusion and GRU

From Fig. 2, it can be found that each part of the data influences the prediction results of the coming time, but the lack of training data will become the decisive factor for this model. By calculating the weight influence between each model, it is easy to obtain a feature matrix with more features by multiplying the matrix, which is conducive to further training of the model. Taking the r layer block as an example, the block output of each independent module is e_h^{r-1}, e_d^{r-1} and $e_w^{r-1} \in R^{N \times T \times \tau}$. The weight values are W_h^i, W_d^i and W_w^i. Interpretation with fusion of recent time modules:

$$a_h^{r-1} = e_h^{r-1} + W_d^i e_d^{r-1} + W_w^i e_d^{r-1} \tag{15}$$

The outputs of the three modules are $a_h^{r-1}, a_d^{r-1}, a_w^{r-1}$. The independence of the three modules does not consider the time series between them. The data of each module is not the same, and the prediction of future traffic has a very important correlation, so simple FC cannot meet the complex correlation in time series. In this part, the variant GRU of LSTM is used to solve this problem. The output obtained from the previous layer is put into the GRU for operation. GRU is defined as follows.

$$z_t = \sigma(W_z X_t + U_z h_{t-1}) \tag{16}$$

$$r_t = \sigma(W_t X_t + U_t h_{t-1}) \tag{17}$$

$$\tilde{h}_t = \sigma(WX_t + U(r_t o h_{t-1})) \tag{18}$$

$$h_t = (1 - z_t) o h_{t-1} + z_t o \tilde{h}_t \tag{19}$$

The model obtains a brand-new value through the GRU and obtains the final predicted value through the loss function.

4 Experimental Analysis

4.1 Dataset

The proposed method is demonstrated on two datasets, PeMSD4 and PeMSD8. Traffic data are aggregated from raw data to every 5-min interval. There are more than 39,000 detectors on highways in California metropolitan areas. Geographical information about sensor stations is recorded in the dataset. Three traffic measurements are considered, including average speed, average occupancy, and total flow. PeMSD4 is the traffic data collected in the San Francisco Bay Area, which contains 3848 detectors on 29 roads from January to February 2018. The data of the first 40 days are the training set, the next 10 days of data are the test set, and the remaining data are the validation set. PeMSD8 Traffic data are collected in San Bernardino from July to August 2016, including 1979 detectors on 8 main roads. The data of the first 40 days are used as the training set, the next 10 days of data are used as the test set and the remaining data are used as the validation set. We set the batch size to 64, set the epoch to 100, and the learning rate to 0.003. Additionally, the kernel size of CNN is 3, and we adjust the time span of data by controlling the stop size of time convolution. Finally, the lengths of the three time windows are set as $T_h = 24$, $T_d = 24$, and $T_w = 24$. Prediction window T_p is set to 12.

4.2 Analysis of Results

We compare our proposed model to several recent traffic flow prediction methods:

- ARIMA [3]: ARIMA is a traditional traffic flow prediction model, that uses the WORD decomposition theorem to perform weakly stationary transformation of the spatiotemporal data and conducts modeling by means of a core variable, which has low computational complexity and is relatively stable.
- VAR [18]: Based on the natural extension of the univariate autoregressive model to dynamic multivariate time series, the analysis of multivariate time series was realized. Not only was VAR successful in the financial field, but VAR also achieved good results in short period traffic flow forecasting and outperformed the univariate sequence methods.
- LSTM [8]: According to the analysis of the temporal characteristics of traffic flow data, a short period traffic flow forecast was performed by filtering which information was useful for future prediction by means of long-short term time series and gating units.
- GRU [19]: GRU uses a new gating unit to analyze the traffic flow data, learn the time series characteristics of the traffic flow data, and predict the future development trend of the traffic flow.
- STGCN [20]: To encode the dynamic graph of node positions in traffic flow, STGCN constructed a multiscale local graph convolution filter composed of local receptive field matrix and signal mapping matrix to recursively operate the structured graph data in the space-time domain.
- GeoMAN [21]: GeoMAN improved the prediction accuracy in two ways: one was to use local attention and global attention to dynamically obtain temporal and spatial

relevance, and the other was to integrate a large number of external factors in the prediction process.

- MSTGCN [22]: MSTGCN carried out traffic flow prediction according to the multi-module fusion method and obtained the final prediction result by defining spatiotemporal convolution blocks, iterating multilayer spatiotemporal blocks, and using residual connections to ensure the characteristics of the original data.
- ASTGCN [6]: ASTGCN added a spatiotemporal attention and spatiotemporal convolution on the basis of the MSTGCN to forecast traffic flow, and the performance was improved to a certain extent.

As seen from the upper part of Table 1, deep learning-based methods generally achieved better performance than traditional time series prediction methods, which indicated that deep learning-based models were more suitable for nonlinear time series analysis. We also considered methods that model both temporal and spatial correlations, such as STGCN, and GeoMAN, which outperformed traditional deep learning models such as LSTM and GRU. GRU could better capture long-short term dependencies, but it still could not capture spatiotemporal dependencies, so the performance improvement was still limited. After adding spatial relationship extraction, the prediction accuracy and stability of the model were greatly improved.

Table 1. Performance of different methods on PeMSD4 and PeMSD8

Model	PeMSD4		PeMSD8	
	RMSE	MAE	RMSE	MAE
ARIMA	68.13	32.11	43.30	24.04
VAR	51.34	33.52	31.11	21.33
LSTM	45.82	29.45	36.96	23.18
GRU	44.92	28.53	35.52	22.13
STGCN	38.29	25.15	27.87	18.88
GeoMAN	37.84	23.64	28.91	17.84
MSTGCN	37.8	24.67	26.15	16.54
ASTGCN	34.56	21.92	29.53	19.45
- Self-attention	37.81	22.28	28.19	18.21
- Side join	35.41	21.22	32.02	19.61
- GRU	33.61	20.39	29.18	18.12
BSAGCRN	32.71	19.87	27.28	17.16

In addition, the performance of GeoMAN was better than that of STGCN, indicating that the multilevel attention model could effectively capture the dynamic correlation of spatiotemporal data. The SAMGCN achieved the highest performance compared with the baseline models through spatiotemporal decoupling and self-attention, demonstrating the superiority of the proposed model in modeling the spatiotemporal features of highway traffic data. To study the contribution of each component of the proposed model, we conducted ablation experiments. The results are reported in the lower part of Table 1. The performance of the model decreased after removing self-attention, side join and GRU. In particular, the performance of the model decreased most significantly after removing self-attention, which proved that self-attention could capture the remote dependency between nodes effectively and bring great gain to our model.

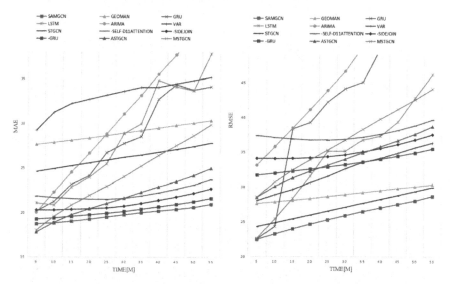

Fig. 4. Prediction results of PeMSD4 by different methods

Comparative experiments were conducted to demonstrate the robustness of our method. The experimental results are reported in Fig. 4 and Fig. 5. The performance of all the models degraded on the two datasets with the extension of time. The decrease rate on PeMSD4 was larger than that of PeMSD8, possibly because there are more monitoring sites on the PeMSD8 dataset and PeMSD8 had more training data than PeMSD4. The methods ARIMA, LSTM and GRU, which only considered temporal correlations, had relatively accurate results in predictions with a duration of 10 min, indicating that temporal correlation has a more important impact on short-term predictions. However, with the increasing of time, their prediction accuracy drops significantly. The BASCGCRN model always achieved the best prediction performance. In particular, the performance of the SAMGCN was much better than that of the other models in long-period forecasting, verifying the robustness of our model.

To further demonstrate the convincingness of our model, single node experiments were conducted. As shown in Fig. 6, we randomly selected four detection sites from all monitoring sites, and randomly selected 6-h real and predicted values at each selected monitoring site for comparison. The horizontal axis of the prediction graph is the duration, and the vertical axis represents the traffic flow value. First, by analyzing the real value, we found that the traffic flow showed a cyclical trend, i.e., the traffic flow would

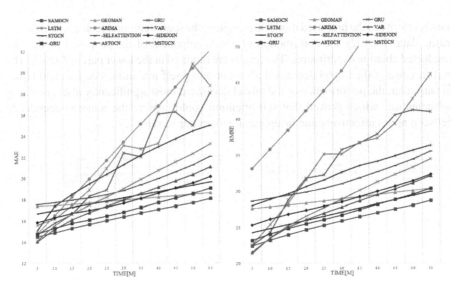

Fig. 5. Prediction results of PeMSD8 by different methods

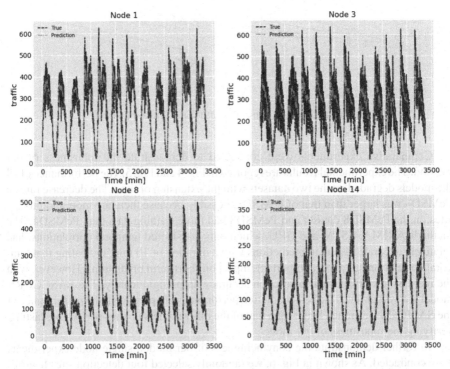

Fig. 6. Predicted and true values of the single node

reach and maintain a brief peak at a certain time, and then decrease rapidly. We found that at the 1600th minute in node 1, there was a sudden change in traffic flow, and this sudden change information is difficult to predict accurately, which is also a challenge for future research. Our model achieved the best results, and can accurately predict the change trend at each node. Although we have not yet perfectly predicted the changes at the mutated nodes, we have obtained reliable predictions.

5   Conclusion

In recent years, the application of deep learning methods to traffic flow prediction has gradually become a trend. In this paper, we propose the SAMGCN model to effectively capture the correlation between individual nodes to improve the performance of traffic flow prediction. The model obtains higher fine-grained spatial and temporal correlation by decoupling the spatiotemporal correlation, prevents its gradient explosion by means of DenseNet, strengthens the transfer of features, and effectively utilizes features. Finally, the time series features are extracted by GRU to promote the prediction performance. Through extensive experimental comparisons, the reliability and practicability of the proposed model are verified. In the future, we will promote our method in the field of air quality prediction.

Acknowledgment. This work was supported by the National Key R&D Program of China under Grant No. 2020YFB1710200, and the National Natural Science Foundation of China under Grant No. 61872105 and No. 62072136.

References

1. Pan, Z., et al.: Spatiotemporal meta learning for urban traffic prediction. IEEE Trans. Knowl. Data Eng. **34**(3), 1462–1476 (2022)
2. Deng, Y., Xiang, J., Ou, Z.: SVR with hybrid chaotic genetic algorithm for short-term traffic flow forecasting. In: Eighth International Conference on Natural Computation, pp. 708–712. IEEE, Chongqing, China, 29–31 May 2012
3. Chen, C., Hu, J., Meng, Q., Zhang, Y.: Short-time traffic flow prediction with ARIMA-GARCH model. In: IEEE Intelligent Vehicles Symposium (IV), pp. 607–612. IEEE, Baden-Baden, Germany, 5–9 June 2011
4. Petrlik, J., Fucik, O., Sekanina, L.: Multiobjective selection of input sensors for SVR applied to road traffic prediction. In: Bartz-Beielstein, T., Branke, J., Filipič, B., Smith, J. (eds.) PPSN 2014. LNCS, vol. 8672, pp. 802–811. Springer, Cham (2014). https://doi.org/10.1007/978-3-319-10762-2_79
5. Cheng, S., Zhou, X.: Network traffic prediction based on BPNN optimized by self-adaptive immune genetic algorithm. In: Proceedings 2013 International Conference on Mechatronic Sciences, Electric Engineering and Computer (MEC), pp. 1030–1033. IEEE (2013)
6. Guo, S., Lin, Y., Feng, N., Song, C., Wan, H.: Attention basedspatial-temporal graph convolutional networks for traffic flow forecasting. In: The Thirty-Third AAAI Conference on Artificial Intelligence, pp. 922–929. AAAI 2019, AAAI Press, Honolulu, Hawaii, USA, 27 Jan–1 Feb 2019

7. Zhao, L., Sun, Q., Ye, J., Chen, F., Lu, C.-T., Ramakrishnan, N.: Multitask learning for spatiotemporal event forecasting. In: Proceedings of the 21th ACM SIGKDD International Conference on Knowledge Discovery and Data Mining, pp. 1503–1512 (2015)

8. Hochreiter, S., Schmidhuber, J.: Long short-term memory. Neural Comput. 9(8), 1735–1780 (1997)

9. Duan, Z., Yang, Y., Zhang, K., Ni, Y., Bajgain, S.: Improved deep hybrid networks for urban traffic flow prediction using trajectory data. IEEE Access 6, 31820–31827 (2018)

10. Zhao, Z., Chen, W., Wu, X., Chen, P.C.Y., Liu, J.: LSTM network: a deep learning approach for short-term traffic forecast. IET Intell. Transp. Syst. 11(2), 68–75 (2017)

11. Wang, S., Zhao, J., Shao, C., Dong, C., Yin, C.: Truck traffic flow prediction based on LSTM and GRU methods with sampled GPS data. IEEE Access 8, 208158–208169 (2020)

12. Bruna, J., Zaremba, W., Szlam, A., LeCun, Y.: Spectral networks and locally connected networks on graphs. In: Bengio, Y., LeCun, Y. (eds.) 2nd International Conference on Learning Representations, ICLR 2014. Banff, AB, Canada, Conference Track Proceedings, 14–16 April 2014

13. Kipf, T.N., Welling, M.: Semisupervised classification with graph convolutional networks. arXiv preprint arXiv:1609.02907 (2016)

14. Jia, X., Li, T., Zhu, R., Wang, Z., Zhang, Z., Wang, J.: Traffic flow prediction based on self-attention mechanism and deep packet residual network. In: ICIT 2019 - The 7th International Conference on Information Technology: IoT and Smart City, pp. 575–580. ACM, Shanghai, China, 20–23 Dec 2019

15. Velickovic, P., Cucurull, G., Casanova, A., Romero, A., Lio, P., Bengio, Y.: Graph attention networks. In: International Conference on Learning Representations, vol. 20, p. 1050 (2017)

16. Fu, X., Zhang, J., Meng, Z., King, I.: MAGNN: metapath aggregated graph neural network for heterogeneous graph embedding. In: Huang, Y., King, I., Liu, T.-Y., van Steen, M. (eds.) WWW'20: The Web Conference 2020, pp. 2331–2341. ACM/IW3C2, Taipei, Taiwan, 20–24 April 2020

17. Hong, H., Guo, H., Lin, Y., Yang, X., Li, Z., Ye, J.: An attention-based graph neural network for heterogeneous structural learning. Proceedings of the AAAI Conference on Artificial Intelligence 34(04), 4132–4139 (2020)

18. Chung, J., Gulcehre, C., Cho, K.H., Bengio, Y.: Empirical evaluation of gated recurrent neural networks on sequence modeling. arXiv preprint arXiv:1412.3555 (2014)

19. Zivot, E., Wang, J.: Vector autoregressive models for multivariate time series. Modeling financial time series with S-PLUS®, pp. 385–429 (2006)

20. Li, C., Cui, Z., Zheng, W., Xu, C., Yang, J. Spatiotemporal graph convolution for skeleton based action recognition. In: Thirty-Second AAAI Conference on Artificial Intelligence (2018)

21. Liang, Y., Ke, S., Zhang, J., Yi, X., Zheng, Y.: Geoman: Multilevel attention networks for geo-sensory time series prediction. In: Lang, J. (ed.) Proceedings of the Twenty-Seventh International Joint Conference on Artificial Intelligence, IJCAI 2018, 13–19 July 2018, Stockholm, Sweden, pp. 3428–3434 (2018). ijcai.org

22. Feng, N., Guo, S., Song, C., Zhu, Q., Wan, H.: Multicomponent spatial-temporal graph convolution networks for traffic flow forecasting. J. Softw. 30(3), 759–769 (2019)

Data Analyses and Parallel Optimization of the Tropical-Cyclone Coupled Numerical Model

Yanqiang Wang[1,2] 🄳, Tianyu Zhang[3](✉), Zhaohui Yin[4], Sai Hao[1], Chenqi Wang[1], and Bo Lin[1]

[1] National Marine Environmental Forecasting Center, Beijing 100081, China
[2] College of Oceanic and Atmospheric Sciences, Ocean University of China, Qingdao 266100, China
[3] Key Laboratory of Climate, Resources and Environment in Continental Shelf Sea and Deep Sea of Department of Education of Guangdong Province, Guangdong Ocean University, Zhanjiang 54008, China
zhaangty@sina.com
[4] Baidu Inc., Beijing 10105, China

Abstract. Tropical cyclones (TCs) are one of the most feared and deadly weather systems in the world. An air-sea coupled numerical model offers a more accurate description of physical processes between atmospheric-ocean fluids. An operational ocean-atmosphere-wave coupled modeling system is employed to improve the prediction accuracy of tropical cyclones in the National Marine Environmental Forecasting Center (NMEFC). Due to the urgent need for operational timeliness, the parallel performance of the operational forecasting system has been analyzed. The parallel algorithm, parallel partitioning grids, and other optimizations were tested after system deployment on the Lenovo cluster of the NMEFC. After optimization, a well-balanced performance of the system is obtained, and computing resources are reasonably utilized, thus laying the foundation for real-time tropical cyclone forecasting.

Keywords: High-performance computing · Tropical cyclones · Model coupling · Parallel partition

1 Introduction

1.1 A Subsection Sample

Tropical cyclones (TCs) are one of the most feared and deadly weather systems in the world. The extreme weather caused by TCs has a tremendous impact on human life, property, and production activities [1]. The storm surges, huge waves, and other marine disasters caused by TCs have brought huge safety hazards and serious economic losses to maritime operations and transportation. The west coast of the Pacific Ocean, where China is located, is one of the areas where most TC activities take place and affects regions in the world. Therefore, the improvement of TC forecasting and warning capability in the Northwest Pacific is of great significance in marine disaster prevention and mitigation, resource development, rights protection, law enforcement, and national security.

Track and intensity prediction is the key to TC warning and disaster prevention. Thanks to improvements in satellite [2] and dropsonde [3–5] observations as well as a better understanding of physical processes that control the motion of TCs [6], considerable progress has been made in TC track forecasts in the past three decades. However, predictions of TC intensity and their change have improved inconspicuously [7]. The principal reason is that a single atmospheric numerical model failed to describe the influence of the intense ocean-atmosphere interaction on the intensity change reasonably. For example, TC surface wind stress induces upwelling and strong turbulent mixing in the upper ocean, which consequently cools the ocean surface due to the entrainment of cold water from the thermocline into the mixed layer [8–10]. The reduction of Sea Surface Temperature (SST) reduces the enthalpy flux from the ocean to the atmosphere, resulting in a decrease in TC intensity. Moreover, strong winds under TC conditions can generate a layer of foam at the air-sea interface [11] and enormous amounts of sea spray into the atmospheric boundary layer. The former impedes the transfer of momentum from the wind to the ocean, and the latter enhances the sensible and latent heat transfer [12–14], resulting in varying TC intensities. In contrast, an air-sea coupled numerical model offers a more accurate description of such physical processes governing the intensity variation.

Based on international advanced atmospheric, oceanic, and ocean wave component models, a coupled numerical prediction system for TCs in the Northwest Pacific was developed [15, 16] and has been employed for operational runs in the National Marine Environmental Forecasting Centers since 2015. A schematic diagram of the coupled model physical concept is shown in Fig. 1. Aiming at better operational application, the prediction system is supposed to spend as little time on the whole forecasting flow as possible. However, it is difficult to reduce the time from the procedure of receiving initial data, pre- and postprocessing. Therefore, parallel optimization for the coupled model is an inevitable choice to achieve the high efficiency and ability of TCs to forecast and pre-alert.

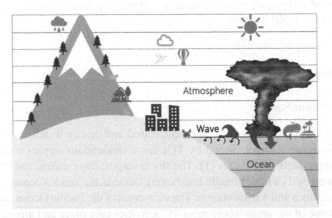

Fig. 1. Schematic diagram of the coupled model physical concept

2   Model Setup

2.1   Atmospheric Model Setup

The Weather Research and Forecasting (WRF) Model version 3.4 was used in this study for the atmospheric component of the Tropical-Cyclone Coupled model. The initial and boundary conditions for the WRF model are initialized on the 5th of July 2018 using 6 hourly Climate Forecast System Reanalysis (CFSR) data with a 1/2 degree spatial resolution. The model domain of WRF in this study covers the northwest Pacific and China's adjacent sea with a horizontal resolution of 1/12 degrees and 890 × 846 grid points. The vertical resolution is employed with 61 sigma levels. This domain includes the area where hurricanes in the west Pacific warm pool are generated and the region by which hurricanes usually pass. The resolution resolved the typhoon with small diameters of several tens of kilometers.

2.2   Hydrodynamic Model Setup

The Regional Ocean Modeling System (ROMS) [17] svn No. 455 was used in this study for the ocean component in the tropical cyclone coupled model. To resolve mesoscale oceanic processes such as mesoscale eddies that might influence typhoon activities, the ROMS horizontal grid number is set to 1339 × 1391. There are 30 layers in the vertical direction.

The ROMS model is initialized on the 5th of July 2018 using the fields of currents, salinity, temperature, and sea surface height from the CFSR ocean dataset with a 1/2 degree spatial resolution. The lateral boundary conditions of the coupled model (including currents, salinity, and temperature) are provided by the same dataset.

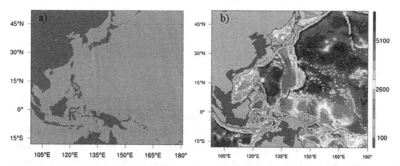

Fig. 2. Model domain in atmospheric model WRF (a) and the bathymetry in ocean model ROMS (b)

2.3   Ocean Wave Model Setup

The Simulating WAves Nearshore (SWAN) version 40.81 developed by Delft University of Technology was used in this study for the ocean wave component in the Tropical-Cyclone Coupled model. For the ocean area, the SWAN used the same grid as the ROMS model. The wind forcing of the SWAN model is the National Centers for Environmental Prediction (NCEP) Final Analysis (FNL) data, with 1/2 degree spatial resolution and 6 h time resolution. The SWAN model was set up to 3600 s for the time step and initialized by a steady state.

2.4 HPC Facilities

Based on the process of the coupled model system, considering the number and performance of the test platform's processors, the system is deployed on the Lenovo cluster of the National Marine Environmental Forecasting Center. The specific configuration is shown in Table 1.

Table 1. HPC facilities used for the tests

System processor	Intel(R) Xeon(R) CPU E5–2680 v4 @ 2.40 GHz
Cores/node	14Core*2
Mem./node	128GB TruDDR4 2400 MHz
MPI network	Infiniband 100Gb/s
Filesystem	IBM GPFS (General parallel file system)
Operating system	Red hat enterprise linux server release 6.8 (Santiago)
Complier	Intel compilers_and_libraries_2017.3.191
Analysis software	Intel parallel_studio_xe_2017.3.053; Paramon 7.2.0;Paratune 7.2.0
Visualization software	NCL 6.4.0; VAPOR 2.6; Python 3.6
Other libraries	NetCDF 4.4.4 etc

2.5 Coupled Variables

We used the Model Coupling Toolkit (MCT) [18, 19] to exchange variables. Figure 3 illustrated coupling variables such as the wind, etc. The WRF atmospheric model provides a 10-m surface wind for the SWAN wave model. The ROMS ocean models receive heat fluxes and momentum fluxes calculated by atmospheric models. The ROMS Ocean models provide SST for the WRF atmospheric models and sea surface currents, sea surface elevations, and ocean bathymetry for the SWAN wave models. The SWAN wave models provide significant wave heights and wavelengths for WRF atmospheric and ROMS oceanic models. The SWAN wave model also provides the ROMS ocean model with the direction of the waves, surface and bottom periods, wave fracture percentages, wave energy dissipation, and bottom orbital velocity.

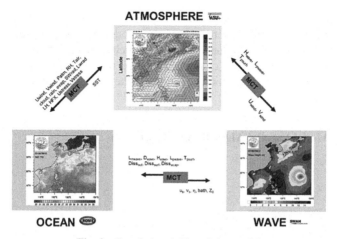

Fig. 3. Coupled variables of the model

3 Scaling Experiments

3.1 Parallel Tests Analysis

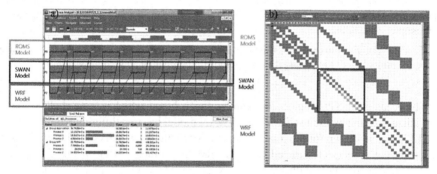

Fig. 4. Intel® Trace Collector data of the coupled model (blue stands for computing and red for MPI communication)

Figure 4a) shows the coupling model ROMS -WRF- SWAN, each using an MPI process. Intel® Trace Collector analysis found that there are two main problems in the parallel computing process of the typhoon-coupled model. First, there is a problem with the internal algorithm of the ocean wave model SWAN, resulting in low parallel computing efficiency. The SWAN ocean wave model only performs two-dimensional operations. In principle, the calculation amount should be much smaller than the atmospheric and ocean models for performing three-dimensional operations. However, as seen from the left figure, the internal operation efficiency of the SWAN model is much lower than that of the WRF atmospheric and ROMS ocean models.

The second problem is that the overall operational efficiency of the load imbalance between the component models of the TC coupled model is low, as shown in Fig. 4b).

Since the operation model of the ocean wave model is much lower than that of the atmosphere and ocean model, the three-component models are assigned the same number of CPU cores in the current operational operation, thereby causing load imbalance among the component models. Before the coupling exchange, the atmosphere and ocean model waited for information exchange for a long time, so the overall operational efficiency of the model was low.

3.2 SWAN Model Parallel Algorithm Optimization

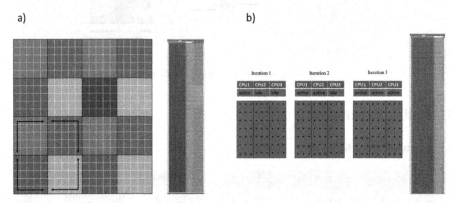

Fig. 5. Parallel algorithm (a) four different colors and (c) block wavefront [20] and Intel® Trace Collector results (b) (d) for (a) and (b) (blue stands for computing and red for waiting)

The ocean wave SWAN model uses an implicit scheme, which is more challenging to parallelize than the explicit scheme. The SWAN model has two different parallel strategies.[20] One is known for the four different color methods, which color the subdomain with different colors, such as red, orange, green and blue, as seen in Fig. 5 a). Each colored subdomain starts with a different order of updates in the same scan. The number of unknowns was replaced in four scans based on the color of the subdomain. The numerical overhead can be reduced, thereby reducing the number of synchronization points with a high degree of parallelism [21, 22].

Another method is known as the block wavefront method [23]. This method breaks down the computation field into many stripes. In each scan, CPU processors belonging to the different stripes communicate. The different unknown numbers of each stripe update are shown in the circle in Fig. 5 c), and the algorithms were as follows.

Step 1: The unknown number N(i, j, l, m) along j = 1 in CPU1 was updated.
Step 2: The unknown number N(i, j, l, m) along j = 2 in CPU1 and j = 1 in CPU2 were updated in parallel. The known number is j = 1 in CPU1.
Step 3: The unknown number N(i, j, l, m) along j = 3 in CPU1, j = 2 in CPU2, and j = 1 in CPU3 will be updated in parallel. The known number is j = 1,2 in CPU1 and j = 1 in CPU2.

Zijlema [24] examines in detail the numerical computational efficiency of the four different colors and block wavefront methods. In this paper, we used the block wavefront method to improve the coupling model operation efficiency. In Fig. 5 b) and d) the block wavefront method reduced the communication time and had a good balance of loading. We obtained an 18.5% performance improvement by SWAN model parallel algorithm optimization. With 196 CPU core simulations for 1 day and the SWAN model, the calculation time was reduced from 250 s to 211 s.

3.3   Ocean Model Grid Optimization

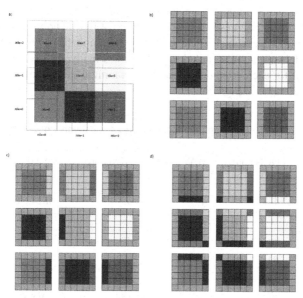

Fig. 6. Ocean model grid with nine tiles reference with https://www.myroms.org/wiki/Paralleli zation

A parallel domain of ROMS with nine tiles is shown in Fig. 6, one color per tile. For MPI jobs, each tile is an MPI process. The number of tiles is set to NtileI and NtileJ in the input file, and the product of both must be equal to the total number of MPI processes. The two mesh points with wide overlapping areas are called ghost points or halo points, which are used for exchange. The exchange in the east–west direction occurred before the north–south direction. Figure 6 b) c) and d) show how the point is updated before and after an update. The interior points are colored with different tiles, and the halo points are colored gray.

The ROMS model is written in Fortran Languages. For the Fortran memory order, the I-element is the rapidly changing index. Therefore, it is advantageous to have more partitions in the J-direction (NtileJ) than in the I-direction (NtileI) to facilitate vectorization. In addition, the coupled model mesh configuration is optimized, and the coupling mode is changed from the same mesh as the WRF model to a different mesh. Under the premise of ensuring that the prediction accuracy is unchanged, the calculation amount of the component model is reduced, and the parallel computing time of the model is further reduced.

4 Parallel Test Results

The parallel algorithm of the SWAN wave model and the parallel partition of the ROMS ocean model were optimized for Sects. 3.2 and 3.3. Then, we take different proportions according to the amount of computation in the model instead of the same number for each different component. Within ~400 CPU cores, there is no significant difference in the results using different numbers for each component of the TC Coupled model. When using processes with more than 400 CPU cores, a better acceleration ratio can be achieved if the number of cores is properly configured. Before optimizing the allocation of cores, the overall operational efficiency for the load imbalance between the different component models of the TC coupled model is low. Finally, we get speedup up to 107 times than serial for each component and MPI communication time was significantly reduced (Table 2, Table 3, Fig. 7).

Table 2. The coupling model uses the same number of cores

Total Cpu cores	OCN cores	WAV cores	ATM cores	Time (s)	Speedup
3	1	1	1	18861.73	
12	4	4	4	4824.64	290.95%
48	16	16	16	1764.85	968.74%
108	36	36	36	759.23	2384.32%
192	64	64	64	416.39	4429.78%
300	100	100	100	278.23	6679.29%
432	144	144	144	218.84	8518.87%
507	169	169	169	215.30	8660.78%
768	256	256	256	240.56	7740.88%

Table 3. The coupling model uses the optimized number of cores

Total Cpu cores	OCN cores	WAV cores	ATM cores	Time (s)	Speedup
3	1	1	1	18861.73	
36	16	16	4	1764.85	968.74%
352	144	144	64	256.40	7256.38%
493	169	196	128	183.46	10181.24%
580	196	256	128	174.58	10703.96%

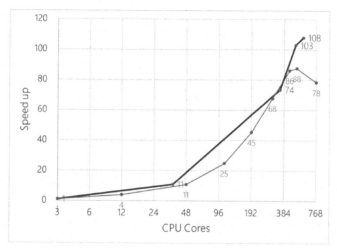

Fig. 7. Speedup of the Tropical-Cyclone Coupled Numerical Model (Data from the Table 2 and Table 3)

5   Model Results Discussion

Figure 2 shows the configuration of the Tropical-Cyclone Coupled Numerical Model domain. The domain of the TC model covers a large area of the western North Pacific, where the components of the ocean and atmosphere are fully interactive. Mesoscale air-sea interaction weather phenomena, such as typhoons, occur frequently in this area.

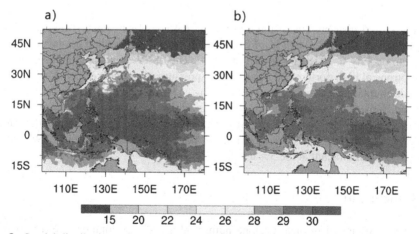

Fig. 8. Spatial distributions of sea surface temperature (°C) at 12 am, July 06, 2018 a) from the coupled model output (tropical-cyclone coupled numerical model) and b) the ERA5 dataset.

Figure 8 shows the Tropical-Cyclone Coupled model results in the simulation of sea surface temperature (SST) at 12 am on July 6, 2018. The features of SST are well captured in the Tropical-Cyclone Coupled Numerical Model. The spatial pattern correlation coefficient (PCC) of SST between the Tropical-Cyclone Coupled model and ERA5 dataset is 0.98, which is statistically significant at the 5% level. The Tropical Cyclone Coupled Numerical Model overestimates the SST over the western Pacific warm pool. The largest bias of SST in the western Pacific warm pool is approximately 2 °C. As shown in Fig. 8, SST biases are common biases in the western Pacific region in ocean–atmosphere coupled models [25–28].

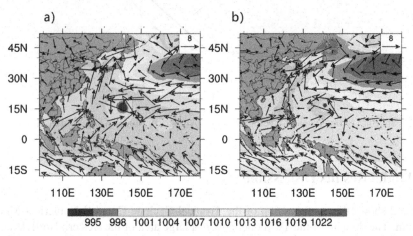

Fig. 9. Spatial distributions of 10 m wind (vectors, units: m/s) and sea-level pressure (shading, units: hPa) at 12 am, July 06, 2018 a) from the coupled model output (tropical-cyclone coupled numerical model) and b) from the ERA5 dataset.

The spatial distributions of the observed (ERA5) and simulated (tropical cyclone coupled numerical model) low-level winds at 10 m and the associated sea-level pressure at 12 am on July 6, 2018, are shown in Fig. 9. The observations are characterized by the subtropical high north of 20°N and a low-pressure center near 17°N, 140°E with a cyclonic circulation anomaly. These features are captured well by the Tropical Cyclones Coupled Numerical Model but with lower pressure in the western North Pacific, which may be caused by the warm SST biases. The spatial pattern correlation coefficient of sea-level pressure between the coupled model output and ERA5 dataset is 0.98, which is statistically significant at the 5% level.

6   Conclusion

The current trends of tropical cyclone numerical model development are higher resolution, higher accuracy and multicomponent coupling. We realized the operational Tropical Cyclones real-time coupling model of the ocean, atmosphere, and wave model using the ROMS – WRF – SWAN model. Through optimization of the SWAN model parallel algorithm, ROMS grid partition and allocation of cores of the coupled model, we obtained speed-ups up to 108 times than serial for each component. The acceleration ratio was significantly improved, and a well-balanced performance of the coupled system was obtained.

Acknowledgments. We thank Dr. Yunfei Zhang and Dr. Xiang Li from the National Marine Environmental Forecasting Center for setting up the coupling modeling system and their valuable suggestions on this work. This research is supported by the National Natural Science Foundation of China (41976200) and the project of Guangdong Ocean University (060302032106). We acknowledge the comments of three anonymous reviewers.

References

1. Yan, D., Zhang, T.: Research progress on tropical cyclone parametric wind field models and their application. Reg. Stud. Mar. Sci. **51**, 102207 (2022)
2. Soden, B.J., Velden, C.S., Tuleya, R.E.: The impact of satellite winds on experimental GFDL hurricane model forecasts. Mon. Wea. Rev. **129**, 835–852 (2001)
3. Aberson, S.D., Franklin, J.L.: Impact on hurricane track and intensity forecasts of GPS dropwindsonde observations from the first-season flights of the NOAA gulfstream-IV jet aircraft. Bull. Amer. Meteor. Soc. **80**, 421–427 (1999)
4. Burpee, R.W., Aberson, S.D., Franklin, J.L., Lord, S.J., Tuleya, R.E.: The impact of omega dropwindsondes on operational hurricane track forecast models. Bull. Amer. Meteor. Soc. **77**, 925–933 (1996)
5. Tuleya, R.E., Lord, S.J.: The impact of dropwindsonde data on GFDL hurricane model forecasts using global analyses. Weather Forecast. **12**, 307 (1997)
6. Emanuel, K.A.: Thermodynamic control of hurricane intensity. Nature **401**, 665–669 (1999)
7. McAdie, C.J., Lawrence, M.B.: Improvements in tropical cyclone track forecasting in the atlantic basin, 1970–98. Bull. Amer. Meteor. Soc. **81**, 989–997 (2000)
8. Shay, L.K., Black, P.G., Mariano, A.J., Hawkins, J.D., Elsberry, R.L.: Upper ocean response to Hurricane Gilbert. J. Geophys. Res. **97**, 20227 (1992)
9. Sanford, T.B., Black, P.G., Haustein, J.R., Feeney, J.W., Forristall, G.Z., Price, J.F.: Ocean response to a hurricane. Part I: observations. J. Phys. Oceanogr. **17**, 2065–2083 (1987)
10. Lin, I.I.: Satellite observations of modulation of surface winds by typhoon-induced upper ocean cooling. Geophys. Res. Lett. **30**, 1131 (2003)
11. Powell, M.D., Vickery, P.J., Reinhold, T.A.: Reduced drag coefficient for high wind speeds in tropical cyclones. Nature **422**, 279–283 (2003)
12. Andreas, E.L., Emanuel, K.A.: Effects of sea spray on tropical cyclone Intensity. J. Atmos. Sci. **58**, 3741–3751 (2001)
13. Bao, J., Wilczak, J.M., Choi, J., Kantha, L.H.: Numerical simulations of air–sea interaction under high wind conditions using a coupled model: a study of hurricane development. Mon. Wea. Rev. **128**, 2190–2210 (2000)
14. Wang, Y., Wu, C.C.: Current understanding of tropical cyclone structure and intensity changes - a review. Meteorol. Atmos. Phys. **87**, 257–278 (2004)
15. Liu, N., Ling, T., Wang, H., Zhang, Y., Gao, Z., Wang, Y.: Numerical simulation of Typhoon Muifa (2011) using a coupled ocean-atmosphere-wave-sediment transport (COAWST) modeling system. J. Ocean Univ. China **14**, 199–209 (2015)
16. Warner, J.C., Armstrong, B., He, R., Zambon, J.B.: Development of a coupled ocean–atmosphere–wave–sediment transport (COAWST) modeling system. Ocean Model. **35**, 230–244 (2010)

17. Shchepetkin, A.F., McWilliams, J.C.: The regional oceanic modeling system (ROMS): a split-explicit, free-surface, topography-following-coordinate oceanic model. Ocean Model. **9**, 347–404 (2005)

18. Jacob, R., Larson, J., Ong, E.: M × N communication and parallel interpolation in community climate system model version 3 using the model coupling toolkit. The Int. J. High Perform. Comput. Appl. **19**(3), 293–307 (2005)

19. Larson, J., Jacob, R., Ong, E.: The model coupling toolkit: a new fortran90 toolkit for building multiphysics parallel coupled models. The Int. J. High Perform. Comput. Appl. **19**(3), 277–292 (2005)

20. *The SWAN Team*: SWAN Scientific and Technical Documentation. SWAN Cycle III version 41.31AB. Delft University of Technology, Delft (2020)

21. Meurant, G.: Domain decomposition methods for partial differential equations on parallel computers. The Int. J. Supercomputing Appl. **2**(4), 5–12 (1988)

22. van der Vorst, H.A.: High performance preconditioning. SIAM J. Sci. Stat. Comput. **10**(6), 1174–1185 (1989)

23. Bastian, P., Horton, G.: Parallelization of robust multigrid methods: ILU factorization and frequency decomposition method. SIAM J. Sci. Stat. Comput. **12**(6), 1457–1470 (1991)

24. Zijlema, M., van der Westhuysen, A.J.: On convergence behaviour and numerical accuracy in stationary SWAN simulations of nearshore wind wave spectra. Coast. Eng. **52**(3), 237–256 (2005)

25. Fang, Y., Zhang, Y., Tang, J., Ren, X.: A regional air-sea coupled model and its application over East Asia in the summer of 2000. Adv. Atmos. Sci. **27**(3), 583–593 (2010)

26. Li, T., Zhou, G.Q.: Preliminary results of a regional air-sea coupled model over East Asia. Chin. Sci. Bull. **55**(21), 2295–2305 (2010)

27. Ren, X., Qian, Y.: A coupled regional air-sea model, its performance and climate drift in simulation of the East Asian summer monsoon in 1998. Int. J. Climatol. **25**(5), 679–692 (2005)

28. Zou, L., Zhou, T.: Can a regional ocean–atmosphere coupled model improve the simulation of the interannual variability of the western north pacific summer monsoon? J. Clim. **26**(7), 2353–2367 (2013)

Factorization Machine Based on Bitwise Feature Importance for CTR Prediction

Hao Li, Caimao Li[✉], Yuquan Hou, Hao Lin, and Qiuhong Chen

School of Computer Science and Technology, Hainan University, Haikou 570228, China
{20085400210028,20085400210018,20085400210039,
20085400210006}@hainanu.edu.cn, lcaim@126.com

Abstract. Click-through-rate (CTR) prediction is a crucial task in recommendation systems. The accuracy of CTR prediction is strongly influenced by the precise extraction of essential data and the modeling strategy chosen. The data of the CTR task are often very sparse, and Factorization Machines (FMs) are a class of general predictors working effectively with it. However, the performance of FMs can be limited by the fixed feature representation and the same weight of different features. In this work, we propose an improved Bitwise Feature Importance Factorization Machine (BFIFM) to improve the accuracy. The necessity of learning the degree of effect of the same feature under various situations is learned through the low-order intersection method, and the deep neural network (DNN) in our model is used in parallel to study high-order intersections. According to the final results obtained, the BFIFM model significantly outperforms other state-of-the-art models.

Keywords: Factorization machines · Deep learning · Recommendation · Sparse data

1 Introduction

Accurate feature expression is very important for click-through-rate (CTR) prediction. At present, there are many models that learn the expression of features, such as NFM [9], AFM, [11] and DeepFM [12]. The basis of this series of models is to convert features into k-dimensional embedding vectors. However, the above models will only give the same weight to the same feature in different situations. Someone introduced a model named IFM to handle this problem, which may account for various degrees of effect of the same feature in different conditions, but the core FEN of the IFM model is to rescale the embedding vector in vectorwise [16]. To extract more information from feature interactions, the Bitwise Feature Importance Network (BFIN) part in this paper can rescale each bit of the embedding vector, which improves the effect of CTR prediction. To filter different important features, we also add a bitwise Concat and Excitation Network (CENET) after embedding the data, which was inspired by FiBiNET [17]. This structure will rescale the input vector and reweight each bit of the vector, which also improves the CTR prediction precision. Inspired by DeepFM, in our model, there is a parallel deep

© The Author(s), under exclusive license to Springer Nature Singapore Pte Ltd. 2022
Y. Wang et al. (Eds.): ICPCSEE 2022, CCIS 1628, pp. 29–40, 2022.
https://doi.org/10.1007/978-981-19-5194-7_3

neural network (DNN), inputting the reweighted vector output by CENET and using DNN to extract high-order feature intersections. Combine the output of DNN and BFIN to obtain the final recommendation result.

2 Related Work

FMs model second-order feature interactions, and the results are rather promising. To further extract the interaction information, Field-aware FM considers the concept of field. Each feature has multiple embedding vectors when combined with the features of other fields.

With the rapid development of deep learning and its successful applications in different fields in recent years, feature interaction based on deep learning has been proposed [19, 20, 27]. The FNN concatenates the FM with the MLP and uses the FM to pretrain the embedding layer [4, 6]. Since the initialization of the FNN embedding weights is FM pretrained, it is not an end-to-end training process. NFM utilizes bi-interacton pooling to obtain the elementwise product of pairwise feature interactions and then stacks multiple nonlinear layers over the bilinear interaction layer to capture the high-order feature interactions [9]. Wide&Deep integrates wide and deep parts; the wide part simulates low-order linear feature interactions, and the deep part simulates high-order feature interactions [7]. DeepFM replaces a wide part in Wide&Deep with FM, thus eliminating the need for feature engineering [12]. The DCN model uses a cross network and applies feature crossing at each layer, which efficiently learns the bounded degree combination features and does not require manual feature engineering [13]. xDeepFM models the low-order and high-order feature interactions explicitly by proposing a novel CIN component [14].

In addition, attention models were applied to machine translation [10], and they can be used in CTR prediction [8, 26]. AFM uses it to extract the key information in the second-order feature interaction [11]. IFM utilizes the FEN for different situations [16]. AutoInt proposes a new method based on self-attention neural networks [15]. The FiBiNET model introduces the SENET network between the embedding layer and the DNN, calculates the importance of each feature, and proposes a bilinear-interaction layer to dynamically learn the relationship between different feature combinations during the training process [17].

However, IFM is vectorwise, and the SENET module in FiBiNET is also vectorwise. We present the BFIFM model, in which CENET can extract the significance of the embedding vector. It is able to extract the information of each bit in the feature vector, and BFIN can extract the weight of the second-order feature interaction in different situations in bitwise, parallel access DNN extracts higher-order feature interactions.

3 Our Approach

FMs achieve better performance than linear regression by second-order interaction [2]. The valued feature is $\mathbf{x} \in \mathbb{R}^n$, where n is the number of features and the majority of elements in $\mathbf{x}$ are 0. The goal of the FM model is to approximate the intersection.

The attention mechanism could also be used in recommendation systems. AFM introduced the attention mechanism in FM. Therefore, the model can effectively learn the weights of second-order feature interactions [11]. We apply CENET to extract the first-order feature weights of the input data after feature embedding in the process of learning the weights of the first-order features. Additionally, depending on the scenario, the impact of the same second-order feature intersection should differ. As a result, extracting the influence components of the second-order feature intersection in various conditions can make progress. The BFIN part in BFIFM can extract the weight of the second-order feature intersection in bitwise. We also add a DNN part to extract high-order interactions. We will go through the suggested BFIFM model in depth in this part. Figure 1 shows the architecture of the BFIFM network.

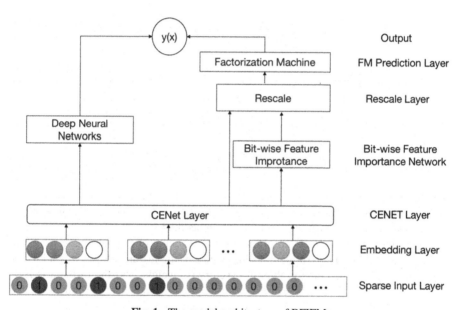

Fig. 1. The model architecture of BFIFM

3.1 Embedding Layer

The primary coverage of the BFIFM model is similar to that of FM, which calculates the feature intersection. The target of the FM model is to simulate the feature intersection. The resulting formulas of BFIFM prediction are as follows:

$$\hat{y} = \text{sigmoid}(y_{BFIN} + y_{DNN}) \tag{1}$$

$$y_{BFIN}(\mathbf{x}) = w_0 + \sum_{i=1}^{n} w_{\mathbf{x},i} x_i + \sum_{i=1}^{n} \sum_{j=i+1}^{n} \langle \mathbf{v}_{\mathbf{x},i}, \mathbf{v}_{\mathbf{x},j} \rangle x_i x_j \tag{2}$$

$$y_{DNN} = \sigma(W^{|H|+1} \cdot a^H + b^{|H|+1}) \tag{3}$$

where $\hat{y} \in (0,1)$ denotes the result of CTR prediction, y_{BFIN} and y_{DNN} are the output of BFIN and DNN respectively.

Embedding Layer. Since the input data to the model are very sparse, it is necessary to embed the input feature data to make it relatively dense [25]. The i-th embedding vector, which is k-dimensional, is usually denoted as $v_i \in \mathbb{R}^k$.

Concat Estimating Network. A CENET layer is added to study first-order weights. The CENET layer includes three steps: concat, excitation, and rescale. Figure 2 shows the CENET network architecture. The three formula expressions in Fig. 2 denote the three steps of CENET, and the different colors indicate different weights after excitation.

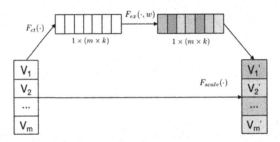

Fig. 2. The network architecture of CENET

Concat multiple input embedding vectors into a new vector, the formula is $C = F_{ct}(V_i)$, where the dimension of C is $1 \times (m \times k)$, $m \times k$ denoted as f. The new vector from the output records each bit of all feature embedding vectors.

The excitation part is made up of two FC with the aim of evaluating the weight of each input vector bit to determine the significance of the features. Since it is a bitwise weight, it has more information than the vectorwise weight, and the formula is $E = F_{ex}(C) = \sigma_2(W_2\sigma_1(W_1C))$, where $E \in R^{1 \times f}$, σ_1 and σ_2 are activation functions, and the parameters to be learned are $W_1 \in R^{f \times \frac{f}{r}}$ and $W_2 \in R^{\frac{f}{r} \times f}$. r is the reduction ratio parameter. The output of this part is the weight of each bit of all feature vectors.

To apply the bitwise weight to each bit of the embedding vector to complete the rescale of the original vector, the formula is $S = F_{scale}(C,E) = [c_1 \cdot e_1, \ldots, c_f \cdot e_f]$. Divide the vector according to the k-dimension on average. The vectors output at this time are the weighted vectors $[V_1', V_2' \ldots V_m']$.

Bitwise Feature Importance Network. Our model, which is inspired by IFM, extracts the various degrees of effect of the same feature in various contexts in a bitwise manner.

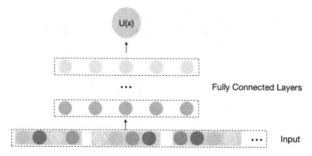

Fig. 3. The network architecture of BFIN

To accomplish the above goal, we first feed the CENET feature vectors V_x into the Bitwise Feature Importance Network (BFIN). After BFIN training, we can obtain U_x. The BFIN network structure is shown in Fig. 3 above. The network is made up of multiple FC that can extract feature information from input data. W_l and b_l are the weight matrix and the bias vector, respectively. The following are the formulas:

$$\mathbf{a}_1 = \sigma_1(\mathbf{W}_1\mathbf{V}_\mathbf{x} + \mathbf{b}_1) \tag{4}$$

$$\mathbf{U}_\mathbf{x} = \mathbf{a}_L = \sigma_L(\mathbf{W}_L\mathbf{a}_{L-1} + \mathbf{b}_L) \tag{5}$$

where σ_l and $\mathbf{a}_l$ denote the activation function and the output of the l-th layer, respectively. U_x is the output of the last layer, which is used to represent the influence degree of every bit of each input vector. The formulas are as follows:

$$\mathbf{m}'_\mathbf{x} = \mathbf{Q}\mathbf{U}_\mathbf{x}\mathbf{P}, \mathbf{P} \in \mathbb{R}^{t \times h}, \mathbf{Q} \in \mathbb{R}^{k \times 1} \tag{6}$$

$$m_{\mathbf{x},(i,j)} = \sum_{i=0}^{k} h \times \frac{\exp(m'_{\mathbf{x},(i,d)})}{\sum_{j=1}^{h} \exp(m'_{\mathbf{x},(i,j)})}, x_{i,j} \neq 0 \tag{7}$$

where h indicates the number of 1 in the sample data and the number of neurons in the final layer of the hidden layers is represented by t. $m'_x \in \mathbb{R}^{k \times h}$ and $d \in [1,h]$, the softmax function is used to normalize m'_x to m_x.

Rescale Layer. The output obtained from BFIN is used to rescale the degree of influence of the feature. The input data are $\mathbf{v}_\mathbf{x}$, and each bit of the perceptual weight of the sample is $m_{\mathbf{x},(i,j)}$. The perceptual weight is combined with the first-order weight $w_{\mathbf{x},i}$ and the embedding vector $\mathbf{v}_i$. The formulas are:

$$w_{\mathbf{x},i} = \sum_{j=1}^{k} m_{\mathbf{x},(i,j)}w_i \tag{8}$$

$$\mathbf{v}_{\mathbf{x},i} = m_{\mathbf{x},i}\mathbf{v}_i \tag{9}$$

FM Prediction Layer. In this part, we import the first-order weight $w_{\mathbf{x},i}$ and embedding vector $\mathbf{v}_{\mathbf{x},i}$ to the FM. We can reformulate Eq. (3) to reduce the runtime, and the prediction result is as follows:

$$\sum_{i=1}^{n} \sum_{j=i+1}^{n} (\mathbf{v}_{\mathbf{x},i}, \mathbf{v}_{\mathbf{x},j}) x_i x_j = \frac{1}{2} \sum_{f=1}^{k} \left[\left(\sum_{j=1}^{n} v_{\mathbf{x},j,f} x_j \right)^2 - \sum_{j=1}^{n} v_{\mathbf{x},j,f}^2 x_j^2 \right]$$

(10)

Deep Component. The DNN part is used to learn the high-order feature intersection, and the dense vector $\mathbf{V}_x$ output by CENET is imported to the DNN. The calculation formula is $a^{(l+1)} = \sigma(\mathbf{W}^{(l)} a^{(l)} + b^{(l)})$, where l and σ denote the layer and activation functions, respectively. After the DNN, the predicted value will be calculated through sigmoid, and the formula is Eq. (3).

In the BFIFM model, the BFIN learns the second-order intersection, and the DNN learns the high-order intersection. The two are combined and share the same embedding vector output by the CENET part. Experiments have demonstrated that our strategy produces great outcomes. The parameters of our model are $\{w_0, w_i, \mathbf{v}_i, \mathbf{W}_l, \mathbf{b}_l, \mathbf{Q}, \mathbf{P}\}$; compared with FM, the extra parameters are $\{\mathbf{W}_l, \mathbf{b}_l, \mathbf{Q}, \mathbf{P}\}$. The parameter $\mathbf{Q}$ is mainly used to rescale the impact of the input vectors bitwise. The time complexity is $O(kn)$.

3.2 Learning

The BFIFM is an enhancement of FMs, so it can be used to implement regression, classification, and ranking tasks. We utilize the Logloss as the loss function:

$$\mathcal{L} = -\frac{1}{N} \sum_{i=1}^{N} (y_i \log(\sigma(\hat{y}_i)) + (1 - y_i) \log(1 - \sigma(\hat{y}_i)))$$

(11)

where $y_i \in \{0,1\}$ denotes the ground truth of the i-th instance and $\sigma(\hat{y}_i) \in (0,1)$ is the prediction CTR. N is the number of training instances.

4 Experiments

4.1 Experimental Settings

Datasets. We test the BFIFM model on two datasets, Criteo and Avazu, both of which are public datasets, and many papers use these two datasets for training and testing. Criteo has 50 million click records, which includes one month ad click logs, and each click record has 39 features[28]. Avazu has more than 40 million click records, and each click record has 23 features[29]. We need training, validation, and test sets. Therefore, we partition the datasets at a ratio of 7:1:2.

Evaluation Metrics. In this paper, the evaluation metrics are AUC and Logloss, as in most CTR prediction models. We aim at CTR prediction, so it is not necessary to evaluate our model in terms of recall, precision and F1.

The formula expression of AUC is as follows:

$$AUC = \frac{\sum (p_i, n_j)_{p_i > n_j}}{P * N} \tag{12}$$

where P is the number of positive samples, N is the number of negative samples, p_i is the predicted score of positive samples, and n_i is the predicted score of negative samples. The magnitude of the AUC is proportional to the performance quality of the model. The formula expression of Logloss is Eq. (11).

Baselines. The BFIFM model was compared against the 9 competitive models below, several of which are state-of-the-art models.

- LR: Logistic Regression models with only linear interactions.
- FM: Modeling Second-Order Interactions with factorization Techniques.
- AFM: On the basis of FM, AFM adds the attention mechanism to the second-order interaction.
- DeepFM: integrates the FM part and the deep network and sets three layers of MLP with layer sizes of 256, 256, and 256.
- xDeepFM: Considering explicit and implicit higher-order interactions, implemented by DNN and CIN.
- AutoInt: Automatic feature learning with multi-head self-attention, and the number of multi-head is set to 2, as recommended by the paper.
- IFM: The hidden dimension in the FEN of the model is set to 256, 256, and 256, as recommended by the paper.
- DIFM: The deep network layer sizes and the number of multi-head self-attention networks in the Dual-FEN of the model are set to 256, 256, 256 and 16, respectively, as recommended by the paper.
- FiBiNET: The number of FC layers in SENet is 2, and the layer sizes of the DNN part are set as recommended in this paper.

Parameter Settings We test all baselines by TensorFlow. For a fair comparison, all models utilize the Adam optimizer to update the Logloss (learning rate 0.001). For the Criteo and Avazu datasets, the embedding sizes are 128 and 256, respectively. The batch size of both datasets is 512. By default, each layer of the neural network in each model has 256 neurons.

4.2 Hyperparameter Study

We research the influence of hyperparameters in this part. The hyperparameters contain the depth of BFIN, dropout ratio and activation function.

Depth of Network. The batch size of both datasets is 512. By default, each layer of the neural network in each model has 256 neurons. Figure 4 above illustrates the effect in BFIN. We can see that as the depth of layers rises, the performance of the BFIFM model keeps increasing. Nevertheless, the model effectiveness gradually deteriorates when it is larger than 2.

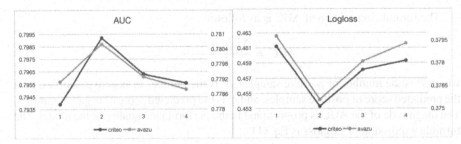

Fig. 4. Impact of hidden layers on AUC and Logloss performance

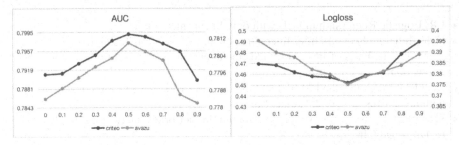

Fig. 5. Impact of dropout on AUC and Logloss performance

Dropout Ratio. Dropout is a very effective way for neural networks to avoid overfitting. Figure 5 above shows the effect of dropout on BFIFM. Setting a correct dropout ratio can promote achievement on the two datasets. For the Criteo and Avazu datasets, the optimal dropout ratio is 0.5, which also proves the effectiveness of dropout on BFIFM.

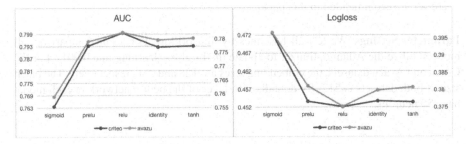

Fig. 6. Impact of the activation function on AUC and Logloss performance

Activation Function. In DNN, we mostly use relu. We use different activation functions in the DNN for comparison. According to Fig. 6, the effect of relu is indeed the best.

4.3 Ablation Study

The BFIFM model uses DNN and Bitwise CENET and BFIN. To explore which part is the most important part of the BFIFM model and whether it is necessary to use these three parts, we conduct ablation experiments on the BFIFM model. In the performance of each part in Table 1, we observed the following:

Table 1. Impact of different components in our BFIFM

Method	Criteo		Avazu	
	AUC	Logloss	AUC	Logloss
FM	0.7832	0.4643	0.7751	0.3883
+ CENET	0.7914	0.4593	0.7781	0.3821
+ CENET &BFIN	0.7983	0.4524	0.7815	0.3761
+ CENET &BFIN &DNN	**0.7995**	**0.4521**	**0.7823**	**0.3752**

- DNN, CENET, and BFIN all have a positive impact on the model. On both datasets, BFIFM performs better than FM after ablation.
- By comparing the results in the Table 1, it can be found that it is necessary to combine DNN, CENET, and BFIN at the same time.
- On the basis of CENET, adding BFIN can improve the model the most.

4.4 Performance Comparison

In the last part, we compare the BFIFM model with the baselines. The comparative performance on the datasets is shown in Table 2, and we can obtain the following conclusions:

- Learning how to interact with features might help your model perform better. LR has the poorest performance of all models because of the lack of feature interaction. Furthermore, learning both low-order and high-order interactions simultaneously can significantly make progress.
- The significance of learning the influence of the features from the input features is also very useful. Table 2 shows that the FM model without learning the importance of the features performs poorly.
- Our proposed BFIFM model performs optimally on both datasets, which also demonstrates the effectiveness of DNN and Bitwise CENET and Bitwise BFIN in SOTA models.

Table 2. Impact of different CTR models

Method	Criteo		Avazu	
	AUC	Logloss	AUC	Logloss
LR	0.7782	0.4795	0.7626	0.3925
FM	0.7832	0.4643	0.7752	0.3883
IFM	0.7980	0.4582	0.7785	0.3771
DIFM	0.7992	0.4573	0.7793	0.3762
AFM	0.7862	0.4584	0.7782	0.3798
DeepFM	0.7983	0.4535	0.7808	0.3766
xDeepFM	0.7986	0.4531	0.7811	0.3761
FiBiNET	0.7970	0.4534	0.7817	0.3758
AutoInt	0.7989	0.4529	0.7810	0.3760
BFIFM	**0.7995**	**0.4521**	**0.7823**	**0.3752**

5   Conclusion

In this paper, we propose a Bitwise Feature Importance Factorization Machine for CTR prediction. Its goal is to use CENET and BFIN to learn input features and representations of different features in second-order interactions, while using DNN to learn high-order interactions. The greatest advantage in BFIFM is that it can effectively learn the relationship of features, which can be used to rescale the degree of influence of features, not only when the feature vector is input but also when the feature is a second-order interaction. Scientific tests on Criteo and Avazu have shown that the BFIFM model is effective. In the two evaluation indicators of Logloss and AUC, BFIFM is superior to the classic LR and other SOTA models, such as DeepFM, IFM, DIFM, AutoInt, etc.

In future work, we will continue to optimize our model from the feature intersection of vectorwise and bitwise and use it not only for CTR prediction but also for recall work.

Acknowledgment. This work is supported by Hainan Province Science and Technology Special Fund, which is Research and Application of Intelligent Recommendation Technology Based on Knowledge Graph and User Portrait (No.ZDYF2020039). Thanks to Professor CaiMao Li, the correspondent of this paper.

References

1. Gortmaker, S.L., Hosmer, D.W., Lemeshow, S.: Applied logistic regression. Contemp. Sociol. **23**(1), 159 (1994)
2. Rendle, S.: Factorization machines. In: ICDM 2010, The 10th IEEE International Conference on Data Mining. IEEE (2010)
3. Vovk, V., et al.: Erratum to: The Fundamental Nature of the Logloss Function (2015)

4. Singh, G., Sachan, M.: Multilayer perceptron (MLP) neural network technique for offline handwritten Gurmukhi character recognition. In: IEEE International Conference on Computational Intelligence Computing Research. IEEE (2015)

5. Juan, Y., Zhuang, Y., Chin, W.-S., Lin, C.-J.: Field-aware factorization machines for CTR prediction. In: Proceedings of the 10th ACM Conference on Recommender Systems, pp. 43–50. ACM (2016)

6. Zhang, W., Du, T., Wang, J.: Deep learning over multi-field categorical data. In: Ferro, N., et al. (eds.) ECIR 2016. LNCS, vol. 9626, pp. 45–57. Springer, Cham (2016). https://doi.org/10.1007/978-3-319-30671-1_4

7. Cheng, H.T., et al.: Wide & Deep Learning for Recommender Systems. In: Proceedings of the 1st Workshop on Deep Learning for Recommender Systems (DLRS 2016), pp. 7–10. Association for Computing Machinery, New York, NY, USA

8. Martins, A., Astudillo, R.F.: From softmax to sparsemax: a sparse model of attention and multilabel classification. In: Proceedings of the 33rd International Conference on International Conference on Machine Learning, vol. 48, pp. 1614–1623 (ICML'16) (2016) JMLR.org

9. He, X., Chua, T.S.: Neural factorization machines for sparse predictive analytics. ACM SIGIR Forum 51(cd), 355–364 (2017)

10. Rivera-Trigueros, I.: Machine translation systems and quality assessment: a systematic review. Lang. Resour. Eval. 56, 593–619 (2021)

11. Xiao, J., Ye, H., He, X., Zhang, H., Wu, F., Chua, T. S.: Attentional factorization machines: learning the weight of feature interactions via attention networks: In: Proceedings of the Twenty-Sixth International Joint Conference on Artificial Intelligence (IJCAI-17), pp. 3119–3125 (2017)

12. Guo, H., Tang, R., Ye, Y., Li, Z., He, X.: DeepFM: A factorization-machine based neural network for CTR prediction. In: Twenty-Sixth International Joint Conference on Artificial Intelligence (2017)

13. Wang, R., Fu, B., Fu, G., Wang, M.: Deepcross network for ad click predictions. In: ACM (2017)

14. Lian, J., Zhou, X., Zhang, F., Chen, Z., Xie, X., Sun, G.: xDeepfm: combining explicit and implicit feature interactions for recommender systems. In: Proceedings of the 24th ACM SIGKDD International Conference on Knowledge Discovery & Data Mining, KDD 2018, London, UK, 19–23 Aug 2018

15. Song, W., Shi, C., Xiao, Z., Duan, Z., Jian, T.: Autoint: automatic feature interaction learning via self-attentive neural networks. In: Proceedings of the 28th ACM International Conference on Information and Knowledge Management (CIKM'19), pp. 1161–1170. Association for Computing Machinery, New York, NY, USA (2018)

16. Yu, Y., Wang, Z., Yuan, B.: An input-aware factorization machine for sparse prediction. In: Twenty-Eighth International Joint Conference on Artificial Intelligence IJCAI-19 (2019)

17. Huang, T., Zhiqi, Z., Junlin, Z.: FiBiNET: combining feature importance and bilinear feature interaction for click-through rate prediction. In: Proceedings of the 13th ACM Conference on Recommender Systems, pp. 169–177 (2019)

18. Lu, W., Yu, Y., Chang, Y., Wang, Z., Yuan, B.: A dual input-aware factorization machine for CTR prediction. In: Twenty-Ninth International Joint Conference on Artificial Intelligence and Seventeenth Pacific Rim International Conference on Artificial Intelligence IJCAI-PRICAI-20 (2020)

19. Ruiqin, W., Zongda, W., Yunliang, J., Jungang, L.: An integrated recommendation model based on two-stage deep learning. J. Comput. Res. Dev. 56, 1661 (2019)

20. Alhijawi, B., Kilani, Y.: The recommender system: a survey. Int. J. Adv. Intell. Paradigms 15(3), 1 (2020)

21. Baldi, P., Sadowski, P.: Understanding dropout. Adv. Neural Inform. Process. Syst. 26(1) (2013)

22. Li, Y., Yuan, Y.: Convergence analysis of two-layer neural networks with ReLU activation (2017)
23. Kingma, D., Ba, J.: Adam: a method for stochastic optimization. Comput. Sci. (2014)
24. Mikolov, T., Chen, K., Corrado, G., Dean, J.: Efficient estimation of word representations in vector space. In: 1st International Conference on Learning Representations, ICLR 2013, Scottsdale, Arizona, USA, Workshop Track Proceedings 2–4 May 2013
25. Barkan, O., Koenigstein, N.: Item2vec: neural item embedding for collaborative filtering. IEEE (2016)
26. Vaswani, A., et al.: Attention is all you need. In: 31st Conference on Neural Information Processing Systems (NIPS 2017), Long Beach, CA, USA (2017)
27. Batmaz, Z., Yurekli, A., Bilge, A., Kaleli, C.: A review on deep learning for recommender systems: challenges and remedies. Artif. Intell. Rev. **52**(1), 1–37 (2018)
28. Kaggle Science Community: Display advertising challenge: Predict click-through rates on display ads. https://www.kaggle.com/c/criteo-display-ad-challeng (2014)
29. Kaggle Science Community: Click-through rate prediction: predict whether a mobile ad will be clicked. https://www.kaggle.com/c/avazu-ctr-prediction (2015)

Focusing on the Importance of Features for CTR Prediction

Yuquan Hou, Caimao Li[✉], Hao Li, Hao Lin, and Qiuhong Chen

School of Computer Science and Technology, Hainan University, Haikou 570228, China
lcaim@126.com

Abstract. Traditional CTR recommendation models have concentrated on how to learn low-order and high-order characteristics. The majority of them make many efforts at combining low-order and high-order functions. However, they ignore the importance of the attention mechanism for learning input features. The ECABiNet model is proposed in this article to enhance the performance of CTR. On the one hand, the ECABiNet model can learn the importance of features dynamically via the LayerNorm and ECANET layers. On the other hand, through the use of a bi-interaction layer and a DNN layer, it is capable of effectively learning the feature interactions. According to the experimental results on two public datasets, the ECABiNet model is more effective than the previous CTR model.

Keywords: CTR Model · ECANET · LayerNorm · ECABiNet

1 Introduction

Currently, the online advertising business has gradually become popular, and improving the CTR of advertisements has become a critical issue. Current CTR models mainly include shallow models and deep models. The article [1] proposes a factorization machine (FM) model that is capable of not only solving problems with enormous sparsity in linear time but also working with any real-valued eigenvectors. This is a typical shallow model, and it is incapable of learning the relationship between higher-order features. Paper [2] proposed the FFM model on the basis of the FM model, which focuses on discrete classification features and enhances the factorization machine algorithm. The classic deep models include the wide and deep model, the DeepFM model, and the FiBiNET model, among others. To achieve memory and generalization capabilities, the wide and deep model [3] combines the linear and deep models. The wide part generates memory for feature interaction through feature intersection, whereas the deep part uses low-dimensional dense features as input to generalize cross-features that do not appear in the training samples. The wide part, on the other hand, necessitates artificial feature engineering, resulting in increased workload.

As a result, the DeepFM model [4] is suggested. Its wide section makes use of the FM model to automatically learn feature intersection, while the deep section continues to make use of the deep neural network DNN, which not only reduces the workload associated with manual feature engineering but also improves recommendation efficiency.

© The Author(s), under exclusive license to Springer Nature Singapore Pte Ltd. 2022
Y. Wang et al. (Eds.): ICPCSEE 2022, CCIS 1628, pp. 41–52, 2022.
https://doi.org/10.1007/978-981-19-5194-7_4

While the preceding models emphasize feature intersection, paper [5] proposes the FiBiNET model, which places a premium on feature importance learning. It employs a SENET layer to determine the relative importance of various features and then generates recommendation results by feeding a bilinear layer into a deep neural network; however, the SENET layer requires a dimensionality reduction operation and is quite complex. By combining the aforementioned issues, this article proposes the ECABiNet model.

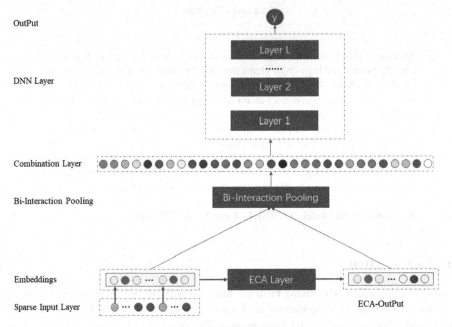

Fig. 1. ECABiNet Model.

2   ECABiNet Model

At first, attention models were applied to machine translation [6] and then to neural network models. The paper [7] proposes the AFM model, which uses an attention mechanism, allowing the model to learn the weights of second-order feature items effectively.

The majority of these established models of attention place a premium on the component of feature intersection. As illustrated in Fig. 1, the ECABiNet model emphasizes the feature embedding portion. The sparse vector input is converted to a feature embedding vector and then normalized before being passed to the ECANET layer via the LayerNorm layer. After pooling the output of the ECANET layer and the feature embedding vector, it is passed to the DNN layer for recommendation.

2.1 Sparse Input and Embedding Layer

The ECABiNet model uses an embedding layer to map the features of the original input into a low-dimensional space and converts discrete variables into continuous vectors. By using an embedding layer, the model not only has the ability to reduce the spatial dimension of discrete variables but also meaningfully represents the features of these original inputs. The output of the embedding layer is $E = [e_1, \ldots, e_n]$.

2.2 Layer Norm

Normalization [8] plays a crucial role in neural networks and can normalize the distribution of data. If the data were not normalized, their distribution would be different, and the distribution of data in each network layer would be constantly changing, which is likely to result in the neural network failing to converge. After training, the eigenvalues of each dimension of the sample are found to be unequal after the traditional machine learning algorithm SVM [9] performs uneven scaling (for example, the eigenvalues of each dimension are multiplied by different coefficients). At the scaling level, we assert that such an algorithm is not immutable. Unless the distribution range of each dimension feature is relatively close, it must be normalized for this type of algorithm.

As illustrated in the following equation, the ECABiNet model performs layer normalization on the feature embedding vector $[e_1, \ldots, e_n]$:

$$y = \frac{x - E[x]}{\sqrt{\text{Var}[x] + \epsilon}} * \gamma + \beta \tag{1}$$

where E and Var denote the mean and variance of all samples in this batch on the k-th feature, respectively. β and γ are used to control the ability of the network to express direct mappings that restore the features previously learned by the LN. The final result is the new feature embedding vector:

$$LN(E) = \text{concat}\big(LN(e_1), \ldots, LN(e_f)\big) \tag{2}$$

where LN denotes the LayerNorm layer and E, e_f denotes the embedding vector.

2.3 ECANET Layer

Since the introduction of SENet [10], the channel attention mechanism has demonstrated considerable promise, and this approach has been shown to be a viable method for improving the overall efficiency of deep convolutional neural networks. SENet is dedicated to developing more complex attention modules to improve performance. For instance, the SENet model makes use of two fully connected layers to enhance its non-linear capability and fit capability. However, it introduces additional parameters and increases the complexity of the model. To resolve the conflict between complexity and performance, a more efficient channel attention (ECANET) module is proposed in [11], as shown in Fig. 2, which not only reduces the complexity of the model but also maintains the performance by avoiding the dimensionality reduction operation of SENet and proper cross-channel interaction.

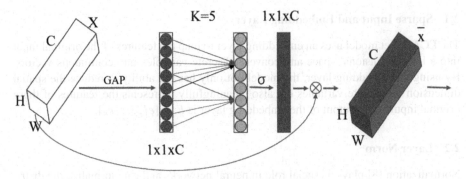

Fig. 2. ECANET model.

The ECANET layer first compresses the embedding vector by the global average pooling method. It compresses the corresponding spatial information on each channel into one value in the corresponding channel; at this time, a pixel represents a channel, and the final dimension becomes $1 \times 1 \times C$. The global average pooling method is shown in the following equation:

$$e'_i = F_{GAP}(e_i) = \frac{1}{k} \sum_{d=1}^{k} e_{id} \tag{3}$$

where k denotes the dimension of the embedding vector and e_i denotes the embedding vector.

After that, the ECANET layer inputs the e_i vector obtained in the previous step into the equation:

$$w_e = \sigma \left(conv1d_k \left(e' \right) \right) \tag{4}$$

The difference of this method is that the size of the convolution kernel used is k, and then to obtain better results, one-dimensional convolution is used to realize the information interaction between channels. $conv1d_k$ denotes one-dimensional convolution, which requires only k parameters, and σ denotes the activation function. The ECANET layer can adaptively select an appropriate value of k, which is proportional to the channel dimension.

The final step is to update the weights. The new embedding vector is obtained by calculating the previously calculated weight factor and the initial embedding vector. The precise formula is depicted by the equation:

$$\begin{aligned} Z &= F_{\text{update}} \left(W_e, E \right) \\ &= \left[w_{e1} \cdot e_1, w_{e2} \cdot e_2, \ldots, w_{ef} \cdot e_f \right] \\ &= \left[v_1, v_2, \ldots, v_f \right] \end{aligned} \tag{5}$$

where E denotes the embedding vector processed by the LayerNorm layer, w_f denotes the corresponding weight embedding vector, and v_f denotes the new embedding vector.

In summary, the ECANET layer has a thorough understanding of the SENet model. The SENet model can learn significant features without adding a large number of parameters and is improved on the basis of the SENet model, which not only reduces complexity but also improves the effect.

2.4 Feature Cross Layer

The ECANET layer learns the new feature embedding vector computed by layerNorm and then computes the weighted feature embedding vector, which is then passed to the bilinear-interaction layer.

Rather than using the conventional inner product (Eq. 6) or Hadamard product (Eq. 7), the bilinear-interaction layer employs a novel bilinear interaction method that combines the two and introduces a new parameter matrix W to learn the feature intersection (Eq. 8).

$$[a_1, a_2, \cdots, a_n] \cdot [b_1, b_2, \cdots, b_n]$$
$$= \sum_{i=1}^{n} a_i b_i \tag{6}$$

$$[a_1, a_2, \cdots, a_n] \odot [b_1, b_2, \cdots, b_n]$$
$$= [a_1 b_1, a_2 b_2, \cdots, a_n b_n] \tag{7}$$

$$p_{i,j} = v_i \cdot W \odot v_j, p_{i,j} \in R^k \tag{8}$$

The intersection vector $p_{i,j}$ can be obtained in three ways:

- All feature groups share a parameter matrix when they are crossed two by two, and the number of extra parameters is $k \times k$, as shown in Eq. 8.
- Each feature group i maintains a parameter matrix W_i with an additional number of parameters $f \times k \times k$, as shown in the following equation:

$$p_{i,j} = v_i \cdot W_i \odot v_j, p_{i,j} \in R^k \tag{9}$$

- Each pair of interaction features $p_{i,j}$ has a parameter matrix W_{ij} with the number of extra parameters $\frac{f \times (f-1)}{2} \times k \times k$, as shown in the following equation:

$$p_{i,j} = v_i \cdot W_{ij} \odot v_j, p_{i,j} \in R^k \tag{10}$$

2.5 DNN Layer

The DNN layer is composed of multiple fully connected layers that are capable of capturing higher-order combinatorial features. Unlike the traditional DeepFM recommendation model, DNN utilizes the output of the combination layer as an embedding vector, which efficiently captures features. As shown in the following equation, the combination layer is primarily responsible for stitching the original embedding vector p with the weight embedding vector q obtained after the ECANET layer:

$$c = F_{concat}(p, q) = [p_1, \cdots, p_n, q_1, \cdots, q_n]$$
$$= [c_1, \cdots, c_{2n}] \tag{11}$$

2.6 Output

The overall formulation of ECABiNet is:

$$\hat{y} = \sigma\left(w_0 + \sum_{i=0}^{k} w_i x_i + y_d\right) \tag{12}$$

where σ represents the sigmoid function, $\hat{y}$ is the predicted result of the ECABiNet, k is the feature size, x is an input and w_i is the $i\text{-}th$ weight of the linear part.

3 Experiment

3.1 Experimental Setup

Datasets This experiment uses two publicly available datasets, and we randomly divided the two datasets into two parts: 90% for training and 10% for testing.

- Criteo. Criteo [12] is one of the most commonly used datasets in the CTR field, and it has 13 continuous characteristics and 26 categorical characteristics. It also has data of more than 40 million user clicks on ads.
- Avazu. Like Criteo, the Avazu [13] dataset is also the most common dataset in the CTR field. It contains more than 40 million user clicks on ads over several days, and it is sorted by time.

Evaluation Criteria AUC and LOGLOSS Are used in this paper to evaluate the model. The AUC value is between 0.5 and 1, with a value closer to 1.0 indicating that the detection method is more authentic. Logloss is the most important probability-based classification metric; the lower the loss score is, the better.

Model Comparison In the experiments, a total of six models are selected for comparison in this paper, namely, FM, AFM, Wide&Deep, DeepFM, FiBiNET and ECABiNet.

Parameter Setting To conduct experiments reasonably, as shown in Table 1, some default hyperparameters are set in this paper.

Table 1. Experimental Hyperparameters.

Parameter name	Parameter value
Dropout	0.5
Optimizer	Adam
Hidden units	(256,128)
Activation	Relu
l2_reg_linear	0.00001
l2_reg_embedding	0.00001

3.2   LayerNorm Effect Comparison

To demonstrate the performance improvement brought by normalizing the embedded feature vector and then performing the feature crossover operation, we will conduct a comparative experiment with the ECABiNet model using LayerNorm and without LayerNorm, where the hyperparameters are as shown in Table 1. The LayerNorm layer is set to the same hidden layer dimension as the ECABiNet model, and the dimension is set to 6 in this experiment. The specific comparison is shown in Fig. 3.

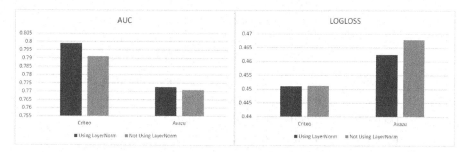

Fig. 3. Comparison of whether to use LayerNorm on the Criteo and Avazu datasets.

From the comparison of AUC and LOGLOSS scores given in Fig. 3, it can be seen that using the LayerNorm layer to perform layer normalization on the feature vector can bring better experimental results.

3.3   Comparison of the Effects of Different Attention Modules

In this part, we compare the effectiveness of the model that uses the ECANET layer as the model to calculate the importance of features and the model that uses the SENET layer to calculate the importance of features through experiments. The results are shown in Fig. 4.

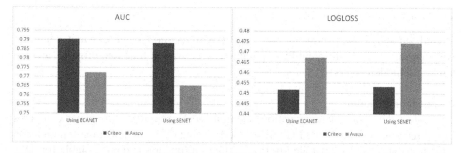

Fig. 4. Comparison of the ECANET and SENET effects on the Criteo and Avazu datasets.

It can be seen from Fig. 4 that the CTR model using the ECANET layer outperforms the SENET layer in both AUC and LOGLOSS.

3.4   Comparison of the Classic Model

To demonstrate the effectiveness of the ECABiNet model, this article compares multiple groups of models using the publicly available Criteo and Avazu datasets. Table 2 illustrates the comparison results:

Table 2. Comparison of forecast results.

	Criteo		Avazu	
	AUC	LOGLOSS	AUC	LOGLOSS
FM	0.7681	0.47831	0.7432	0.41012
AFM	0.7722	0.46217	0.7511	0.39551
Wide&Deep	0.7796	0.46011	0.7545	0.39425
FiBiNet	0.7802	0.45628	0.7598	0.39334
DeepFM	0.7891	0.45243	0.7621	0.39021
ECABiNet	**0.7988**	**0.45102**	**0.7753**	**0.38432**

Being good at using attention modules can sometimes bring good results for CTR models. Comparing the AFM model and the FM model in Table 2, it can be found that the effect of the model can be improved by adding an attention module during feature intersection.

Effectively combining shallow and deep models, that is, learning both high-order and low-order features concurrently, can significantly improve the CTR model. As demonstrated in Table 2, the performance of the deep DeepFM and FiBiNET models is superior to that of the shallow AFM and FM models.

Traditional attention models add attention modules when features intersect, and sometimes adding attention mechanisms to embedded feature vectors works surprisingly well. As shown in Table 2, both ECABiNet and FiBiNET outperform AFM and DeepFM.

3.5   Study HyperParameter

To find suitable hyperparameters for the ECABiNet model, we conduct comparative experiments from different hyperparameters (activation function, dropout, and the number of hidden layers).

– Activation Function
  To select the appropriate activation function more accurately, this paper compares the prediction results of the ECABiNet model using different activation functions on the two datasets. As illustrated in Fig. 5, the relu function is more suitable for deep models.

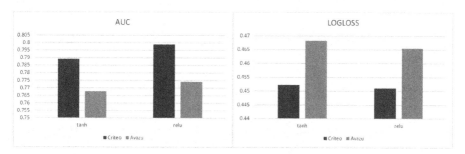

Fig. 5. Comparison of different activation functions on the Criteo and Avazu datasets.

– Dropout
  To improve model performance, dropout randomly drops neural units. To improve the performance of the ECABiNet model, this paper studies the performance of dropout from 0-1, as shown in Fig. 6. The model works best when dropout is 0.9.

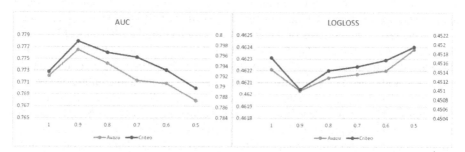

Fig. 6. Comparison of different dropout effects on the Criteo and Avazu datasets.

– Number of hidden layer
  In the DNN layer, the number of different hidden layers also affects the performance of the model. This paper compares the effect of the number of hidden layers from layers 1–7 on two public datasets. As shown in Fig. 7, the results show that ECABiNet works best when the number of hidden layers is 5 or 6.

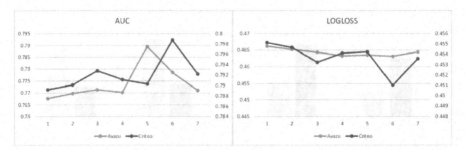

Fig. 7. Comparison of different numbers of hidden layers on the Criteo and Avazu datasets.

4 Related Work

Currently, many CTR models have been proposed in the field of recommender sys-tems. Thanks to the proposal of ResNet [14], paper [15] proposed the deep crossing model, which converts sparse features into low-dimensional dense features by adding an embedding layer and uses a stacking layer, or concat layer, to connect the segment-ed feature vectors. Then, the combination and transformation of features are complet-ed through the multilayer neural network, and finally, the calculation of CTR is com-pleted with the scoring layer. On the basis of the Deep Crossing model, the FNN model [16] uses the hidden layer vector of FM as the embedding of user and item, which avoids training the embedding from random state completely, thereby improving the recommendation effect. The traditional DNN directly completes the intersection and combination of features through multilayer fully connected layers, but this method lacks a certain 'target', so the paper in [17] proposed the PNN model. To balance the memory ability and generalization ability, Google proposed the Wide&Deep model. However, the wide part requires artificial feature engineering, which leads to increased workload. Therefore, the DeepFM model is proposed. To address the issue of the insufficient expression capability of the wide part, Google published the DCN model [18] in the following year. The main idea is to use the cross network to replace the original wide part, which increases the interaction between features. Considering the possibility of improvement in the DNN part, paper [19] proposes the NFM model. From the perspective of modifying the second-order part of the FM, the NFM model replaces the feature intersection part of the FM with a DNN with a Bi-interaction Pooling layer, forming the unique Wide&Deep architecture improves the recommendation effect. User characteristics are important, but introducing the interests of users into the model often brings unexpected effects. Paper [20] proposed the DIEN model, which is not only a further 'evolution' of the DIN model [21], but more importantly, DIEN simulates the process of user interest evolution by introducing the sequence model AUGRU.

5 Conclusions

In the CTR field, new models are constantly being introduced, but few have considered the importance of embedding vector layers. The ECANET network layer is used in this paper to teach the CTR model about the importance of embedding vectors, which improves the performance of the model. Additionally, this paper introduces LayerNorm in the embedded feature vector layer to normalize the features and improve accuracy. Our experimental results also show that the ECABiNet model outperforms existing models on both datasets.

Acknowledgment. This work is supported by Hainan Province Science and Technology Special Fund, which is Research and Application of Intelligent Recommendation Technology Based on Knowledge Graph and User Portrait (No.ZDYF2020039). Thanks to Professor CaiMao Li, the correspondent of this paper.

References

1. Rendle, S.: Factorization machines. In: ICDM 2010, The 10th IEEE International Conference on Data Mining, Sydney, Australia, 14–17 December 2010 (2010)
2. Juan, Y., Zhuang, Y., Chin, W.S., Lin, C.J.: Field-aware factorization machines for ctr prediction. In: Proceedings of the 10th ACM conference on recommender systems, pp. 43–50 (2016)
3. Cheng, H.T., et al.: Wide & deep learning for recommender systems. In: Proceedings of the 1st workshop on deep learning for recommender systems. pp. 7–10 (2016)
4. Guo, H., Tang, R., Ye, Y., Li, Z., He, X.: Deepfm: a factorization-machine based neural network for ctr prediction. In: Twenty-Sixth International Joint Conference on Artificial Intelligence (2017)
5. Huang, T., Zhang, Z., Zhang, J.: Fibinet: combining feature importance and bilinear feature interaction for click-through rate prediction. In: Proceedings of the 13th ACM Conference on Recommender Systems. pp. 169–177 (2019)
6. Rivera-Trigueros, I.: Machine translation systems and quality assessment: a systematic review. Language Resources and Evaluation 1–27 (2021)
7. Xiao, J., Ye, H., He, X., Zhang, H., Wu, F., Chua, T.S.: Attentional factorization machines: Learning the weight of feature interactions via attention networks. arXiv preprint arXiv:1708.04617 (2017)
8. Ba, J.L., Kiros, J.R., Hinton, G.E.: Layer normalization. arXiv preprint arXiv:1607.06450 (2016)
9. Chauhan, V.K., Dahiya, K., Sharma, A.: Problem formulations and solvers in linear svm: a review. Artif. Intell. Rev. **52**(2), 803–855 (2019)
10. Hu J, Shen L, Sun G.: Squeeze-and-excitation networks. In: Proceedings of the IEEE conference on computer vision and pattern recognition. Pp. 7132–7141 (2018)
11. Wang, Q., Wu, B., Zhu, P., Li, P., Hu, Q.: Eca-net: Efficient channel attention for deep convolutional neural networks. In: 2020 IEEE/CVF Conference on Computer Vision and Pattern Recognition (CVPR) (2020)
12. Kaggle Science Community: Display advertising challenge: Predict click-through rates on display ads. https://www.kaggle.com/c/criteo-display-ad-challeng (2014)
13. Kaggle Science Community: Click-Through Rate Prediction: predict whether a mobile ad will be clicked. https://www.kaggle.com/c/avazu-ctr-prediction (2015)
14. He, K., Zhang, X., Ren, S., Sun, J.: Deep residual learning for image recognition. In: Proceedings of the IEEE conference on computer vision and pattern recognition, pp. 770–778 (2016)
15. Shan, Y., Hoens, T.R., Jiao, J., Wang, H., Yu, D., Mao, J.: Deep crossing: webscale modeling without manually crafted combinatorial features. In: Proceedings of the 22nd ACM SIGKDD international conference on knowledge discovery and data mining, pp. 255–262 (2016)

16. Zhang, W., Du, T., Wang, J.: Deep learning over multi-field categorical data. In: European conference on information retrieval, pp. 45–57. Springer (2016)
17. Qu, Y., Cai, H., Ren, K., Zhang, W., Yu, Y., Wen, Y., Wang, J.: Product-based neural networks for user response prediction. In: 2016 IEEE 16th International Conference on Data Mining (ICDM), pp. 1149–1154. IEEE (2016)
18. Wang, R., Fu, B., Fu, G., Wang, M.: Deep & cross network for ad click predictions. In: Proceedings of the ADKDD'17, pp. 1–7 (2017)
19. He, X., Chua, T.S.: Neural factorization machines for sparse predictive analytics. In: Proceedings of the 40th International ACM SIGIR conference on Research and Development in Information Retrieval, pp. 355–364 (2017)
20. Zhou, G., et al.: Deep interest evolution network for click-through rate prediction. In: Proceedings of the AAAI conference on artificial intelligence, vol. 33, pp. 5941–5948 (2019)
21. Zhou, G., et al.: Deep interest network for click-through rate prediction. In: Proceedings of the 24th ACM SIGKDD international conference on knowledge discovery & data mining, pp. 1059–1068 (2018)

Active Anomaly Detection Technology Based on Ensemble Learning

Weiwei Liu[1], Shuya Lei[1(✉)], Liangying Peng[2], Jun Feng[2], Sichen Pan[2], and Meng Gao[3]

[1] Artificial Intelligence On Electric Power System State Grid Corporation Joint Laboratory, State Grid Smart Grid Research Institute Co.Ltd., Beijing 102209, China
oweiwlo@163.com, sxcjdxlsy@163.com
[2] State Grid Zhejiang Information and Telecommunication Branch, Hangzhou 310016, China
[3] Harbin Institute of Technology, Harbin 150001, China

Abstract. Anomaly detection is an important problem in various research and application fields. Researchers design reliable schemes to provide solutions for effectively detecting anomaly points. Most of the existing anomaly detection schemes are unsupervised methods, such as anomaly detection methods based on density, distance and clustering. In total, unsupervised anomaly detection methods have many limitations. For example, they cannot be well combined with prior knowledge in some anomaly detection tasks. For some nonlinear anomaly detection tasks, the modeling is complex and faces dimensional disasters, which are greatly affected by noise. Sometimes it is difficult to find abnormal events that users are interested in, and users need to customize model parameters before detection. With the wide application of deep learning technology, it has a good modeling ability to solve linear and nonlinear data relationships, but the application of deep learning technology in the field of anomaly detection has many challenges. If we regard exceptions as a supervised problem, exceptions are a few, and we usually face the problem of too few labels. To obtain a model that performs well in the anomaly detection task, it requires a high initial training set. Therefore, to solve the above problems, this paper proposes a supervised learning method with manual participation. We introduce the integrated learning model and train a supervised anomaly detection model with strong stability and high accuracy through active learning technology. In addition, this paper adopts certain strategies to maximize the accuracy of anomaly detection and minimize the cost of manual labeling. In the experimental link, we will show that our method is better than some traditional anomaly detection algorithms.

Keywords: Anomaly detection · Ensemble learning · Artificial anomaly detection · Methods to reduce labor cost · Model self-training

1 Introduction

The application of density-based methods to anomaly detection is one of the earliest known anomaly detection methods. The core principle of density-based outlier detection

© The Author(s), under exclusive license to Springer Nature Singapore Pte Ltd. 2022
Y. Wang et al. (Eds.): ICPCSEE 2022, CCIS 1628, pp. 53–66, 2022.
https://doi.org/10.1007/978-981-19-5194-7_5

methods is that outliers can be found in low-density regions, while normal points are assumed to appear in dense regions, i.e., objects that are significantly different from their nearest neighbors. Those that are far away from their nearest neighbors are labeled and always considered outliers, and Breunig et al. [1] proposed a local outlier factor (local outlier factor, LOF) method, which is one of the earliest clustering outlier detection methods based on loose correlation density. This technology uses K nearest neighbors. The algorithm mainly compares the density of each point p and its neighbor points. To judge whether the point is an abnormal point, if the density of point p is lower, it is more likely to be regarded as an abnormal point. Tang et al. [2] improved LOF [1] and simplified LOF [3] and called it a connection-based outlier factor (COF). This method is very similar to the LOF method, and the only difference is how the record density estimate is calculated. In addition, density-based outlier detection methods have been continuously proposed in follow-up, such as RDF [4], GDP + DLC [5], and RDOS [6]. Although some density-based methods have been proven to have better performance, in most cases, they are more complex and computationally expensive than statistical methods [7], they are sensitive to parameter settings, such as determining the size of neighbors.

Statistical-based techniques can be used to detect outliers in supervised, semisupervised, and unsupervised ways. In statistics-based methods, data points can be modeled using a random distribution, some data points can be marked as outliers based on their relationship to the distribution model, and the declaration of outliers and normal points depends on the data distribution model. Statistics-based methods are usually divided into two categories: parametric methods and nonparametric methods. The main difference between the two methods is that the former assumes a potential distribution model in the given data and estimates the distribution model from the known data. parameter. The latter method does not assume any prior knowledge of the distribution model [8]. For the parametric method, Yang et al. [9] introduced a globally optimal Gaussian mixture model (Exemplar-Based GMM) method for unsupervised outlier detection. In their technique, they first implemented a globally optimal expectation-maximization (EM) algorithm to fit a GMM to a given dataset, treating the outlier factor of each data point as the sum of the weighted mixture proportions. Weights represent relationships to other data points. For nonparametric methods, Zhang et al. [10] also studied an adaptive kernel density technique based on Gaussian kernels for detecting anomalies in nonlinear systems, and Qin et al. [11] proposed a new local isolation cluster semantics, which makes good use of KDE to efficiently detect local outliers in data streams. While statistics-based anomaly detection is mathematically interpretable and generally has good performance, the quality of the results produced is unreliable for practical purposes due to the lack of prior knowledge about the underlying distribution and the need to make assumptions about the distribution model. Mostly unreliable for situations and applications. Distance-based methods detect outliers by calculating the distance between points. A point that is far from its nearest data point is considered an outlier.

The most common definition of distance-based outlier detection focuses on local neighborhoods. k-nearest neighbors (KNN) [12] and traditional distance thresholding. Knorr and Ng were among the first scholars to study distance-based outlier computation, a moderately nonparametric approach for distance-based anomaly detection suitable for medium- to high-dimensional large datasets. They tend to have a firmer foundation, are more flexible and are more computationally efficient than statistical techniques. Luan Tran et al. [13] proposed a distance-based real-time data anomaly detection method. This method not only determines the anomaly points based on distance but also designs many pruning rules, which greatly reduce the search space. The method has both time and space. Good performance. However, in terms of high dimensional spaces. However, distance-based anomaly detection has some similar disadvantages to statistical methods and density-based methods because their performance will degrade due to the limitation of dimensionality, and objects in the data usually have discrete proper ties, which makes the definition of these object distances difficult. Most of the existing anomaly detection schemes are unsupervised methods, such as density-based, distance-based, and cluster-based anomaly detection methods, which usually face the following challenges:

- Most unsupervised methods cannot be well combined with prior knowledge in some anomaly detection tasks. For some nonlinear anomaly detection tasks, the modeling is complex and faces dimensional disaster, which is greatly affected by noise.
- Sometimes it is difficult to find abnormal events that users are interested in.
- User-defined model parameters are required before detection.
- Because there are few outliers, supervised anomaly detection methods usually face the problem of too few labels, and it is difficult to train a model that performs well.

With the development of machine learning and deep learning, many people apply it to the problem of anomaly detection as a supervised or semisupervised classification problem and let the model learn the complex features in the data, which can solve the problem of dimensionality. problem, and unlike many unsupervised methods, there are problems that are sensitive to parameters. In addition, traditional anomaly detection methods are usually binary classifications. If it is regarded as a supervised classification problem, anomalies can be divided into multiple categories to achieve multicategory cause detection for anomalies, which is a lot of unsupervised Algorithms are hard to do. However, for model learning, in the problem of anomaly detection, supervised models often lack the labels required for training. Therefore, some people have proposed using active learning technology, but there are still few related methods in this field. How are the existing solutions? Minimizing manpower while improving the accuracy of anomaly detection remains to be studied. The contributions of this paper are as follows:

- we propose a supervised anomaly detection method with human involvement, which addresses the problem of insufficient labels in supervised models by introducing human effort.
- In the design of the model, we adopted an ensemble learning model. In the experimental test part, it is shown that the ensemble learning model has the characteristics of strong stability and high accuracy.
- In addition, we design related manual annotation strategies and query functions to minimize labor costs and maximize anomaly detection accuracy.
- Finally, we adopt a semisupervised method to further improve the model performance.

2 Problem Statement

Given an initial training set X, X contains normal data points and abnormal data points:$X = (X_1, X_2, \ldots, X_N)$. The corresponding labels are $L = (l_1, l_2, l_3 \ldots, l_N)$, where $l_i \in \{0, 1\}$ is the label corresponding to the data point in the training set, 0 represents the corresponding normal data point, and 1 represents the abnormal data point. For large batch unlabeled datasets: $D = (y_1, y_2, y_3 \ldots, y_M)$.

Our goal is to train an ensemble learning model F with p learners. The ensemble learning model relies on the active learning framework to manually perform some specific data points in D according to a certain sample selection strategy with manual participation. label, and retrain model F.

Ultimately, each learner is able to score data points for anomalies so that the score of normal points is as close to 0 as possible, and the score of outliers is as close to 1 as possible while minimizing the cost of manual labeling.

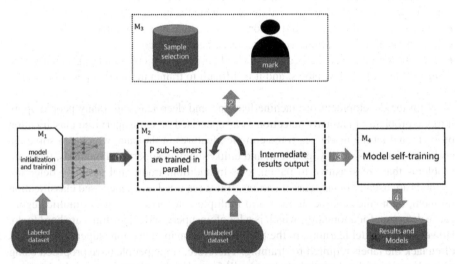

Fig. 1. Model overall architecture (AADEL)

3 Proposed Model

This section introduces our model: Active Anomaly Detection Based on Ensemble Learning (AADEL). In general, we combine the prior knowledge of anomaly detection with the ensemble learning model and introduce artificial participation through active learning to impröve the expressiveness of the model. Finally, we use semisupervised methods to further improve the performance of the model.

This section introduces our model: Active Anomaly Detection Based on Ensemble Learning (AADEL). In general, we combine the prior knowledge of anomaly detection with the ensemble learning model and introduce artificial participation through active learning to improve the expressiveness of the model. Finally, we use semisupervised methods to further improve the performance of the model.

The overall framework method process is shown in Fig. 1. Specifically, first, the M1 module is responsible for the initialization of the learner, generating p sublearners, and then, we will train p sublearners. After each round of training, the integration of the learning model will output intermediate results, corresponding to the M2 module in the figure. At this time, the sample selection algorithm will select some data points according to a certain strategy according to the output results, hand them over to the M3 module in the corresponding figure, and hand these marked points to the M2 module for re-enhancing training. After repeating it a certain number of times, the model will enter the self-training stage M4. Finally, a model with better performance and detection results corresponding to the current dataset is obtained. Next, we will introduce the model in detail in three parts: supervised ensemble learning model, human participation and model self-training (Fig. 2).

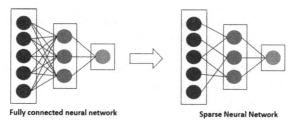

Fully connected neural network   Sparse Neural Network

Fig. 2. Sparse neural network architecture

3.1 Supervised Ensemble Learning Model

In the supervised algorithm of machine learning, our goal is to learn a model that is stable and performs well in all aspects, but the actual situation is often not ideal, and sometimes only multiple models with preference can be obtained. Better in some respects. The emergence of ensemble learning can effectively solve this problem. It combines multiple learners. Even if a weak classifier obtains an incorrect prediction, other weak classifiers can correct the error. The typical ensemble learning framework includes bagging and boosting.

The main idea of bagging is to extract K subtraining sets from the training set, use the K subtraining sets to train to obtain K models, and vote the K models to obtain the classification results. For regression problems, the mean of the models is calculated as the final result. All models are equally important. Boosting is an adaptive algorithm; that is, it can adapt to the respective training error rates of weak learners. The specific process is to first assign the same initial weight to each sample. The weights are adjusted to increase the weights of misclassified samples so that in the next round of learning, misclassified samples receive more attention.

It can be found that in the process of parallel computing, each prediction function of bagging can be generated in parallel, while each prediction function of boosting can only be generated sequentially because the latter model needs the results of the previous round of models.

In this paper, a random sparse neural network is used as the sublearner because the sparse neural network has the characteristics of lower training time complexity than the traditional neural network. Neural networks usually have the characteristics of overfitting due to a large number of parameters. There are problems in training such neural networks in terms of time and accuracy. However, sparse neural networks may also overfit, combining multiple such neural networks. The overall benefit can be improved, and our experimental results also show that the effect of our ensemble learning model is better than that of the traditional deep learning single model. In addition, because our sublearners are independent, we can train p sublearners in parallel, which will greatly reduce the time consumption when training the model.

In terms of implementation, we use a three-layer neural network structure, and the activation function of the neural node uses the SIGMOD function. The sparse neural network is realized by randomly removing part of the connection between the two layers, as shown in the following figure:

For each sparse neural network VE, there are three layers in total, and its output layer is a neuron for outputting abnormal scores. For input and hidden layers: $q = \Psi(x, \Theta_t)$. For hidden and output layers: $\Phi(x, \Theta) = \Theta_s^T q$.

Θ_t is the corresponding sparse matrix, so when each neural network is initialized, the matrix weights between the first layer and the second layer are randomly set to 0 to construct the matrix (Fig. 3).

$$
\begin{array}{ccc}
w_1 & w_2 & w_3 \\
w_4 & w_5 & w_6 \\
w_7 & w_8 & w_9 \\
w_{10} & w_{11} & w_{12} \\
w_{13} & w_{14} & w_{15}
\end{array}
\implies
\begin{array}{ccc}
w_1 & 0 & 0 \\
0 & w_5 & 0 \\
w_7 & 0 & 0 \\
0 & 0 & w_{12} \\
0 & w_{14} & 0
\end{array}
$$

Fig. 3. The construction process of the parameter matrix

It is worth noting that in the process of backpropagation of the neural network, different from the traditional BP neural network, the new neural network structure does not need to update the zero elements in the parameter matrix when the weight is updated in the backpropagation. which saves time and ensures that the element in the corresponding position of the parameter matrix is always zero.

3.2 Human Participation

The previous section introduced the machine learning model part of the overall process of the algorithm. In the classification task of machine learning, the more training data there are, the better, but in the problem of anomaly detection, there are few anomalies, and it is difficult for us to obtain sufficient anomalies. The training set of data is provided to the ensemble learning model for training. Therefore, we consider the combination of active learning technology and a semisupervised method to improve the model. In this section, we first introduce active learning technology with human participation and focus on sample selection and manual annotation.

3.2.1 Sample Selection

Sample selection is also called the function of inquiry in active learning technology. In the process of active learning, one or a batch of the most useful sample points is selected and handed over to humans for labeling.

Currently, the most widely used uncertainty query is the uncertainty query. For uncertainty, we can use the concept of information entropy to understand it. The greater the information entropy is, the greater the uncertainty and the richer the information contained. In fact, many uncertainty-based active learning query functions are designed using information entropy, so the uncertainty strategy is to find ways to find samples with high uncertainty because the rich information contained in these samples is useful for model training.

In the previous section, we integrated the base classifier for anomaly detection work. Based on ensemble learning to combine the characteristics of multiple classifiers, we used entropy query-by-bagging (EQB) to select samples for manual labeling.

The basic idea of EBC is to use the available set of labeled samples to train a set of classifiers. This set of classifiers constitutes a committee, and then the committee classifies the unlabeled samples in the candidate sample set and selects the "least inconsistent" samples classified by the committee. In EQB, the basis for judging the"least inconsistency" is voting entropy, as shown in Fig. 4. The specific calculation process of EQB is given below:

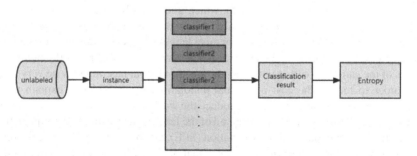

Fig. 4. EQB query step diagram

1. First, K training sets are selected by bagging from the initial training set, and then this training set is used to train p classification models. This group of models constitutes a committee.
2. The classifier in the committee is used to predict each sample in the unlabeled sample set, and each sample is labeled according to the predicted class so that each sample has K labels.

3. EQB uses these labels labels to calculate the entropy of the samples:

$$x^{EQB} = \arg\max_{x_i \in U}\{\frac{H^{BAG}(x_i)}{\log(N_i)}\} \tag{1}$$

where:

$$H^{BAG}(x_i) = \sum_{w=1}^{N_i} p(y_i^* = w|x_i) \log[p(y_i^* = w|x_i)] \tag{2}$$

where $p(y_i^* = w|x_i)$ represents the probability that sample xi is predicted to be w by the p models, that is, the number of votes that sample xi is predicted to be w is divided by p, and Ni is the total number of categories. By observing the above formula, we find that when all the classifiers in the committee predict that the category of the sample consistently H^{BAG} is 0, which indicates that for the current classification model, the category of the sample is almost certain. Handing samples to manual labeling generally cannot improve the model, and a large H^{BAG} generally means that the greater the divergence in the classification of sample points, the greater the amount of information contained. It is possible to hand over such sample points to manual labeling to improve the ability of the model more effectively.

3.2.2 Manual Annotation

When we hand over some sample points to manual annotation, the annotator checks the information of each feature of the current point, determines whether the point is an abnormal point, and feeds it back to the model, but there may be hundreds or thousands of features due to multidimensional data, which undoubtedly creates a huge challenge for the labelers. Even if the data points have only dozens of features, it is not a small challenge. That is, in the active learning manual labeling stage, the model selects a batch of sample points through sample selection. When manually labeling, if the number of features of the sample points is too large, the cost of manual labeling may be very large.

In order to reduce the cost of manual annotation, we consider to explain to the annotator why the anomaly is caused to reduce the work of the labeler. With these explanations, the labeler can reduce the workload of the labeling according to the explanation. We introduce a sequential feature explanation (SFE), given a data point, where the order of features is the order of feature importance that distinguishes this data point as a normal or an anomaly; that is, at the top, the features have the greatest impact on the model anomaly discrimination.

For example, an SFE of length k is represented as follows:

$$E = (e_1, e_2, \ldots, e_k) ei \in \{1 \ldots, n\}$$

Among them, the earliest feature is the feature that plays the most important role in model discrimination. If a data point x is given, then Xe1 is the most important feature.

Given the SFE E for the data point X, present this point X_{e1} to the annotator. If the annotator can judge by the current feature information, then we have completed the annotation of this point. Otherwise, the next feature will be provided to the annotator, which means that the annotator can then look at X_{e2} and make an analysis and judgment. The process of incrementally adding features to the presented information set will continue until the analyst can make a decision. The process may also be terminated early due to time constraints.

Next, we focus on how to generate sequential feature explanations (SFE), giving the definition of the problem:

Given a trained machine learning model F and a data point X, generate a sequential feature explanation SFE for it, where the SFE is ranked according to its influence on the model discrimination.

Before sorting the SFEs, we first analyze the neural network. The forward propagation of the neural network can be represented by the parameter matrix:

Assuming that the forward data points are $X = \{x^1, x^2, \ldots, x^t\}$, the number of layers of the neural network is three, the data points have a total of t dimensions, and the number of neurons in the middle layer hidden layer is h, then the matrix from the input layer to the hidden layer is represented as follows. Here, we only show the weight parameters and ignore the bias parameters: Input layer to hidden layer:

$$[x^1, x^2, \ldots, x^t] \begin{bmatrix} w_{11} & w_{21} & \ldots & w_{h1} \\ w_{12} & w_{22} & \ldots & w_{h2} \\ \ldots & \ldots & \ldots & \ldots \\ w_{1t} & w_{2t} & \ldots & w_{ht} \end{bmatrix} = [W_1 X, W_2 X, \ldots, W_h X] \tag{3}$$

Hidden layer to output layer:

$$[W_1 X, W_2 X, \ldots, W_h X] \begin{bmatrix} w_{o1} \\ w_{o2} \\ \ldots \\ w_{oh} \end{bmatrix} = w_{o1} W_1 X + w_{o2} W_2 X + \ldots + w_h W_h X$$
$$= (w_{o1} W_1 + w_{o2} W_2 + \ldots + w_h W_h) X \tag{4}$$

Assume:

$$O = w_{o1} W_1 + w_{o2} W_2 + \ldots + w_h W_h \tag{5}$$

Among them, O is a matrix of size $1 \times t$, and each column of the matrix corresponds to each eigenvalue of X, which exists as its parameter, The larger the absolute value of the parameter is, the more important the feature corresponding to the parameter is to the prediction of the current model result. Therefore, the feature ranking corresponding to the absolute value of the parameter is taken and fed back to the experts. Considering the ensemble learning model, the feature ranking of each sub-learner may be different. Therefore, it is necessary to superimpose the rankings of the p sub-learner features to recommend users in order.

3.3 Model Self-training

In the previous section, we introduced manual participation, which to a certain extent alleviated the problem of poor model performance caused by the lack of ensemble learning model labels. In this section, we use semisupervised algorithms to further improve the anomaly detection ability of ensemble learning.

The traditional self-training algorithm completes the expansion of the training dataset by selecting unlabeled data samples with high confidence, but there is a problem that if the predetermined category is wrong, it is obvious that in the process of algorithm iteration, each iteration will cause the accumulation of errors to worsen the performance of the model. For the above problems, we use a self-training method with repeated labels, which can effectively solve the accumulation of errors and improve the performance of the classifier.

The specific idea of the self-training check labeling method is to use the classifier obtained at the i-1th iteration to repredict all unlabeled samples at the ith iteration of the algorithm and then select a larger amount of prediction data to expand the next set of training samples used in an iteration. That is, the unlabeled data samples selected in each iteration will be predicted again in each iteration to ensure that errors can be corrected in the subsequent model training process. The following is the pseudocode of the improved self-training repeated labeling algorithm:

Algorithm 1. Self-training repeat labeling algorithm

---

input : Tagged training data X_{train} ; Unlabeled dataset $\tilde{X}$;

      Initial value of prediction probability threshold α ; Change step of prediction probability threshold γ And lower limit β $(\alpha > \beta)$

output : Trained classifier F' ;

1   $U_1 \leftarrow \emptyset, U_2 \leftarrow \emptyset$

2   **while** $Len(U_2) < Len(\tilde{X})$ *and* $\alpha > \beta$ **do**

3     $F \leftarrow train\_model(X_{train} \cup U_2)$;

4   **for** $\tilde{X}_i$ *in* $\tilde{X}$ **do**

5   **if** $\max F(\tilde{X}_i) > \alpha$ **then**

6         $U_1 \leftarrow U_1 \cup \{(\tilde{X}_i, \tilde{y}_i)\}$;

7   **if** U_2 *is not empty and* $Len(U_1) < Len(U_2)$ **then**

8       $\alpha \leftarrow \alpha - \gamma$;

9   **else do**

10    $U_2 \leftarrow U_1$;

11   $U_1 \leftarrow \emptyset$

12   $F' \leftarrow F$

13   **return** F';

First initialize the set U_1 and the set U_2, in the process of each iteration, the data points that meet the requirements in X_e, that is, the predicted probability greater than the threshold α, are added to U_1 (line 4–6 of the algorithm 1), and U_2 will be used for each iteration. The training samples that meet the requirements are added to the sample set. If the number of points with a prediction probability greater than α in the previous round is larger than that in the current round, then reduce the threshold α according to the step size γ (line 7–8 of the algorithm 1) and repeat the process. Prediction, it is worth noting that the threshold α is always required to be greater than β, and the iteration will stop if α is less than β.

3.4 Experiment

3.4.1 Experimental Setup

In the experimental testing phase, we used three common anomaly detection datasets, including disease diagnosis, human activity, and forest coverage datasets (Tables 1 and 2).

Table 1. Dataset description

dataset	describe	data link
Annthyroid	Disease diagnosis	https://www.openml.org/d/40497
HAR	Human activity	https://www.openml.org/d/ 1478
Covertype	Forest cover rate	https://archive.ics.uci.edu/ml/datasets/covertype

Table 2. Dataset dimension and data type information

dataset	Dimension	Type
Annthyroid	22	float
HAR	562	float
Covertype	54	float

For the division of datasets, 10,000 data points were extracted from each dataset, 10 percent of which were used as training data to simulate prior knowledge, and the test set accounted for 90 percent. For the number of outliers, the number of outliers in the training set is 60, and the number of outliers in the test set is 540. For both the training set and the test set, the proportion of outliers is 6 percent. The experimental parameters are as follows: the number of sparse neural networks $p = 10$, the optimizer is gradient descent, the learning rate is 0.01, and the number of training rounds is 10000.

3.4.2 Experimental Results of the Ensemble Learning Model

We conducted accuracy experiments and compared them with the biased neural network deep learning anomaly detection DADDN. On a test set of 9000 data points (Table 3):

Table 3. Results under three datasets

dataset	Ensemble learning model	DADDN
Annthyroid	7654	6154
HAR	8190	8312
Covertype	7542	7235

From the perspective of the experimental results, the characteristics of the integrated learning model determine that it has a higher accuracy, and we use a sparse neural network for each sublearner to allow the model to overcome the problem of overfitting and make the model more stable and has good transferability on various datasets (Fig. 5).

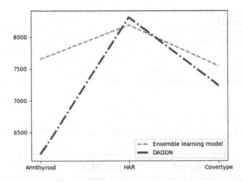

Fig. 5. Comparison of test results between the two methods

3.4.3 Experimental Results of AADEL

To verify the performance of the overall model, on the basis of the above experimental settings, a total of 100 points were selected for manual labeling during each iteration of the training process, and the number of labels accounted for 1 percent of the total test set.

For comparative experiments, we choose the unsupervised method LOF and the previously introduced supervised method DADDN.

By observing the experimental results in Fig. 6, it can be found that our method has improved the accuracy, and the anomaly detection accuracy is higher and more stable than the traditional supervised and unsupervised anomaly detection methods. Compared with unsupervised methods, our method can better mine the abnormal patterns that users are interested in due to the combination of domain knowledge, while unsupervised methods may capture some boring noise and lead to high false positives. Therefore, the accuracy of our method is higher than that of classical unsupervised methods such as lof (Table 4).

Table 4. Experimental results of AADEL and other methods

dataset	AADEL	DADDN	LOF
Annthyroid	7931	6154	4532
HAR	8614	8312	3769
Covertype	7601	7235	2301

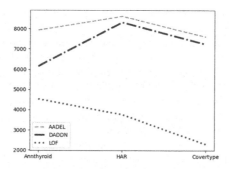

Fig. 6. Comparison of AADEL and other methods

3.5 Conclusion

This paper proposes an integrated learning anomaly detection technology with manual partition. We use the integrated sparse neural network. This integrated learning model has good stability and high accuracy, and the sparse neural network structure can ensure that it will not overfit and has good mobility on many datasets. Then, we use active learning technology to further improve the model because the supervised model may have high requirements for the training set, and in the anomaly detection problem, it is difficult to obtain a training set that meets the requirements. Finally, we propose a model self-training method, which is helped to mine new abnormal patterns and support the continuous addition method, which is helped to mine new abnormal patterns and support the continuous addition of time series. In the final experiment, we prove the effectiveness of the model.

Acknowledgements. The project is supported by the State Grid Research Project "Study on Intelligent Analysis Technology of Abnormal Power Data Quality based on Rule Mining" (5700-202119176A-0-0-00).

References

1. Breunig, M., Kriegel, H., Ng, R.T., Sander, J.: LOF: Identifying density-based local outliers. ACM SIGMOD Rec. **29**(2), 93–104 (2000)
2. Tang, J., Chen, Z., Fu, A., Cheung, D.: Enhancing effectiveness of outlier detections for low density patterns. In: Advances in Knowledge Discovery and Data Mining, pp. 535–548. Springer, Berlin, Germany (2002)
3. Schubert, E., Zimek, A., Kriegel, H.-P.: Local outlier detection reconsidered: A generalized view on locality with applications to spatial, video, and network outlier detection. Data Mining Knowl. Discovery **28**(1), 190–237 (2014)
4. Ren, D., Wang, B., Perrizo, W.: RDF: a density-based outlier detection method using vertical data representation. In: Proc. Int. Conf. DataMining, pp. 503–506 (Nov 2004)
5. Bai, M., Wang, X., Xin, J., Wang, G.: An Efficient algorithm for distributed density-based outlier detection on big data. Neurocomputing **181**, 19–28 (2016). Mar.
6. Tang, B., He, H.: A local density-based approach for outlier detection. Neurocomputing **241**, 171–180 (2017). Jun.
7. Kriegel, H., Kröger, P., Zimek, A.: Outlier detection techniques. In: Proc. Tutorial KDD, pp. 1–10 (2009)
8. Eskin, E.: 'Anomaly detection over noisy data using learned probability distributions. In: Proc. 17th Int. Conf. Mach. Learn. (ICML), pp. 255–262 (Jul. 2000)
9. Yang, X., Latecki, L.J., Pokrajac, D.: Outlier detection with globally optimal exemplar-based GMM. In: Proc. SIAM Int. Conf. on Mining(SDM), pp. 145–154 (Apr. 2009)
10. Zhang, L., Lin, J., Karim, R.: Adaptive kernel density-based anomaly detection for nonlinear systems. Knowl. -Based Syst. **139**, 50–63 (2018). Jan.
11. Qin, X., Cao, L., Rundensteiner, E.A., Madden, S.: Scalable kernel density estimation-based local outlier detection over large data streams. In: Proc. EDBT, pp. 421–432 (2019)
12. Dang, T.T., Ngan, H.Y.T., Liu, W.: Distance-based k-nearest neighbors outlier detection method in large-scale traffic data. In: Proc. IEEE
13. Luan, T., Min, Y.M., Shahabi, C.: Real-time distance-based outlier detection in data streams. Proceed

Automatic Generation of Graduation Thesis Comments Based on Multilevel Analysis

Yiwen Zhu, Zhaoyi Li, Yanzi Li, and Yanqing Wang[✉]

School of Information Engineering, Nanjing Xiaozhuang University, NanJing 211171, China
wyq0325@126.com

Abstract. In the evaluation of graduation theses, teachers' evaluation criteria for graduation theses are inconsistent, subjective and not completely reasonable and fair. This paper proposes using the BERT model to analyze the existing graduation papers in colleges and universities and make quantitatively evaluate students' graduation projects according to the given relevant parameters. The purpose of this method is to use standards to make comprehensive, systematic and accurate evaluations and avoid the phenomenon of high repetition and similarity caused by a large number of teachers' comments. This can not only effectively improve the efficiency of graduation design evaluation but also improve the fairness of evaluation. In this paper, changing the review work of the graduation thesis from pure manual operation to machine review combined with manual operation can not only reduce manpower consumption but also make the review work more objective and fair, making it more objective on the basis of traditional subjective review.

Keywords: Multilevel analysis · Reviews the automatically generated · BERT model

1 Introduction

Although there are many studies on the realization of automatic comment generation systems, there is no mature product on the market. With the development of artificial intelligence, more progress will be made in natural language processing. At present, the market has been suitable for middle school below the automatic generation of comments software, but the graduation project evaluation system for college students is still vacant.

There are few formal studies on the semantics of natural language in domestic linguistics and computational linguistics, and there is still a gap in the research on the comprehensive semantic description of Chinese. In recent years, the volume of text data on the Internet, news articles, scientific papers and online product reviews has grown dramatically [1]. Some units in the domestic industry have developed a text generation system based on templates. For example, Xinhua News Agency has developed a system to generate corporate financial reports from financial data. Based on manual templates, the system fills the required data into the written templates to generate financial reports.

© The Author(s), under exclusive license to Springer Nature Singapore Pte Ltd. 2022
Y. Wang et al. (Eds.): ICPCSEE 2022, CCIS 1628, pp. 67–79, 2022.
https://doi.org/10.1007/978-981-19-5194-7_6

Because the template used is relatively fixed, the financial statements and annual reports generated for different enterprises are relatively similar, and the language is not vivid.

Writing automatic evaluation system development began in the 1960s in the United States and is the earliest automatic evaluation system. In the mid-1990s, with the development of artificial intelligence technology, the development of automatic writing evaluation systems, especially natural language processing and intelligent language teaching systems, also made progress. In foreign countries, the development of automatic writing scoring technology has become increasingly mature. Automatic writing scoring system has been 60 years of history. Although these foreign scoring systems have the advantages of being reliable and objective, they are mainly aimed at native English learners.

Graduation thesis comment automatic generation system technology can bring greater convenience to college graduation thesis comment work. Graduation theses are an important basis for each class of graduates to successfully graduate from school. The quality of the graduation thesis is not only an important basis for examining whether graduates have preliminary research ability but also an important symbol of examining whether graduates have the ability to comprehensively use the knowledge learned to solve practical problems [2]. It is an important mechanism to improve the quality of graduation theses to determine a reasonable and scientific evaluation index paper and evaluation method, which can guide graduates and teachers to form a positive and serious attitude and design a scientific research program.

Graduation thesis review work from pure manual operation to machine review combined with manual operation can not only reduce manpower consumption but also make the paper review work more objective and fair. Using this paper to make the teaching work to the network, intelligent management era further.

2   Technical Principle

2.1   BERT Model Introduced

The BERT model is a bidirectional coding model represented by multiple stacked transformer coding layers. It plays an important role in many NLP tasks and achieves the best results. It is a milestone in natural language processing.

The BERT model has the following advantages. The BERT model adds a self-attention mechanism, which can solve the problem of semantic disappearance of long sequence data extremely well. Second, the BERT model is an improvement of two-way transmission based on the transformer model. The referenced model has a good effect, and semantic relevance has been improved. Finally, the underlying architecture of the BERT model can be well applied to different fields of natural language processing so that the cost of pretraining is very small. There are two BERT pretraining models.

The transformer number refers to the number of stacked transformer modules used by the BERT model, the dimension refers to the dimension vector specified for encoding the data of the input model and converting the data into the dimension vector corresponding to the input when the transformer transmits data, and the multiattention number refers to the stack number of the attention mechanism. In addition, it is stacked in parallel to ensure that the processing time is not increased in the process of data transmission and that more semantic relevance can be retained.

2.2 Basic Structure of the BERT Model

The main structure of the BERT model is a transformer. The specific BERT model structure is shown in Fig. 1. It can be seen from the figure that the most critical step of the BERT model is the multiattentional mechanism to extract semantic information. The left side of the figure is the encoding stage, and the right side is the decoding stage. The whole BERT model uses the self-attention mechanism to replace the recurrent neural network structure used in the previous sequential-to-sequence model. In the process of text processing, the semantic related information can be retained by using the attention mechanism training, and the interdependent relationship can be obtained from it.

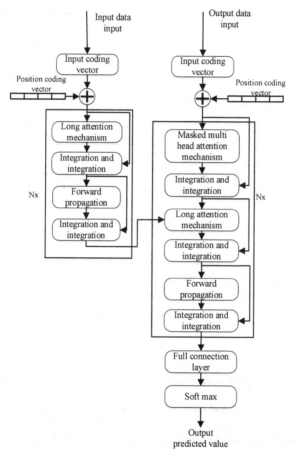

Fig. 1. Structure of the BERT model.

The BERT architecture model is obvious in the coding phase. First the input text vector data and the code, as a result of the position of the words have an influence on semantics, so before the incoming model training location information must be added to the text vector of data after long attention mechanism, the dependent relationship between words is first obtained, and the weight vectoris updated. Then, the updated data are combined with the incoming data for domestication processing. At this time, the vector obtained is added with an attention mechanism. Next, the data are propagated forward through the feedforward neural network, that is, the full-connection layer training, and then normalized processing. This process is known as the transformer model, where after N modules and a final normalization, the final output vector can be viewed as a text vector representation of the input text. In the decoding stage, the basic principle is similar to that in the encoding stage. Different from the encoder, the multiattentional mechanism has a mask and an extra operation of the multiattentional mechanism. This time, in addition to the original target document vector, the input of the encoding stage is also passed in the output text vector to affect the decoding state.

In N modules of the decoding stage, the output text vector is passed each time. Finally, the prediction words of the target output are obtained through the full connection layer and softmax layer.

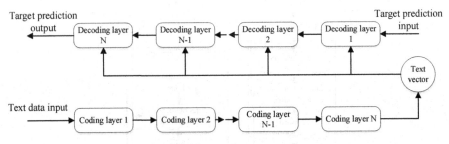

Fig. 2. Data flow in BERT.

The main data transfer is roughly the same as the traditional Seq2Seq, which is a two-stage processing mode. Stack, many coding and decoding layers, each coding and decoding layer of internal structure as shown in Fig. 2, and not every module in the decoding stage is output a prediction, but in the final output overall predictive value, this is because the inside has many parallel computing network, allowing model to deal with many data at the same time [3].

2.3 Comparison with Other Algorithms

This paper adopts the current and more effective BERT model. Table 1 is the comparison of the advantages and disadvantages of the existing partial models, where the transformer model is the multiple duplex superposition of the BERT model [4].

Table 1. Advantages and disadvantages of some existing models.

Existing partial models	Advantages and disadvantages
Seq2Seq	advantage: Able to deal with the unequal length of input and output sequences
	insufficient: There are exposuure deviations and inconsistencies between training and test measurements
VAE	advantage: Have the natural ability to solve the single problem of text generation
	insufficient: There is a problem that the inherent KL divergence disappears
GAN	advantage: Generation discriminant model can generate false data throuth multiple iterations
	insufficient: Discriminator difficult to train
Transformer	advantage: Based on attention mechanism,it can better capture the relationship between data and generate high-quality data
	insufficient: The complexity of the model is high,especially when dealing with long sequencedata

3 Project Analysis

3.1 Technical Route

The technical route of the project is studied, first confirming and training the neural network, writing evaluation categories and rules into the database and generating evaluation scheme data.Overall individual student evaluation and specific characteristic evaluation were generated according to the quantitative value read from the database.Then output according to the category weight of the student paper comments, exit page (Fig. 3).

(1) First, enter the main page to establish the graduation thesis ontology of comments. The existing graduation thesis of Nanjing Xiaozhuang University, contains the theme, structure, language and other information of the paper. This paper constructs a graduation thesis generating comment ontology.

(2) After this paper analyzes the graduation thesis, all the main information contained in it is extracted to form a neural network framework. In this paper, the evaluation category, planning and so on are written to the database.

(3) The source text of comments is preprocessed to generate comment scheme data and input into the database synchronously. The quantized value is read in the database, and the initial total student evaluation is generated with grammar analysis. There are several main categories in the operation database. Keywords of ontology relation concepts are allocated according to categories, and teachers can analyze the characteristics according to the specific situation.

(4) Match the results of the preprocessing module with the keywords of the key category editing module to obtain comments and output comments. Finally, the matching

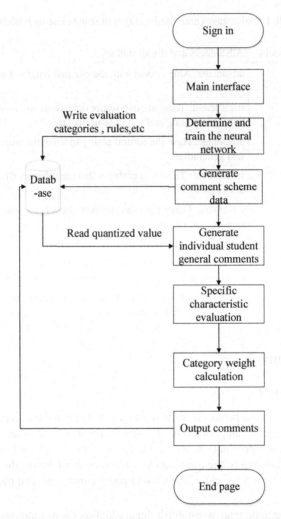

Fig. 3. Flow chart of automatic generation paper of comments based on multilevel analysis.

results are fed back to the ontology framework of the existing database to form a closed loop. Through this process, a mature paper is formed.

3.2 Technical Analysis

The technical principle of this paper mentioned that the existing natural language at home and abroad still has many problems, such as low accuracy, long testing time, and excessive language repetition. Aiming at these problems, therefore proposing the corresponding solutions, the BERT model is used to analyze the thesis reviews that are automatically generated and corresponding to the corresponding requirements, set weight, establish corresponding evaluation architecture, add teachers modify links, and make the automatically generated comments on fluency and fairness have further ascension.

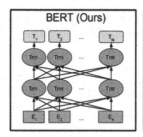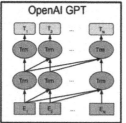

Fig. 4. BERT and GPT structure drawing.

BERT is a bidirectional overlay of multiple transformers, with each circle in Fig. 4 being a transformer, so the common GPT equivalent has to be proposed. From the perspective of innovation, BERT does not have many structural innovations. Both BERT and GPT use a transformer structure. Compared with GPT, it has a bidirectional structure, while GPT is unidirectional. For this paper, only one output layer needs to be added for fine-tuning.

The experimental results show that the training results of the BERT model are good, and the test rate is also good in the development set. To more intuitively display the comparison between BERT model and GPT model, this paper lists their situations under different conditions, as shown in Table 2.

(1) GPT training on BooksCorpus(8 million words); BERT's training was BooksCorpus (8 million words) and Wikipedia (25 million words).
(2) GPT uses sentence separators ([SEP]) and classifier tokens ([CLS]), which are only introduced for fine tuning; BERT learned the embedding of [SEP], [CLS] and sentences A/B during pretraining.
(3) GPT received 1M steps of training and a batch of 32,000 words; BERT went through 1M steps of training to produce a batch of 128,000 words.
(4) GPT used the same 5E-5 learning rate for all fine-tuning experiments; BERT chooses the task-specific fine-tuning learning rate that performs best on the development set.

The BERT model is more suitable for language automatic generation modules. Compared with GPT algorithm, it performs better in the number rate of operations and further improves the system's fluency and stability.

4 Project Implementation

This paper uses a neural network to generate comments automatically. Teachers often do repetitive work when writing comments, but what they truly need to do is to determine the scores of all aspects of students' writing, namely weight, and the specific process of generating comments can be completed by the computer [5].

According to the method to generate similar to the teacher comments for comments, the conditions need to establish a corpus, and then modify the database and the second input student information, and then based on the input information of students for each evaluation point corresponding neural network training, finally produce the current student project corresponding evaluation content of each evaluation, the following figure (Fig. 5).

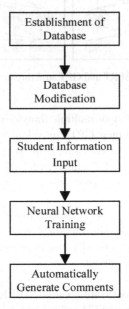

Fig. 5. Computer generated comments process.

4.1 Database Established Modification

The user information database is used to store user information. Other databases include storage of evaluation angles, evaluation points, evaluation levels, etc. At present, the teaching management department has introduced some graduation thesis quality evaluation standards, such as from the selected topic, literature review, foreign language translation, academic level and practical ability and comprehensive ability of applying basic theory and basic skills, words and graphics quality "several indexes such as the assessment of the quality of graduation thesis, and to determine the weights of each indicator [6]. To establish a comprehensive and fair evaluation system, the construction of the database, this paper uses a large number of existing graduation theses, as well as a complete set of comment templates, to obtain the initial database and the multi-level framework of graduation thesis comment text. This paper establishes a two-level comprehensive evaluation index system by taking the evaluation angle database of a graduation thesis evaluation automatic generation system as an example.

The modification part of the database mainly addresses adding and deleting the evaluation angle, evaluation point, evaluation grade, specific grade and evaluation content and adjusting the order among their respective contents. There are many factors affecting graduation theses, different regions, different schools, and different departments will lead to final graduation thesis scoring, and comments produce differences. Let alone in the hands of different teachers who will have different views, there will be personal subjective factors, which is inevitable. At this time, the graduation thesis weight ratio can solve this problem to a certain extent. The same weight can weaken the teacher's own subjective color to a certain extent, so that the graduation thesis comments are more fairer.

Table 2. Graduation thesis comments automatic generation system processing source data.

(1)	The structure of the thesis is reasonable and scientific with clear logic accurate expression of viewpoints and fluent language
(2)	Topics of social hot issues comprehensive discussion simple language
(3)	Content close to the professional requirements to write, full withdrawal of professional characteristics
(4)	Study----for the title, fully reflect the characteristics of The Times, the thesis has reasonable structure, clear thinking and obvious viewpoint
(5)	This artical conforms to the professional requirements,reflects the hot social issues, the structre is appropriate, the thinking is cear, the point of view is asically correct
(6)	......

Because the tutor is a very important part of the students' graduation design, it can make a modification to the paper, but also as a personalized improvement. For example, students cannot show the quality of the graduation thesis, but can also show this link to add, delete, change and other operations. The revision of comments is also an indispensable part of the graduation thesis.

4.2 Student Information Input

This section provides an interface for entering various aspects of the currently selected student's graduation performance. To obtain a comprehensive comment text, we need to map each student's graduation thesis to the user and instructor usage page, structure the comment text, and write it into the database table of the corresponding student's graduation thesis [7] (Fig. 6).

Taking the scoring rules of a university graduation thesis as an example, the scoring criteria of its graduation thesis are show in Table 3. As shown in the table, for graduation thesis specific scoring rules, grade judgment is a complete system. The recent success of deep neural network techniques in natural language processing largely depends on what is called the distribution hypothesis. This can be understood as a simplified version of the classical structuralist hypothesis. The paper then reinterprets the structuralist scheme using insights from evidence theory. In this way, by establishing a connection between computational logic and classical structuralism, more accurate results can be obtained [8]. Partition according to the rules shown in the table, fill in the corresponding module of the system, and then perform quantitative scoring.

Table 3. A college graduation thesis quantitative scoring rules part table

Graduation thesis grading criteria

Name			Student number	Class		
				Standard for evaluation		
			Good (86%-100%)	Middle (70%-85%)	Pass (60%-69%)	Fail (<60%)
1	Goal1	20	1. The topic selection is closely combined with the engineering practice and the workload is large; Be able to put forward the problems to be solved according to the assignment, the current situation of research and development at home and abroad and the analysis of relevant technical data; 2. The questions raised correctly cover the main work of the subject; 3. Be able to correctly use the principles of natural science and engineering to analyze problems.	1. The topic selection is closely combined with the engineering practice and the workload is moderate; Be able to put forward the problems to be solved according to the research and development status at home and abroad and the analysis of relevant technical data; 2. The questions raised correctly cover the core work of the subject; 3. Be able to use natural science and engineering principles to analyze problems.	1. The topic selection is closely combined with the actual project and the workload is general; Be able to put forward problems to be solved according to the assignment; 2. The questions raised are basically correct, but not comprehensive enough; 3. Be able to use natural science and engineering principles to analyze problems, but not comprehensive enough.	1. The topic selection is closely combined with the engineering practice and the workload is insufficient; Unable to put forward the problems to be solved according to the assignment; 2. The questions raised are incorrect and not comprehensive enough; 3. Did not use the principles of natural science and engineering to analyze the problem.
2	Goal2	20	Be able to widely consult relevant technical materials, fully, correctly and in detail analyze the current situation of relevant research work and technological development at home and abroad, and have a broad international vision.	Be able to consult relevant technical materials and analyze the the current situation of relevant research work and technological development at home and abroad, with a certain international perspective.	Be able to consult relevant technical data, analyze the current situation of relevant research work and technological development at home and abroad, and have a general international perspective.	The access to relevant technical data is not comprehensive, the analysis of relevant research work and technological development at home and abroad is incorrect, and there is no international vision.
3	Goal3	10	For the problems to be solved, a variety of solutions can be proposed and described in detail through investigation and research, literature review and other methods;	Be able to propose a variety of solutions to the problems to be solved;	Be able to propose solutions to the problems to be solved;	No substantive solution is proposed;

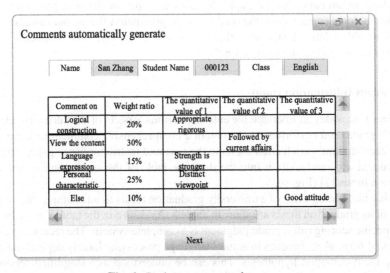

Fig. 6. Student assessment data page.

4.3 Neural Network Training

According to the BERT model, the neural network corresponding to each evaluation point is trained according to the samples input by the teacher through preprocessing. In the link of calling comments, the basic work to be done is to combine the evaluation words and sentences corresponding to the selected students' characteristic attributes into smooth comment sentences under the corresponding templates and strategies. The evaluation words and sentences are selected from excellent comments and stored in the database according to their characteristics. The evaluation editor is used to automatically generate comments. According to the evaluation function, the system characteristic parameter defines the following statement: the full opening (combination of words and sentences by the students' personality characteristics), the full text conclusion (can be made of a compliment, motivation, hope, advice and other properties of sentences), and the content description (may conduct evaluation subjects, learning evaluation, analysis, assessment of ability, learning advice words combination).

This paper uses university graduation thesis 5000. After systematic training, the similarity between the quantitative score and the teacher's manual score was 78.65%. The results are sufficient to prove that to a certain extent, the paper works for teachers when students are naering graduates. It not only reduces the pressure on teachers but also has certain reference value. That is, the evaluation result fairness and objectivity of this paper.

4.4 Automatically Generate Comments

After the teacher sets the scoring standard, the weight proportion, and then the neural network training, can calculate the score point of the current student's graduation project and automatically generate comments.

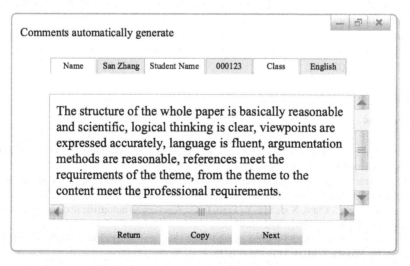

Fig. 7. Comments automatically generate pages.

As shown in Fig. 7, the system can output comments briefly, greatly reducing the repetition of evaluation writing by teachers, making it more intelligent and more objective than the manual evaluation.

5 Conclusion

There are many mechanisms to improve the quality of undergraduate graduation thesess, among which the most important is to determine the scientific and effective evaluation standard and methods of undergraduate graduation thesis quality. Existing on the market are used to establish the following comments automatic generation of software, but college students' graduation design review paper is still vacant. This is the advantage of this paper, and developing a system can be used for the automatic generation of college students' graduation design reviews to help college teachers from the onerous boring design review. Through the machine to review the graduation thesis and give the appropriate comments, in the future era of artificial intelligence, graduation thesis comment generation technology is bound to have broad development.

After this paper, it can assist teachers in evaluating students' graduation projects. The teacher determines the weight and can conduct a quantitative evaluation of the student's graduation thesis. The system is simple to operate and can be judged comprehensively, systematically and accurately. It can not only avoid the phenomenon of high repetition and similarity caused by a large number of teachers' reviews, but also effectively improve the work efficiency of graduation project reviews. It is fairer and more objective and can truly achieve the intelligent management of teaching work.

Thanks. This article is supported by the scientific research project of Nanjing Xiaozhuang University (Project name: Graduation paper "Evaluation" -- Evaluation Automatic generation system based on multilevel analysis, No.2020XSKY001).

This article is supported by the project (Project name: Research and development of self generating remote controller for intelligent toilet, No.2021320108001139).

This article is supported by the project(Project name: Modeling of energy industry based on artificial intelligence holographic interactive atlas system, No.2021320108002080).

References

1. Hima Bindu Sri, S., Dutta Sushma, R.: A survey on automatic text summarization techniques. J. Physics. Conference Series **2040**(1), 1742–6588 (2021)
2. Sun, C.: Research on design and implementation of college graduation thesis management system. Information Record Material **22**(11), 175–176 (2021)
3. Li, J.-P., Zhang, C., Chen, X.-J., Hu, Y., Liao, P.-C.: A review of automatic text summarization. Comp. Res. Develop. **58**(01), 1–21 (2021)
4. Zhang, C., Du, Y.: Research Progress of automatic short text generation technology. Data and Computing Frontiers **3**(03), 111–125 (2021)
5. Wang, R.: Research on automatic generation of student comments based on Vector machine. Fujian Computer **34**(10), 129–131 and 142 (2018)

6. Chen, Y.: Study on countermeasures of improving the quality of graduation thesis of chemistry and Chemical Engineering specialty in provincial universities. Guangdong Chemical Industry **49**(04), 213–215 (2022)
7. Yin, H.: The comprehensive evaluation of text automatic generation system based on ontology research. Computers and Telecommunications, 47–49 (2014)
8. Gastaldi, J.L., Pellissier, L.: The calculus of language: explicit representation of emergent linguistic structure through type-theoretical paradigms. Interdisciplinary Science Reviews **46**(4), 0308–0188 (2021)

A Survey of Malware Classification Methods Based on Data Flow Graph

Tingting Jiang[1], Lingling Cui[2(✉)], Zedong Lin[1], and Faming Lu[1(✉)]

[1] Shandong University of Science and Technology, Qingdao 266590, Shandong, China
fm_lu@163.com
[2] Shandong Water Conservancy Vocational College, Weifang 276826, Shandong, China
349119513@qq.com

Abstract. Malware is emerging day by day. To evade detection, many malware obfuscation techniques have emerged. Dynamic malware detection methods based on data flow graphs have attracted much attention since they can deal with the obfuscation problem to a certain extent. Many malware classification methods based on data flow graphs have been proposed. Some of them are based on user-defined features or graph similarity of data flow graphs. Graph neural networks have also recently been used to implement malware classification recently. This paper provides an overview of current data flow graph-based malware classification methods. Their respective advantages and disadvantages are summarized as well. In addition, the future trend of the data flow graph-based malware classification method is analyzed, which is of great significance for promoting the development of malware detection technology.

Keywords: Malware detection · Malware classification · Data flow graph · Graph neural network

1 Introduction

The amount of malware is growing rapidly. According to the Av-Test statistics report [1], the cumulative number of malicious programs in 2021 is 1.31264 billion, an increase of 173.4 million compared with 2020. The data show that malicious pro-grams are growing fast. Malware endangers network security seriously and poses a huge threat to our lives. In this regard, various malware detection tools have been widely developed. To avoid detection, many malware obfuscation technologies, including packing and deformation technology, have emerged in recent years [2]. In this case, effectively detecting malware becomes a challenge.

Malware detection can be achieved through software automatic classification technology. Current malware classification methods can be divided into two categories: static analysis methods and dynamic analysis methods [3]. The static analysis method identifies malware by analyzing static features without executing the malware program. Static features include byte sequences, symbolic signatures, file header information, etc. [4]. They are usually simple and efficient. However, they are easily circumvented by some code self-protection methods, such as encrypted and polymorphic methods [5, 6]. In contrast, the dynamic analysis method places a malware in a controllable virtual environment for execution, captures the behavioral trajectory of malware, and detects malware by analyzing the behavioral characteristics from the trajectory. Usually, the sequence of API calls, or the interaction information among system entities, is used as the basis for malware classification [7]. Because the behavioral characteristics of malware have a certain stability, dynamic analysis methods can deal with the obfuscation problem to a certain extent.

Some dynamic analysis methods obtain the behavior characteristics of malware by analyzing their system API call sequence. They cannot deal with obfuscation techniques such as call sequence rearrangement, false call sequence injecting, and equivalent semantic replacement [8]. However, the interaction information among system entities is not easily confused by the above methods. Many studies have constructed data flow graphs to model the interaction information, and many malware classifications based on data flow graphs have been developed. For example, Tobias Wüchner et al. [9] presented a quantitative data flow graph construction method, which expresses the data flow relationship in the form of a graph structure. Yang Pin et al. [10] constructed a data flow graph with attributes to express the behavioral characteristics of malware. Attributes include times attributes and quantity attributes. Malware detection methods based on data flow graphs can overcome the problems caused by some obfuscation methods, including call sequence rearrangement, invalid API insertion, and semantic equivalent substitution. For example, references [9] and [11] define a series of feature attributes for malware data flow graphs and present a malware classification method based on them. In addition, reference [12] proposes a method to compute the subgraph similarity of data flow graphs and then develops a malware detection method based on the similarity index. It is more efficient than graph-matching-based malware detection. Taking a data flow graph as input, reference [10] uses graph progressive neural networks for malware classification. Based on the GraphSAGE framework (graph sample and aggregate) [13], the node aggregation algorithm is modified, and a graph convolutional network ADGCN (attributed data-flow graph convolutional network) is constructed. The ADGCN is suitable for directed weighted graphs, and a pooling layer with an attention mechanism is added to obtain graph-level features.

In view of the widespread attention of data flow graphs, as well as their advantages in dealing with malware obfuscation, this paper surveys of malware classification methods based on data flow graphs. After analyzing the advantages and disadvantages of different data flow graph variants and their corresponding malware classification methods, this paper points out the future development trend of data flow graph-based malware detection methods.

2 Data Flow Graph

A data flow graph is a graph structure used to represent the data flow between system entities generated by calling system API functions. It can effectively extract the behavior characteristics of malware. Unlike API call sequences, data flow graphs and their corresponding malware detection methods are robust to many obfuscation techniques. This section summarizes the basic concepts, construction methods and various variants of data flow graphs.

2.1 Basic Concepts of the Data Flow Graph

Reference [9] proposed a construction method for malware data flow graphs. In a data flow graph, each node represents a system entity. Each edge represents a data flow event between system entities. An edge can be associated with some attributes. For example, Fig. 1 presents an instance of data flow graphs. A directed edge of the process P_1 pointing to the registry R_1 represents the data flow that occurs when the process P_1 writes the registry R_1. The attribute on the edge indicates that process P_1 writes registry R_1 at time t_2, and the total amount of data written is s_2.

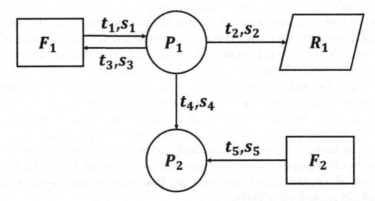

Fig. 1. Data flow graph example.

Formally speaking, a data flow graph can be represented by a 3-tuple $DFG = (V, E, A, \lambda)$, where $V = \{v_1, v_2, \ldots, v_n\}$ is the set of nodes representing entities. An entity is usually a file, a network socket, a registry key, or a process. $E = \{(v_{src}, v_{dst})|v_{src}, v_{dst} \in V\}$ is the set of directed edges representing data flow events among entities. A represents attributes associated with a data flow, which can be the size of a data flow, the occurrence time of flow, or many other attributes of a data flow. The labeling function λ retrieves the attribute value assigned to a node or edge.

During the running process of a malware, many system API functions are called, which generates a corresponding event. It usually causes data flow between system entities, according to which we can construct a data flow graph. The source node of a data flow is set to be a precursor, and the destination node is set to be a successor.

A directed edge is added between them, which represents the data flow relation. Each time a new event is captured, we first check whether the related entity nodes and their corresponding edges have already existed in the data flow graph. If they exist, we only need to update the attributes associated with the edge. Otherwise, we add a new node for each nonexistent entity, add an edge if necessary, and initialize the edge attribute values. This process can be expressed as Formula (1). In formula (1), E is the set of directed edges representing data flow events among entities and e represents a directed edge.

$$update(DFG, src, dst, A) = \begin{cases} \begin{bmatrix} V \\ E \\ \lambda[(e,A) \leftarrow (e, A_{new})] \end{bmatrix}, & if\ e \in E \\ \begin{bmatrix} V \cup \{src, dst\} \\ E \cup \{e\} \\ \lambda[(e,A) \leftarrow (e_{new}, A_{new})] \end{bmatrix}, & otherwise \end{cases} \quad (1)$$

2.2 Data Flow Graph Corresponding to Common APIs

For some common API functions of Windows operating systems, this section gives their corresponding data flow graph fragment.

File System Operations

- ReadFile: This function is called to read bytes from a file into memory. When capturing such an event, we take the target file as a data flow source node and the calling process as a data flow destination node. The number of bytes read by this call, its timestamp, or the total read times can be set as the attributes associated with the data flow;
- WriteFile: This function is called to write bytes to a file. When capturing such an event, we take the calling process as a data flow source node and the target file as a data a destination node. The number of bytes read by this call, its timestamp, or the total read times can be set as the attributes associated with the data flow.

Socket Operation

- Recv: This function is called to read bytes from a network socket. When capturing such an event, we take the target network socket as a data flow source node and the calling process as a data flow destination node. The number of bytes read by this call, its timestamp, or the total read times can be set as the attributes associated with the data flow;
- Send: This function is called to write bytes to a network socket. When capturing such an event, we take the calling process as a data flow source node and the target network socket as a destination node. The number of bytes read by this call, its timestamp, or the total read times can be set as the attributes associated with the data flow.

Registry Operation

- RegQueryValue: This function is called to read bytes from a registry. When capturing such an event, we take the target registry as a data flow source node and the calling process as a data flow destination node. The number of bytes read by this call, its timestamp, or the total read times can be set as the attributes associated with the data flow;
- RegSetValue: This function is called to write bytes to a registry. When capturing such an event, we take the calling process as a data flow source node, and we take the target registry as a data flow destination node. The number of bytes read by this call, its timestamp, or the total read times can be set as the attributes associated with the data flow.

Process Operation

- CreateProcess: This function is called to create a new process. When capturing such an event, we take the calling process as a data flow source node and the new process data flow destination node. The number of bytes read by this call, its timestamp, or the total read times can be set as the attributes associated with the data flow;
- ReadProcessMemory: This function is called to read bytes from another process. When capturing such an event, we take the target process as a data flow source node and the calling process data flow destination node. The number of bytes read by this call, its timestamp, or the total read times can be set as the attributes associated with the data flow;
- WriteProcessMemory: This function is called to write bytes to another process. When capturing such an event, we take the calling process as a data flow source node and take the target process as a data flow destination. The number of bytes read by this call, its timestamp, or the total read times can be set as the attributes associated with the data flow.

Related examples are shown in Table 1.

2.3 Extension of Data Flow Graphs

To capture more malware behaviors, some references further extend the information on edge properties of data flow graphs. We call this the extended data flow graph. References [9, 11, 12] take both the data flow size and the timestamp as the edge attributes in a data flow graph.

While s represents the size of the data flow, t represents the timestamp. Events between the same pair of entities are merged instead of creating their own events. If the data flow from node A to node B occurs multiple times in the data flow graph, for example, $e_1(A,B,s_1,t_1)$ and $e_2(A,B,s_2,t_2)$, they will be merged into $e(A,B,s_1 + s_2, min(t_1 + t_2))$. An example is shown in Fig. 2. P_1 represents a process node and

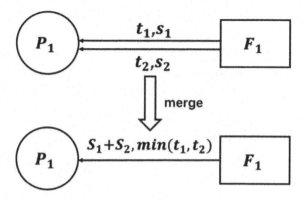

Fig. 2. Quantitative data flow diagram example of data flow size and occurrence time

F_1 represents a file node. P_1 reads F_1 at time t_1 and time t_2, and the data flow sizes are s_1 and s_2, respectively. The generation process is shown in formula (2), and the meaning of the symbols in formula (2) is shown in Table 2.

$$update(G,(src,dst,s,t)) = \begin{cases} \begin{bmatrix} N \\ E \\ \lambda \begin{bmatrix} (e,size) \leftarrow (e,size) + s \\ (e,time) \leftarrow (min(\lambda(e,time),t) \end{bmatrix} \end{bmatrix}, & if \ e \in E \\ \begin{bmatrix} N \cup \{src,dst\} \\ E \cup \{e\} \\ \lambda \begin{bmatrix} (e,size) \leftarrow s \\ (e,time) \leftarrow t \end{bmatrix} \end{bmatrix}, & otherwise \end{cases} \qquad (2)$$

The attributes used in reference [10] are times and quantity. The times attribute records the data flowing times of a directed edge, and the quantity attribute records the total amount of data flowing on the directed edge. As shown in the example shown in Fig. 3, data flows from file F_1 to process P_1 s times, and the data amount is l in total.

The generation algorithm of a data flow graph is as follows, where G = (V, E) represents the data flow graph, V represents the finite set of nodes, and E represents the set of directed edges.

Table 1. Example of event data flow diagram

Action category	API function	Sample graph
File system operations	ReadFile	Calling Process ←t,s— Source File
	WriteFile	Calling Process —t,s→ Destination File
Socket operation	Recv	Calling Process ←t,s— Source Address
	Send	Calling Process —t,s→ Destination Address
Registry operations	RegQueryValue	Calling Process ←t,s— Source Key
	RegSetValue	Calling Process —t,s→ Destination Key
Process operation	CreateProcess	Calling Process —t,s→ Callee Process
	ReadProcessMemory	Calling Process ←t,s— Source Process
	WriteProcessMemory	Calling Process —t,s→ Destination Process

Table 2. Some symbolic meanings

Symbol	Meaning
G	Data flow graph
N	A node set consisting of system entities
E	An edge set consisting of data flows between system entities
e	An edge that represents the data flow
src	Originating system entity
dst	Destination system entity
s	Data flow size
t	Timestamp

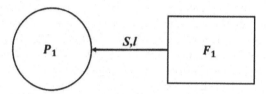

Fig. 3. Quantity flow size, times attribute data flow diagram example

---

Algorithm 1. Generation algorithm of data flow graphs

Input: API sequence $S=\{api_1, api_2, ..., api_n\}$;

Output: data flow diagram $G = \{V, E\}$;

1. FOR api in S DO
2. $event \leftarrow parse(api)$;
3. $v_{src} \leftarrow event.src$;
4. $v_{dst} \leftarrow event.dst$;
5. $e \leftarrow (v_{src}, v_{dst}, event.size)$;
6.   IF v_{src}, v_{dst}, e NOT in G
7. $G.add\,Nodes(v_{src}, v_{dst})$;
8. $G.add\,Edges(e)$;
9.   ELSE
10. $G.update(e)$;
11.   END IF
12. END FOR

3 Malware Classification Based on Data Flow Graph

The data flow graph models data flow relationships generated between a series of events during calling API functions by malware. The data flow graphs can effectively represent the behavior of malware without interference from obfuscation techniques such as API sequence rearrangement and insertion invalid API functions. Malware classification methods combined with data flow graphs also have good performance. The existing data flow graph-based malware classification methods can be classified into three types: combining with user-defined features, combining with graph similarity calculation, and combining with graph neural networks.

3.1 User-Defined Data Flow Graph Feature-based Malware Classification

Reference [9] proposed a heuristic method to determine the behavior expressed by the data flow graph of malware. The basic idea of the heuristic method is to determine the data flow and related attributes corresponding to the typical high-level behavior of malware. The heuristic methods are classified into three types: replication heuristics, manipulation heuristics, and quantitative heuristics. Replication heuristics can trigger the activity of malware characterized by replication behavior to infect local or remote systems. There are four types of replication heuristics: local replication, network replication, binary infection, download and launch payload.

- Local replication: malicious behavior of malware replicating itself by generating cloned processes from its own binary image file.
- Network Replication: malicious behavior of malware replicates itself by sending its binary image over the network to infect remote systems.
- Binary Infection: malicious behavior of malware infecting a target binary by appending its own code to benign binaries.
- Download and Start Payload: malicious behavior of malware reloading and executing other malicious payloads from the Internet.

   Manipulation heuristics target certain data flows that contain specific high-level semantics related to manipulation of the system by specific activities such as integrity or data confidentiality. The three main types of manipulation heuristics include leaking cached Internet data, a Spawn shell, and a deploying system driver.

- Leak Cached Internet Data: malicious behavior in which malware attempts to steal sensitive data from a dedicated cached Internet file by sending it to a remote network location.
- Spawn Shell: malicious behaviors of malware pasting a command shell into a socket to open a back door.
- Deploy System Driver: malicious behavior resulting from malware attempting to deploy and load malicious system drivers into the Windows kernel to inject its rootkit functionality.

   Quantitative heuristics trigger typical quantitative data flow properties of malware. Quantitative heuristics include both single-hop data entropy and registry call ratio.

- Single-hop Data Amount Entropy: A malicious behavioral data flow signature that triggers malware to replicate itself by infecting other binaries or injecting code into other system processes.
- Registry Call Ratio: malicious behavior with a relatively high rate of registry-related streams since malware typically performs many registry queries to probe the state of the infected system before starting the actual malicious activity.

Reference [11] proposed a malware detection method based on similarity metrics. The method proposes several metrics based on the local and global features of malware behavior. The local features include entropy and dataflow proportion.

- By calculating normalized entropies of edge features, such as dataflow size and event count, features such as the uniformity of dataflow distribution and the relative sensitivity of node outgoing edges are captured. The entropy criterion can be used to detect malware that infects other executables or processes by injecting or appending its own binary image.
- Dataflow proportion calculates the proportion of a certain type of output data flow from a node to all outgoing data flows from that node. The dataflow proportion calculates different proportions of traffic to distinguish malware from benign malware. Usually malware visits a node with a high percentage of outgoing edges.

The global features include closeness and betweenness.

- Closeness is the inverse of the average distance from a node to other nodes in the same graph. A high closeness indicates that the node is closely connected to all other nodes in the graph. Closeness can be used to detect malware that replicates by infecting other processes and binaries.
- Betweenness represents the relative fraction of all shortest paths between all possible pairs of nodes through a particular node. A high betweenness indicates that the node communicates communications among many nodes. This metric captures how often a process mediates partial data flow between other system resources. It can be used to identify malware for man-in-the-middle attacks.

3.2 Data Flow Graph Similarity-Based Malware Classification

The graph similarity-based malicious program detection method uses machine learning techniques to achieve the identification of malicious codes by designing graph similarity measures [14]. The method will model the behavior of malware through data flow graphs and build a library of existing data flow graphs of malware. Based on the data flow graph, the similarity between the unknown code and the malicious code in the library is calculated to classify the unknown code. Reference [15] is the first to propose behavioral subgraphs that exist only in malicious programs. Reference [16] proposed extracting malware behavior features based on HotPath such as the maximum common subgraph. Reference [17] proposed constructing frequent subgraphs to represent common behaviors of malware of the same family. Reference [18] first converts the malware's API call sequence into a call graph through dynamic analysis. Then, the abstract node converts the call graph into a semantic signature. Finally, all subgraphs are extracted and the behavioral similarity of the two malware types is analyzed by subgraph similarity.

A subgraph similarity-based malware detection method is proposed in the literature [12]. The method expresses malicious features by constructing a dataflow graph representing malicious behaviors of malware and extracts feature subgraphs of the dataflow graph, which reduces the structural complexity of the graph while covering all malicious behaviors. Using the inverse topology to map subgraphs into strings makes matching simpler and faster. After the above steps, a feature subgraph library is constructed and similarity learning is performed to obtain subgraph embedding. In the training model phase, feature extraction is first performed on the samples in the training set. Based on this, graph vectors are obtained using a feature subgraph library. After that, the training sample kernel matrix is constructed by the graph similarity function. Finally, the classifier is constructed by the kernel matrix for malware detection. In the detection phase, after feature extraction of the unknown code, the similarity between the graph vector and the training sample vector is calculated to classify the malware.

3.3   Graph Neural Network-Based Malware Classification

As a kind of non-Euclidean data, a graph has an irregular and disordered spatial structure and stores a large amount of graphical information. A graph mainly consists of objects and relationships between objects. Graph neural networks can be applied to graph structures where traditional machine learning cannot work effectively. The data flow graph of malware is a typical graph, which consists of entity nodes and the relationships between nodes. Using graph neural networks, the malware data flow graph is modeled effectively to obtain the relation features of the data flow graph [19]. The Graph Convolutional Neural Network(GCN) [20] can extract the features of the graph structure more effectively. Reference [21] proposed a novel behavioral malware detection method based on Deep Graph Convolutional Neural Networks (DGCNNs) to learn directly from API call sequences and their associated behavioral graphs.

A malware family classification method based on attribute data flow graph is proposed in reference [10]. The method executes PE files of malicious codes in a sandbox and extracts the system API functions called during the malicious code run to obtain the API call sequences. The attribute dataflow graph is constructed based on the sequence, and the dataflow graph is classified using a graph convolutional neural network. The attribute dataflow graph classifies nodes into four types: files, processes, registries and sockets. Attributes consider the size and number of data flows between nodes. The improved GraphSAGE node aggregation algorithm is proposed to be applicable to directed weighted graphs by adding an attention mechanism in the pooling layer. The nodes in the attribute data flow graph are mapped to meaningful vectors by node embedding. The graph is expressed by graph embedding and the malware is classified by graph embedding (Fig. 4).

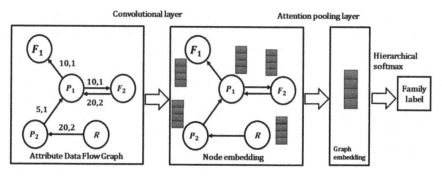

Fig. 4. Graph convolutional neural network ADGCN

4 Discussion

The existing data flow graph-based malware detection model has good detection performance in all aspects and can effectively detect various obfuscation techniques, such as the insertion of useless API sequences and API call rearrangement. The data flow graph-based model that considers the event occurrence time as well as the dataflow can capture the time of malware operation and obtain the temporal feature of nodes. However, some methods update events by simply summing the dataflow, ignoring the number of events that occurred. Other methods consider the number of events as well as the dataflow in the malware data flow model but ignore the temporal information of nodes.

The heuristic-based methods express malicious behaviors through data flow graphs to identify malware with high accuracy and low false positives, and have high performance in terms of time-space efficiency. The heuristic-based approach, although it can detect using multiple behavioral descriptions of different malware families, still has difficulty coping with the speed of malware updates.

The metric-based methods construct a feature set including centrality, etc. for each node in the data flow graph and uses machine learning techniques to train a classifier to achieve malware identification. These methods have good robustness and significantly improve detection accuracy.

The subgraph similarity-based methods can effectively avoid the computational inefficiency caused by graph matching. Considering the relationship between subgraphs, the problem that the traditional graph matching method ignores the relationship between subgraphs is solved. These methods have high malware detection accuracy. However, the study did not evaluate the data volume of the subgraph.

The graph neural network-based methods reduce malware detection time by reducing the complexity of constructing graphs. These methods have high detection accuracy but have limited effect on malicious code that can simulate the detection environment and hide malicious behavior.

In addition, the various types of data flow graph-based approaches described above do not model the order between instructions in API calls, resulting in the loss of sequential information. These approaches have advantages and disadvantages. The good is that they can circumvent the negative impact of obfuscation techniques such as instruction disorder on malware detection. The disadvantage is that API call order is actually one of the key features of malware. Discarding API call order completely tends to degrade malware detection performance. To this end, we will investigate the scheme of incorporating api order information in the dataflow graph to improve the accuracy of malware detection. Finally, the abovementioned related studies have not graded the node granularity when constructing the data flow graph and selecting the nodes of data flow graph entities. In our future research, we will propose a data flow graph construction method with grained nodes for different types of viruses.

5   Conclusion

Existing methods based on data flow graphs alleviate the limitations of API sequence-based malware detection, such as the inability to effectively detect API sequence rearrangement and insertion of invalid sequences, and achieve better performance in the detection of malware. The problems faced by the existing methods and the improvement directions are as follows.

1. The call sequence order feature is completely discarded, and future work can add API call order information to the data flow graph.
2. In constructing the data flow graph, the addition of edge attributes is considered comprehensively to further improve the data flow graph.
3. The node granularity of the dataflow graph is not graded, and the node granularity can be graded according to the malware type in the future.

References

1. AV-TEST anti-virus testing agency static analysis report. https://www.avtest.org/de/statistiken/malweare
2. Yan, L.: Detection and classification of malicious programs based on deep learning. Xidian University
3. Damodaran, A., Troia, F.D., Visaggio, C.A., Austin, T.H., Stamp, M.: A comparison of static, dynamic, and hybrid analysis for malware detection. J. Comput. Virol. Hack. Tech. **13**, 1–12 (2017)
4. Jialai, W., Chao, Z., Xuyan, Q., et al.: Overview of intelligent detection of malware on windows platform. J. Comput. Res. Dev. **58**(5), 18 (2021)
5. Bat-Erdene, M., Park, H., Li, H., Lee, H., Choi, M.S.: Entropy analysis to classify unknown packing algorithms for malware detection. Int. J. Inf. Secur. **16**(3), 1–22 (2016)
6. Cesare, S., Xiang, Y.: A fast flowgraph based classification system for packed and polymorphic malware on the Endhost. In: IEEE International Conference on Advanced Information Networking & Applications. IEEE (2010)
7. Zhou, Y., Lu, T., Du, Y., Guo, R., Bao, Y., Li, Mo: Detection and analysis of windows malicious code based on thread fusion features. Comput. Eng. Appl. 1–11 (2020)
8. Min, X., Tianfu, Z.: Malicious code detection method based on behavioral features. Netw. Inform. **6**, 14–16 (2009)
9. Wüchner, T., Ochoa, M., Pretschner, A.: Malware detection with quantitative data flow graphs. In: Proceedings of the 9th ACM symposium on Information, computer and communications security (ASIA CCS'14), pp. 271–282. Association for Computing Machinery, New York, NY, USA (2014)

10. Pin, Y., Yue, Z., Lei, Z.: Malware code family classification based on attribute data flow graph. Inform. Secur. Res. **6**(3), 7 (2020)
11. Wüchner, T., Ochoa, M., Pretschner, A.: Robust and effective malware detection through quantitative data flow graph metrics. In: Almgren, M., Gulisano, V., Maggi, F. (eds.) Detection of Intrusions and Malware, and Vulnerability Assessment, pp. 98–118. Springer International Publishing, Cham (2015). https://doi.org/10.1007/978-3-319-20550-2_6
12. Jie, W., Changqing, W.: Malware detection method based on subgraph similarity. J. Softw. **31**(11), 12 (2020)
13. Hamilton, W., Ying, Z., Leskovee, J.: Inductive representation learning on large graphs. In: Advance in Neural Information Processing Systems, pp. 1024–1034. MIT Press, Cambridge, MA (2017)
14. Wang, Z., Shen, H., Cao, G., Cheng, X.: A review of graph classification research. J. Softw. **33**(1), 171–192 (2022)
15. Christodorescu, M., Jha, S., Kruegel, C.: Mining specifications of malicious behavior. In: Proc. of the 1st India Software Engineering Conf., pp. 5–14. ACM, New York (2008)
16. Park, Y., Reeves, D.S., Stamp, M.: Deriving common malware behavior through graph clustering. In: Proc. of the 6th ACM Symp. on Information, Computer and Communications Security, pp. 497–502. ACM, New York (2011)
17. Fan, M., et al.: Frequent subgraph based familial classification of android malware. In: IEEE International Symposium on Software Reliability Engineering. IEEE (2016)
18. Kwon, J.-H, Lee, J.H., Jeong, H.-C., Lee, H.: Metamorphic malware detection using subgraph matching. J. Korea Inst. Inform. Secur. Cryptology 21(2) (2011)
19. Wu, Z., Pan, S., Chen, F., Long, G., Zhang, C., Yu, P.S.: A comprehensive survey on graph neural networks. IEEE Trans. Neural Netw. Learn. Syst. **32**, 4–24 (2019)
20. Kip, F.T.N., Welling, M.: Semi-supervised classification with graph convolutional networks. IEEE Trans. Neural Netw. Learn. Syst. **32**, 4–24 (2016)
21. Schranko de Oliveira, A., Sassi, R.J.: Behavioral malware detection using deep graph convolutional neural networks. TechRxiv. Preprint (2019). https://doi.org/10.36227/techrxiv.100 43099.v1

Anomaly Detection of Multivariate Time Series Based on Metric Learning

Hongkai Wang[1], Jun Feng[2], Liangying Peng[1($\boxtimes$)], Sichen Pan[1], Shuai Zhao[2], and Helin Jin[3]

[1] State Grid Zhejiang Information and Telecommunication Branch,
Hangzhou, China
peng_liangying@zj.sgcc.com.cn
[2] Grid Zhejiang Information and Telecommunication Branch, Hangzhou, China
[3] Harbin Institute of Technology, Harbin, China

Abstract. Most of the current methods for anomaly detection in time series are unsupervised. However, unsupervised learning assumes the distribution of the data and cannot obtain satisfactory results in some scenarios. In this paper, we design a semisupervised time series anomaly detection algorithm based on metric learning. The algorithm model mines the features in the time series from the perspectives of the time domain and frequency domain. Furthermore, we design a loss function for anomaly detection. Different from the two-class loss function, in the scenario of the loss function we designed, the normal data will be clustered and distributed in the embedding space, and the abnormal data will be far from the normal data distribution. Furthermore, we extend our designed metric learning model to a semisupervised learning model, extending the labeled dataset with the unlabeled dataset by setting different confidence levels. We conduct experiments on different public datasets and compare them with commonly used time series anomaly detection algorithms. The results show that our model has a good effect. At the same time the semisupervised setting does improve the accuracy of model detection.

1 Introduction

In the scenario of anomaly detection of multivariate time series, many existing algorithms cannot achieve good results [8]. In particular, the distance-based anomaly detection algorithm, usually considers that a location with a low distribution density of points is more likely to be an anomaly. However, this kind of algorithm is very dependent on the definition of distance. The usual distance definition methods include Euclidean distance and Editing distance. However, this kind of distance definition method becomes inapplicable in the case of high-dimensional time series data. Euclidean distance does not consider the relationship between dimensions, and edit distance can only measure the distance of a single dimension, and then expand to multiple dimensions. These methods have poor performance in high-dimensional scenarios. Extending to high-dimensional

Y. Wang et al. (Eds.): ICPCSEE 2022, CCIS 1628, pp. 94–110, 2022.
https://doi.org/10.1007/978-981-19-5194-7_8

time series data, the situation becomes more complicated. In addition to individual data points, we also need to consider the temporary dependence that exists between data points. Therefore, anomaly detection in multidimensional time series is very challenging.

Today, popular anomaly detection models are usually unsupervised. In the time series of real scenes, label data are more difficult to obtain [2]. However, these unsupervised learning methods assume the distribution of the data and consider those points in the data where the distribution gathers as normal points. However, in practical scenarios, such a simple assumption can lead to many misjudgments. For example, there may also be some clustered outliers in the clustered distribution points. For example, in the data generated by wind turbines, anomalies include sparse outliers and stacked outliers. Stacked outliers are abnormal points in the aggregated distribution. Using traditional unsupervised anomaly detection methods will misjudge these data as normal points. However, at this time, we know the anomaly of this part of the point, and the unsupervised learning method cannot use this knowledge to help detect the anomaly.

Supervised learning can solve problems where data labels cannot be exploited. In real scenarios, prior knowledge can also be converted into labeled data to utilize this knowledge. The supervised learning anomaly detection algorithm has been researched in network intrusion detection, but less research has been conducted in other fields. On the one hand, data labels are difficult to obtain in real-world scenarios, and the cost of labeling datasets is too high; on the other hand, even if we have labeled datasets, it is difficult for us to obtain all types of anomalies and abnormal randomness. The label data required for supervised learning anomaly detection algorithms can be very large. Therefore, it is unrealistic to use supervised learning algorithms to solve the problem of anomaly detection in real scenarios [3]. Therefore, in summary, the challenges faced by anomaly detection in high-dimensional time series data are as follows:

- Currently, the dimension of time series is high and the number is large, and it is difficult for traditional anomaly detection methods to obtain good detection results.
- The unsupervised anomaly detection algorithm has a strong assumption about the data distribution, and considers that the points of the cluster distribution are normal points. However, the actual situation may be more complicated than this, and some abnormal points are not discretely distributed.
- Supervised learning anomaly detection algorithms can utilize knowledge in different fields, but the reality is that we cannot obtain such a large and complete data label, so supervised learning anomaly detection algorithms are not practical.

Based on these challenges, we propose a time series anomaly detection model based on metric learning, and propose a new loss function adapted to deep metric learning anomaly detection. Combining the advantages of CNN and LSTM, our model is a model that can effectively extract features from high-dimensional time series data and embed them into low-dimensional space. The model uses random dimension permutation and short-time Fourier for feature extraction of raw time series. Among them, random dimension permutation can effectively mine feature associations between different dimensions, and add randomness to improve the robustness of the model. Short-time Fourier techniques can mine sequence-related information in the frequency domain. It also includes raw time series input into the LSTM module to mine short-term dependencies in time series data to form the final model. To better allocate the influence of the vectors of the embedding space of different dimensions on the center, we add an attention module to optimize the solution method of the category center in the place where the model finds the normal category center. Moreover, we propose a metric learning loss function suitable for anomaly detection, and experimentally verify that it is superior to loss functions such as cross entropy.

Deep metric learning is generally used to solve classification tasks [11]. There are many metric learning methods for classification tasks, but the basic idea is to make the distance between the data of the same category in the metric space as close as possible, and the distance between the data of different categories as far as possible. Such as the use of the prototype network [10]. However, this is not suitable for anomaly detection scenarios. Usually, anomaly detection can be regarded as a binary classification problem. However, there may be many types and causes of anomalies, so it is unreasonable to directly regard anomalies as one category. A point of view we agree with is that anomaly detection is a one-class classification problem [6], that is, judging whether the data are of a normal category. If the data are not of a normal class then it is turned into an anomaly. One-class SVM is an algorithm that thinks like this. Therefore, in the process of anomaly detection by metric learning, we only need to save a "center" of a normal category. The normal samples are as close as possible to this center, and the anomalies are as close as possible to this center [1]. The loss function we designed is based on this idea.

In addition, our model is also able to utilize a large amount of unlabeled data, extending to a semisupervised learning model. We can use a small amount of data to first learn the parameters of the model and the center of the class, and then use the unlabeled data to augment our dataset. At the same time, users can select different thresholds according to their needs to obtain data with different confidence levels. The final model only needs to use a small amount of data to achieve good results. In summary, the contributions of our paper can be summarized as follows:

- We propose a model that can extract the features of high-dimensional time series, and jointly mine data information from three perspectives: high-dimensional, time-series, and frequency domains.
- Based on the proposed network model, we propose a metric learning loss function suitable for anomaly detection. It can provide a low-dimensional distance space representation for time series. In this distance space, normal categories will be clustered together, and abnormal data will be far away from normal data clusters.

- We extend the model to semisupervised learning. The model can get good performance using only a small amount of labeled data and some unlabeled data.
- We conduct experiments on different public datasets and also compare supervised and semisupervised learning. The results show that both our model and the semisupervised setting have good performance.

The content of the following article is summarized as follows. First, a basic introduction and definition of some definitions, such as time series, anomaly scoring, etc. Then we mainly introduce our high-dimensional time series feature extraction model, which includes and processing modules, including segmentation, short-time Fourier transform on segments and random dimension permutation. Then the convolutional neural network, the recurrent neural network and the attention mechanism added to the center are assembled to form the final network model. Then we introduce the loss function of the model and the form of anomaly scoring, and extend the model to the semisupervised learning module. Finally, experiments are carried out on public datasets to verify the effectiveness and accuracy of the model.

2 Preliminaries

Time Series

A time series $Q = <\mathbf{q_1}, \mathbf{q_2}, \ldots, \mathbf{q_n}>$ is a chronologically organized sequence of vectors, each of which $\mathbf{q_i} = (s_i^{(1)}, s_i^{(2)}, \ldots, s_i^{(k)})$ represents the data generated at time t_i, where $1 \leq i \leq n$. And the vector sequence is arranged in chronological order, that is, when $i < j$, we have $t_i < t_j$. When $k = 1$, the time series is *univarate*, and when $k > 1$, the time series is *multivariate*. Here we mainly study the challenges brought by the current high-dimensional time series data, that is, the scenario of $k > 1$.

Anomalies and Anomaly Scoring in Time Series

Given a time series $Q = <\mathbf{q_1}, \mathbf{q_2}, \ldots, \mathbf{q_n}>$, our goal is to find those data points that are incorrect. To achieve this, for each point $\mathbf{q_i}$ in the sequence, we can calculate the anomaly score $OS(\mathbf{q_j})$ of it. The higher $OS(\mathbf{q_j})$ is, the more likely $\mathbf{q_i}$ is an anomaly.

Semisupervised Time Series Anomaly Detection

The current mainstream anomaly detection algorithms are unsupervised learning, and there are also a small number of supervised learning anomaly detection examples. However, semisupervised learning anomaly detection [9] is more suitable for use in real scenarios. In the context of semisupervised learning for multidimensional time series anomaly detection, we believe that the labels used in training the model are not sufficient. That is, only a portion of the data has labels. Therefore, we plan to use our model to label unlabeled data so that these data can also be used in supervised learning multidimensional time series anomaly detection algorithm. Therefore, the dataset used in the training process is a set of tuples, $((\mathcal{Q}, \boldsymbol{y}), \widetilde{\mathcal{Q}}))$, where $\mathcal{Q}$ has the label set $\boldsymbol{y}$, while $\widetilde{\mathcal{Q}}$ has no label. In order to solve the problem of insufficient datasets, the usual semisupervised learning multidimensional sequence anomaly detection problem first uses the labeled dataset $(\mathcal{Q}, \boldsymbol{y})$ to train a classifier $\tilde{f}_Q \mapsto y$, and then use this classifier to label our unlabeled dataset $\widetilde{\mathcal{Q}}$, and finally get an expanded dataset. The problem can then be treated as a supervised learning task.

3   Proposed Model

This section introduces our model. The first is data preprocessing, including segmentation, short-time Fourier transform and random dimension permutation. The preprocessing part allows the model to better extract features from the data. Then there is our proposed network model, which combines different inputs through convolution and LSTM and other structures to extract features at different levels, and finally combines them to form the final embedding vector. Then comes the loss function part of the model, in which we propose a loss function based on metric learning for anomaly detection. Then combined with the loss function, it illustrates how the model scores anomalies. Finally, the model is extended to the form of semisupervised learning, making it effective in detecting anomalies even with a small number of samples. The overall flow of the model is shown in Fig. 1.

3.1   Preprocessing

The first step is to segment the data. Using average segmentation similar to PAA [4] will further reduce our limited data, and at the same time, it is not good enough for continuous features of time series data. Therefore, to better obtain data from small labeled samples, and at the same time to allow the segmented data to retain the continuity in the data as much as possible, we consider direct coverage between segments during segmentation. If the specified segment size is s, we define a coverage rate of τ, and unlike the average segment, we let two adjacent segments have $\lceil \tau \times s \rceil$ is identical. For example, the sequence $<1, 2, 3, 4, 5, 6, 7, 8>$. If the segment size is 4, then the result of the segment is $<\{1, 2, 3, 4\}, \{3, 4, 5, 6\}, \{5, 6, 7, 8\}>$. After segmenting, we obtain a matrix of $n \times s \times d$, where n is the number of segments, s is the segment length, and d is the dimension of the data point.

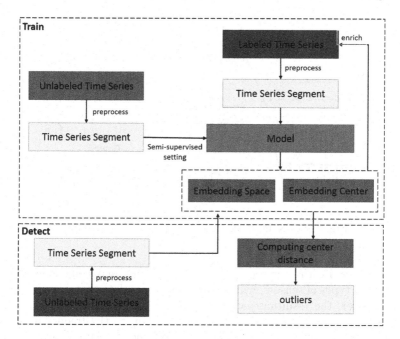

Fig. 1. Overall flow

We preprocess the obtained sequence of data segments, we preprocess it to extract features from the data. The ones used here include random dimension permutation and short-time Fourier transform. Random dimension permutation is a method that can mine the correlation between different dimensions of multi-dimensional time series. The short-time Fourier transform can obtain the frequency domain signal corresponding to the data segment, which can provide a new perspective for anomaly detection in many cases. The specific method is as follows:

Random dimension permutation can mine the associations between different dimensions in multi-dimensional time series, and can also effectively mine the patterns contained in some dimensions. We extend it again so that it can accommodate time-series data segments. Assuming that our time series dimension is k, we want to divide each data segment into g groups. Then we can calculate the dimension size of the data in each group as $\varphi = \left\lfloor \frac{m \cdot \alpha}{g} \right\rfloor$, where φ is a parameter that controls the size of the group, and $\lfloor \cdot \rfloor$ is the symbol for rounding down. Each random dimension permutation process needs to randomly arrange the dimensions of the data, and then select the first φ dimensions as the result obtained this time. In the data segment scenario, we only need to randomly arrange each dimension of the data segment according to the same arrangement rule, and finally for each data segment, we can obtain a $g \times s \times \varphi$ size matrix. An example of random dimension permutation is shown in Fig. 2.

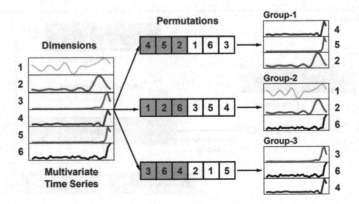

Fig. 2. An example of random dimension permutation

When we deal with time series, the data we obtain may come from different locations and types of sensors, which makes us obtain data that are not synchronized. However, the resulting pattern will be embodied in a certain "shape", when we transform the data from the time domain to the frequency domain, which will give the model a new perspective to mine the data. Here we use the short-time Fourier transform [5]. The research of Li et al. shows that the short-time Fourier transform can effectively extract the features of time series data in the time domain, and then substitute it into the neural network as the preprocessed data for the next step of training. For the time series segment obtained by the previous processing, the mth data in the nth data can be converted into:

$$STFT^{(\tau,s)}\{\mathbf{x}\}(m,n) = \sum_{t=1}^{T} \mathbf{x}(t) \cdot \mathbf{w}(t - s \cdot m) \cdot e^{-j\frac{2\pi n}{\tau}(t - s \cdot m)}$$

where $STFT^{(\tau,s)}$ represents our short-time Fourier transform function with parameters τ and s. The difference between the Fourier transform and the short-time Fourier transform is that the short-time Fourier transform performs the Fourier transform on the local segments after the data are segmented. In our data preprocessing process, the data has been segmented, so only the Fourier transform on the segment is equivalent to the short-time Fourier transform.

The two preprocessing processes are from different aspects, one is to extract features between different dimensions in high-dimensional data, and the other is to mine more information in the frequency domain in single-dimensional time series data, and consider substitute the original data into the model, and finally form all the inputs of the model.

3.2   Encoder for High-Dimensional Time Series Data

The main purpose of this part of the encoder is to make a low-dimensional representation of a high-dimensional time series data segment for further processing and use. TapNet is an effective high-dimensional time series feature extraction method [12], unlike theirs, our model is calculated in segments, which is very good for data scale expansion capability. Moreover, the short-time Fourier transform module is added to our module, which can extract more information from the frequency domain.

The network model is mainly divided into three parts, one of which is to flatten the raw time series segments, and then the module for feature extraction through LSTM. One part is to perform a short-time Fourier transform on time series data and then perform a one-dimensional convolution to extract features. The last part is a process of randomly extracting time series data, that is, random dimension permutation, and then performing one-dimensional convolution in different dimensions. Among them, the LSTM module is proposed to better extract time series features in the time series. The purpose of flattening the original time series segments is to make the data fit the input of the LSTM module. In this part, our input is $X \in \mathcal{R}^{s \times d}$, and the size of the output is the size of the latent space of the LSTM module, i.e. $X_{lstm} \in \mathcal{R}^{s \times h_{lstm}}$.

Then, methods based on short-time Fourier transform and random dimension extraction are both based on convolutional neural networks. Because of the particularity of time series, the convolutional neural networks we use are all one-dimensional. Moreover, batch norm and ReLU activation functions are added after each convolutional layer to better extract features, and finally a maxpool module is used for pooling. The short-time Fourier module changes the thinking and extracts the features of the time series in the frequency domain. Research shows that such feature extraction is effective. Then, after performing the short-time Fourier transform module on the time series segment, the convolution module can also obtain an output $X_{SFTF} \in \mathcal{R}^{s' \times h_{sftfconv}}$.

The random dimension permutation module is designed to randomly extract some dimensions in the data multiple times, which can not only randomly learn the information interaction between different dimensions, but also can improve the robustness of the model. The specific operation process is similar to TapNet. For each decimation, we obtain an output $X^{(i)_{randomconv}} \in \mathcal{R}^{d_f \times 1}$. Finally, the results of these three parts are expanded and spliced to obtain a synthetic long vector. Then this long vector is merged and dimensionally reduced through a fully connected layer to obtain the final embedding space vector. Then it can be trained or predicted according to different scenarios (Fig. 3).

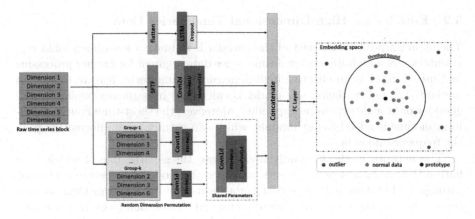

Fig. 3. Model

3.3   Attentional Center Learning

The main goal of our learning is the parameters of the network model, and the center of the normal data in the embedding space. Because we can only work with limited data, there may be an inductive bias in the learning of the network model [7]. Therefore, we added an attention module to the last center learning, that is, attentional center learning. This attention mainly solves that the normal data and abnormal data may focus on different embedding vectors when calculating, that is, the contribution of different dimensions to the center calculation in the embedding space may be different, and each dimension may not be the same. Want to wait for the weight.

The idea of the attention module is as follows: when calculating the center $\mathbf{c}$, attention can be added to each dimension, which is formulated as $\mathbf{c} = \sum_i A_i \cdot Hi$, where A is the attention vector of the embedding dimension of the center, and H is all datasets of the normal category. A is the parameter to be learned through learning, $A = softmax(\mathbf{w}^T tanh(VH^T))$, where w and V are both learned during the learning process model parameters. If this attention module is added, after the embedding vector is finally obtained, an attention operation must be performed on the vector to obtain the final embedding vector.

3.4   Loss Function

Intuitively, the anomaly detection problem can be regarded as a binary classification problem, but it is somewhat different from the traditional binary classification problem. However, there are only two types of result labels for anomaly detection, due to the randomness and diversity of outliers. Therefore, we do not solve it as a binary classification problem, but an "either-or" single-class detection. We only need to learn the category features of normal data as much as possible. If the data have a large deviation from the normal category, we can treat it as an anomaly. Therefore, our loss function is formulated as follows:

$$loss = \frac{1}{|Q_p| + |Q_o|}[\sum_{\mathbf{x} \in Q_p} d(f_\phi(\mathbf{x}), \mathbf{c}_s) + \alpha \sum_{\mathbf{x} \in Q_o} (md(f_\phi(\mathbf{x}), \mathbf{c}_s))]$$

where $\mathbf{c}_s = \frac{1}{|S_p|} \sum_{(\mathbf{x},y) \in S_p} f_\phi(\mathbf{x})$, S_p is a sample; Q_s is a query sample dataset of normal data, Q_o is a query sample dataset of abnormal data, $f_\phi : \mathcal{R}^D \rightarrow \mathcal{R}^M$ is the embedding function learned by the neural network, ϕ is the parameter set to be learned by the neural network; α is the parameter used to balance the two distance weights, $d : \mathcal{R}^M \times \mathcal{R}^M \rightarrow [0, +\infty)$ is a distance method to measure the similarity of two vectors. This is the Euclidean distance, because the specific features are learned by the network model, so the form of the definition is not important. The meaning of the final loss function definition is to make the normal data as small as possible from the class center, and make the abnormal data as large as possible from the class center. The parameters finally learned by the model are the parameter set ϕ learned by the neural network and the category center $\mathbf{c}_s$ of the normal data.

In addition, the parameters of the model and the center of the embedding space are updated. Because the data brought in by our method for each training are the data segment at this moment, the data segment at the next moment cannot be brought into the center of the data segment completely as the center of the whole, which is easily destroyed due to anomalies. The stability of the model. For the center $\mathbf{c}_{t-1}$ obtained at the previous moment, and the center $\mathbf{c}'_t$ obtained from the normal data after the training of the current data segment, the calculation method of the final center is: $\mathbf{c}_t = \mathbf{c}_{t-1} + \beta \mathbf{c}'_t$, that is, the calculation of the center is incremental. Therefore, our model can also be easily extended to the mode of online algorithms. This is also one of the strengths of our model. The model training process can be shown as Algorithm 1.

Algorithm 1. The process of training the model, where $SUBSERIES$ $(Q, start, end)$ represents the subsequence of the sequence Q from $start$ to end

---

Require: Training dataset $Q =< (\mathbf{q_1}, y_1), (\mathbf{q_2}, y_2), \ldots, (\mathbf{q_n}, y_n) >$, where $y_i \in 0, 1$, 0 means normal data, 1 means abnormal data, window size s, coverage rate τ, class center obtained in the previous round $\lrcorner_{t-1}$
Ensure: Backpropagate the result of the loss trained on the current t batch and obtain the new class center c_t
$\quad Q_t \leftarrow SUBSERIES(Q, s + (1 - \tau)s(t - 1), 2s + (1 - \tau)s(t - 1))$
$\quad J \leftarrow 0$
$\quad$ **for** $(\mathbf{x}, y) \in Q_t$ **do**
$\quad\quad J = J + (1 - y)d(f_\phi(\mathbf{x}), \mathbf{c}_{t-1}) + \alpha y(m - d(f_\phi(\mathbf{x}), \mathbf{c}_{t-1}))$
$\quad$ **end for**
$\quad J \leftarrow \frac{J}{|Q_t|}$
$\quad$ Backpropagation J
$\quad$ **for** $(\mathbf{x}, y) \in Q_t and y = 0$ **do**
$\quad\quad \mathbf{c}_t = \mathbf{c}_{t-1} + \beta f_\phi(x)$
$\quad$ **end for**

During the detection process, we judge the degree of abnormality of the data by scoring the abnormality of the data. The higher the abnormality score of a data segment is, the higher the probability that the data are abnormal. After training the neural network, we obtain the embedding function f_ϕ we need and the class center c_s of the normal data. The logic of the model we finally obtain is as follows: if the data segment is relatively close to the category center c_s after passing through the encoder, then the possibility of being abnormal is relatively small, otherwise it means that the possibility of being abnormal is relatively larger. Based on this, the final use of f_ϕ to calculate the abnormal score is mixed with the distance measurement method used in the previous training $d: \mathcal{R}^M \times \mathcal{R}^M \to [0, +\infty)$, if the defined method is Euclidean distance $d(x, y) = ||xy||_2^2$, then the abnormal score of data segment x is $OS(x) = ||f_\phi(x) - c_t||_2^2$.

3.5   Semisupervised Learning

If the number of labels provided is extremely low, our algorithm may repeatedly resample on small datasets, which may lead to overfitting problems. Some methods to solve the problem of overfitting due to few samples, such as data augmentation and adding noise to the data, have good results in the field of image processing. However, the time series data we deal with are often relatively simple real-valued data. Although the dimension is high, the data between different dimensions have their own unique meaning. These methods are often not applicable to these data. However, considering that our data often come from sensor networks, sensor networks will bring us much unlabeled data. We can generate some pseudolabels on these data to enrich our training dataset to prevent overfitting. Purpose. At this time, the small sample problem is transferred from the original only small sample data to a method similar to semisupervised learning. Let us introduce our simple semisupervised learning method using self-training based on the model learned from small samples.

For unlabeled datasets, the set U is obtained using the previous data preprocessing method, and then we calculate a confidence value for each data through the embedding function f_ϕ that has been learned earlier. Because the size of the anomaly score can reflect the probability that the test data are anomalous, we can simply use $confidnce(x) = \frac{1}{OS(x)}$ as the confidence value of the data segment. Then define a ratio γ, and add these data segments with high confidence as normal data to the training dataset to continue training. The specific algorithm for expanding the training dataset is shown in Algorithm 2.

Here, we consider that in the semisupervised setting in the anomaly detection scenario, most of the labeled samples are normal samples, and because there are more normal samples, the correct rate of identifying normal samples is higher than that of abnormal samples, that is, the model is more sensitive to normal categories. The sample recall is high, and the recall for anomalous categories is low. Therefore, we have reason to believe that if we identify an unlabeled sample as an anomaly in a semisupervised setting, then it should be an anomaly with a high probability, because we have seen a lot of normal data during training, in a sense, the judgment of normal data is in a state of "overfitting". Therefore,

---

Algorithm 2. Augment the dataset with self-training

$getTopK(L, k)$ is to get the first k elements in the list L
Require: Unlabeled dataset U, confidence ratio γ
Ensure: The augmented dataset D

$Confidence \leftarrow \frac{1}{OS(U)}$
$Confidence \leftarrow sort(Confidence, order = descending)$
$threshod \leftarrow min(getTopK(confidence, \gamma|U|))$
$D \leftarrow D \cup \{x \in U | \frac{1}{OS(X)} < threshold\}$

---

the data obtained in such a scenario is still abnormal with a high probability of being abnormal, so we can give it a higher weight in training. Because our model is judged by anomaly scoring, and there is no clear boundary between categories, this idea can be implemented in another way: set two thresholds, one is the non-abnormal threshold, and the normalized range is $0 - \gamma_{normal}$, the other is the abnormal threshold, the range after normalization is $\gamma_{outlier} - 1$, where the value of γ_{normal} can be set to be harsher (as small as possible), because we have enough normal data, so setting the threshold can ensure that the model learns the correct normal data; $\gamma_{outlier}$ can be set looser, such as 0.6 and 0.7, because we have learned enough The normal data can still obtain a large abnormal score under the condition of such category imbalance, indicating that it is very likely to be abnormal data. The overall improved process is shown in Algorithm 3.

Algorithm 3. Augment the dataset with improved self-training

$getTopK(L, k)$ is to get the first k elements in the list L, $getLastK(L, k)$ is to get the last k elements in the list L
Require: Unlabeled dataset U, confidence ratios γ_{normal} and $\gamma_{outlier}$
Ensure: The augmented dataset D

$Confidence \leftarrow \frac{1}{OS(U)}$
$Confidence \leftarrow sort(Confidence, order = ascending)$
$threshod_{normal} \leftarrow min(getTopK(confidence, \gamma_{normal}|U|))$
$threshod_{outlier} \leftarrow min(getLastK(confidence, \gamma_{outlier}|U|))$
$D \leftarrow D \cup \{x \in U | \frac{1}{OS(X)} < threshod_{normal}\} \cup \{x \in U | \frac{1}{OS(X)} > threshod_{outlier}\}$

---

4   Experiments

4.1   Dataset

The datasets used here are the oil chromatography datasets provided by the State Grid and some public datasets. There is no abnormality in the oil chromatography data, only different data states, so here we consider inserting some abnormal intervals into the data. These abnormal intervals increase or decrease their values on the basis of the original data, and then mark them as abnormal.

There are abnormal and normal data labels in public datasets. Here, some high-dimensional and single-dimensional time series anomaly detection datasets are mainly selected to suit the scenarios of our method. Included here are the server machine dataset and the ECG Dataset. Among them, server machine dataset is a high-dimensional time series dataset, which is the data collected by the author in multiple scenarios. The ECG dataset is a dataset in the field of electrocardiography, most of which are single-dimensional datasets, which are also in the field of time series anomaly detection. Commonly used datasets.

4.2   Setup

Evaluation Metrics

A measure of the accuracy of the results was obtained by calculating the ROC-AUC and PR-AUC. In the anomaly detection scenario, because the anomalies in the data are usually few, the results obtained by directly using the accuracy rate are meaningless. Using the AUC value can comprehensively consider the precision and recall rate, which is more practical. This measurement method comprehensively considers TP, FP, TN, FN in the binary classification problem, and sets various thresholds.

Hyperparameter Settings

The settings of hyperparameters are mainly distributed on the related settings of the neural network and the related settings of preprocessing. First, in our experimental setting, the coverage rate $\tau = 0.5$, the random dimension permutation is set to three groups, and each group randomly selects half of the currently used dataset dimensions. The window size is set to 20. In the settings related to the neural network, the vector dimension of the finally obtained embedding space is 64.

4.3   Result

Accuracy

First, the experimental results on the dataset provided by the grid are presented. Because there are no anomalies in the dataset, we insert a certain percentage of anomalies into the dataset to conduct experiments. Detect one or more of these segments as anomalies by numerically increasing them. The first experiment is the effect of different modules on the experimental results. We remove different modules to test the accuracy. The results of ROC-AUC are shown in Table 1, and the results of PR-AUC are shown in Table 2.

Table 1. ROC-AUC results on the grid dataset

Abnormal increase rate	ROC-AUC			
	Our model	No preprocessing	Use CNN only	Use LSTM only
12%	1.00	1.00	1.00	1.00
10%	1.00	0.972	0.987	0.982
7%	0.999	0.953	0.954	0.962
5%	0.995	0.948	0.940	0.953
2%	0.550	0.562	0.463	0.623

Table 2. PR-AUC results on the grid dataset

Abnormal increase rate	PR-AUC			
	Our model	No preprocessing	Use CNN only	Use LSTM only
12%	0.999	0.999	0.999	0.999
10%	0.999	0.955	0.967	0.962
7%	0.996	0.932	0.962	0.982
5%	0.958	0.962	0.951	0.942
2%	0.149	0.253	0.213	0.153

In addition, we also verified the effect of our loss function, which is mainly compared with cross entropy. As a classic binary classification loss function, cross entropy is well represented. The results are shown in Table 3.

Table 3. Comparison of loss functions on power grid data

Abnormal increase rate	ROC-AUC		PR-AUC	
	Our model	Cross entropy	Our model	Cross entropy
12%	1.00	0.999	0.999	0.999
10%	1.00	0.983	0.999	0.965
7%	0.999	0.981	0.996	0.921
5%	0.995	0.965	0.958	0.915
2%	0.550	0.532	0.149	0.135

Then, we also validate the effectiveness of our method on public datasets. The datasets used are server machine Dataset and ECG Dataset. Similarly, we first verify the effectiveness of different modules on public datasets for different modules, and the results are shown in Table 4 and Table 5.

Table 4. ROC-AUC results on the public dataset

Dataset	Our model	No preprocessing	Use CNN only	Use LSTM only
SMD 1	0.862	0.832	0.852	0.823
SMD 2	0.936	0.895	0.871	0.891
ECG1	0.968	0.952	0.912	0.935
ECG2	0.996	0.992	0.992	0.996

Table 5. ROC-AUC results on the public dataset

Dataset	Our model	No preprocessing	Use CNN only	Use LSTM only
SMD 1	0.776	0.786	0.723	0.767
SMD 2	0.852	0.811	0.843	0.821
ECG1	0.891	0.812	0.863	0.827
ECG2	0.683	0.694	0.563	0.593

In addition, we compare our method with other methods on public datasets, including the deep learning time series anomaly detection model LSTM autoencoder, the classic anomaly detection method isolation forest and one-class SVM. The results are shown in Table 6 and Table 7.

Table 6. Comparative experiments on public datasets(ROC-AUC)

Dataset	Our model	Isolated forest	LSTM-AE	oc-SVM
SMD 1	0.862	0.847	0.842	0.755
SMD 2	0.936	0.863	0.925	0.879
ECG1	0.968	0.935	0.963	0.924
ECG2	0.996	0.912	0.872	0.885

Table 7. Comparative experiments on public datasets(PR-AUC)

Dataset	Our model	Isolated forest	LSTM-AE	oc-SVM
SMD 1	0.776	0.426	0.538	0.688
SMD 2	0.852	0.723	0.785	0.256
ECG1	0.891	0.912	0.852	0.798
ECG2	0.683	0.523	0.292	0.463

As mentioned earlier, semisupervised learning algorithms can effectively use unlabeled data in the data to improve the ability of the model. Here we mainly discuss whether the use of semisupervised learning algorithms has a positive impact on the results. Because labeled data are a small part of the data, we use a 2:8 ratio to divide labeled data and unlabeled data, and then compare the metric learning using only labeled data with our semisupervised learning algorithm using unlabeled data, respectively. The results are shown in Table 8.

Table 8. Comparative experiments on public datasets(PR-AUC), where SL indicates supervised learning and UL indicates unsupervised learning

Abnormal increase rate	ROC-AUC		PR-AUC	
	SL	UL	SL	UL
12%	1.00	1.00	0.999	0.999
10%	0.992	1.00	0.999	0.999
7%	0.972	0.999	0.954	0.996
5%	0.952	0.995	0.921	0.958
2%	0.425	0.550	0.135	0.149

5   Conclusion

For the high-dimensional time series anomaly detection scenario, our paper designs a semisupervised time series anomaly detection algorithm based on metric learning. First, unique preprocessing is performed for high-dimensional time series. It includes feature extraction of time series data, correlation extraction before different dimensions of high-dimensional data, and conversion to frequency domain mining of time series data features. Then, different from binary classification, we design a new loss function suitable for anomaly detection. In the feature space trained by this loss function, normal samples will be clustered and distributed, and abnormal data will be scattered and distributed far away from normal samples. We also extended the model to semisupervised learning. Only a small number of labeled and unlabeled samples are required to obtain good results. We have conducted experiments on power grid datasets and public datasets, including the comparison of the functions of different modules within the model and the comparison of different methods. The results show that our method can achieve good anomaly detection results.

Future work may continue to use metric learning models, extending the models to online learning. Additionally, research on anomaly detection algorithms for small sample time series will be considered. The data augmentation method mentioned in the paper is a solution, but few-shot learning encounters more problems.

Acknowledgments. The project is supported by State Grid Research Project "Study on Intelligent Analysis Technology of Abnormal Power Data Quality based on Rule Mining" (5700-202119176A-0-0-00).

References

1. Booth, B.G., Sijbers, J., Keijsers, N.L.W.: Outlier detection for foot complaint diagnosis: modeling confounding factors using metric learning. IEEE Intell. Syst. **36**(3), 41–49 (2021)
2. Chandola, V., Banerjee, A., Kumar, V.: Anomaly detection: a survey. ACM Comput. Surv. (CSUR) **41**(3), 15 (2009)
3. Ezeme, O.M., Mahmoud, Q.H., Azim, A.: A framework for anomaly detection in time-driven and event-driven processes using kernel traces. IEEE Trans. Knowl. Data Eng. **34**(1), 1–14 (2022)
4. Fotso, V.S.S., Nguifo, E.M., Vaslin, P.: Grasp heuristic for time series compression with piecewise aggregate approximation. RAIRO Oper. Res. **53**(1), 243–259 (2019)
5. Li, S., Hong, D., Wang, H.: Relation inference among sensor time series in smart buildings with metric learning. In: The Thirty-Fourth AAAI Conference on Artificial Intelligence, AAAI 2020, The Thirty-Second Innovative Applications of Artificial Intelligence Conference, IAAI 2020, The Tenth AAAI Symposium on Educational Advances in Artificial Intelligence, EAAI 2020, New York, NY, USA, 7–12 February 2020, pp. 4683–4690. AAAI Press (2020)
6. Manevitz, L.M., Yousef, M.: One-class SVMs for document classification. J. Mach. Learn. Res. **2**(Dec), 139–154 (2001)
7. Neyshabur, B., Tomioka, R., Srebro, N.: In search of the real inductive bias: on the role of implicit regularization in deep learning. In: Bengio, Y., LeCun, Y., (eds.) 3rd International Conference on Learning Representations, ICLR 2015, San Diego, CA, USA, 7–9 May 2015. Workshop Track Proceedings (2015)
8. Riffo, V., Mery, D.: Automated detection of threat objects using adapted implicit shape model. IEEE Trans. Syst. Man Cybern.: Syst. **46**(4), 472–482 (2015)
9. Ruff, L., et al.: Deep semi-supervised anomaly detection. In: ICML Workshop on Uncertainty Robustness in Deep Learning (2019)
10. Snell, J., Swersky, K., Zemel, R.S.: Prototypical networks for few-shot learning. In: Guyon, I., et al. (eds.) Advances in Neural Information Processing Systems 30: Annual Conference on Neural Information Processing Systems 2017, Long Beach, CA, USA, 4–9 December 2017, pp. 4077–4087 (2017)
11. Sun, P., Yang, L.: Low-rank supervised and semi-supervised multi-metric learning for classification. Knowl. Based Syst. **236**, 107787 (2022)
12. Zhang, X., Gao, Y., Lin, J., Lu, C.T.: TapNet: multivariate time series classification with attentional prototypical network. In: The Thirty-Fourth AAAI Conference on Artificial Intelligence, AAAI 2020, The Thirty-Second Innovative Applications of Artificial Intelligence Conference, IAAI 2020, The Tenth AAAI Symposium on Educational Advances in Artificial Intelligence, EAAI 2020, New York, NY, USA, 7–12 February 2020, pp. 6845–6852. AAAI Press (2020)

Social Network Analysis of Coauthor Networks in Inclusive Finance in China

Jiamin Yan[1], Fenjing An[1], Ruiqi Wang[1], Ling Chen[2], Xi Yu[1(✉)], and Mingsen Deng[1,2(✉)] (iD)

[1] School of Information, Guizhou University of Finance and Economics, Guiyang 550025, China
{yuxi,msdeng}@mail.gufe.edu.cn

[2] Guizhou Provincial Key Laboratory of Computational Nano-Material Science, Guizhou Education University, Guiyang 550018, China

Abstract. The proposal and innovation of inclusive finance provide a very valuable pathway to realize social equity and eliminate poverty, which has attracted extensive attention, especially from developing countries. Based on the papers on inclusive finance published in the Chinese journal database CNKI from 2014 to 2018, we constructed an undirected weighted coauthor network 2154 authors. By employing social network analysis, we found that the number of authors in the field of inclusive finance increased rapidly. Although the cooperation between them was still very low and the cooperation authors were relatively fixed, the scale of cooperation was rapidly expanding. Although no scholar could always be at the center position in the coauthor network, the knowledge transfer path was significantly reduced. Financial universities and some financial institutions were the most important promoters of inclusive finance. Knowledge discovery in this field was promoted alternately by several center authors and cooperation by many scholars 2014-2018. We believe these discoveries are of great significance to promote knowledge sharing and innovation in the academic community of inclusive finance.

Keywords: Co-author network · Inclusive finance · Average weighted degree · Author centrality · Cohesive subgroups

1 Introduction and Motivation

With the rapid development of the world economy and the improvement of the level of civilization, it is a practical need to eliminate poverty and achieve social equity to benefit more groups from the achievements of economic development. The United Nations introduced the concept of inclusive finance in 2005, which aims to provide appropriate and effective financial services to all social strata and groups in need of financial services at an affordable cost based on the principles of equal opportunity and business sustainability [1]. Countries around the world have proposed different measures to promote the inclusive financial system, and these measures have promoted the growth of the population inclusive of inclusive finance year by year [2, 3]. However, the problems faced by inclusive finance are also very practical. For lower-income groups and small and microenterprises, it is usually difficult to obtain loans due to the lack of collateral required by banks. Therefore, research on inclusive finance and various methods to reduce credit risk and other financial risks has received extensive attention from researchers all over the world, especially in developing countries [4–6].

With the in-depth understanding of inclusive finance and the various factors that limit it, many issues often exceed the research capabilities of a single researcher or a single research team. As a result, scientists from various backgrounds have come together to form many research teams. This kind of cooperation between researchers of different backgrounds has greatly promoted the implementation of the concept of inclusive finance. The research papers they copublished were a very concrete embodiment of this collaboration. To better evaluate some new ideas and technological innovations in the field of inclusive finance, it is necessary to systematically analyze these research papers to obtain different research groups and different views they hold.

Regardless of the field of scientific research, finding outstanding researchers is the key to achieving scientific innovation, technological development and policy evolution [7]. Very recently, innovation project risks in companies have been evaluated by social network analysis [8]. Finding promising young researchers and fostering them is an important issue for sustainable development in the field. It is a common practice to achieve the above two goals through the coauthor networks formed by the author of the papers. Based on social network analysis (SNA) methods, coauthor networks in many fields have been studied, such as physics, biology, computer science, library and information service. Based on the Medline, arXiv and databases maintained by the journal Mathematical Reviews, M. E. J. Newman constructed a complex network with scientists as nodes in biology, physics, and mathematics. This answered a broad variety of questions about collaboration patterns of scientific research and the change in collaboration patterns between subjects over time [9]. A. Alireza et al. found that betweenness centrality is the preferential attachment driver in the evolution of collaboration networks of scientists [10]. Through social network analysis, Rong et al. concluded that scientists in the field of library and information science need to strengthen communication with each other [11]. Sun et al. proposed a method to find the most influential authors and papers through the collaborator network with network density and betweenness centrality [12]. Daud and Fujita et al. proposed the possibility of looking for promising researchers in coauthor networks by employing StarRank and network centrality [13, 14], respectively. By using the social network analysis method to study the authors' cooperation network of the journal affiliated with the China Computer Federation (CCF), Fu et al. pointed out that coauthorship in computer science was very different from the other fields and eight key persons and the group teams where the key persons were located [15]. Some online databases, such as the ISI Web of Science, CNKI and DBLP, usually contain a large amount of publication information, including the title, author, abstract, publication year, and digital object identifier (DOI) of the papers, which provide rich data resources and the chance for understanding the cooperation patterns of scientists.

Therefore, to better find the cooperative relationship between researchers in the field of inclusive finance in China, we built a coauthor network based on the papers in the field of inclusive finance published extracted from the CNKI database in this paper. Through the analysis of the general characteristics, author centrality and 2-core calculations of the coauthor network with 2154 authors, we found that the undirected weighted network changed remarkably with increasing years. Although the overall cooperation was still not high and the coauthorship was relatively fixed, the scholars and the cooperation between them increased rapidly, and the path of knowledge transfer was significantly shortened. Finally, by studying the maximal subgroup, we found that innovation and knowledge discovery in inclusive finance were promoted alternately by several center scholars and the cooperation between many authors.

2 Data Collection and Preprocessing

We have extracted the dataset including only journal papers from the CNKI database that contain the keyword "Inclusive Finance" for the five years from 2014 to 2018. CNKI is a database of Chinese academic literature, including various resources such as journal papers, dissertations, newspapers, conference proceedings, and yearbooks. It provides search, navigation, online reading and download services for these studies. The raw data are a series of papers on inclusive finance published by Chinese scholars, including the title, authors, affiliation unit, publication year, and abstract. We removed papers without authors and published in the name of the editorial board. The authors with the same name were distinguished according to their institutions. The effective data we obtained were 3540 papers, including 400 papers in 2014, 562 papers in 2015, 710 papers in 2016, 865 papers in 2017, and 1003 papers in 2018. The number of papers published has grown significantly over time.

To establish a coauthor network in inclusive finance, we analyzed the above 3540 papers and found that 1053 papers were copublished papers with more than one author. That is, only 29.74% of the 3540 papers were coauthorship, while more than 2/3 papers were published independently. There are 2154 authors of these papers, and the degree of cooperation, which is the ratio of total authors and total papers, is only 1.1 person per paper. This means that scientific research cooperation in the field of inclusive finance is at a very low level, which is lower than most scientific fields. In the field of high energy theory, computer science and journal papers affiliated with CCF, the average authors of each paper are 1.99, 2.22 and 1.35 [15, 16], respectively. The author's coauthorship and the evolution trend are shown in Fig. 1(a). The trend of coauthorship papers is not stable, with an increased ratio per year. The highest rate of coauthorship papers, which is the total number of coauthorship papers divided by the total number of published papers, is 31.67% in 2015. It is worth pointing out that although the proportion of coauthorship papers in the field of inclusive finance is relatively low, the total number of coauthorship papers still shows an obvious upward trend. The cooperative publishing of papers in the field of inclusive finance has also received significant attention.

In the above copublished papers, we conducted a statistical analysis of the cooperating institutions. As shown in Fig. 1(b), most scholars choose to cooperate in the same institution. Most of them are members of a team of the four major banks in China, teachers and students from the same institution. Although the cooperation of such a single institution is very stable, it is not as conducive to the overall innovation and exchange in the field of inclusive finance. It is worth noting that the scale of cooperating institutions

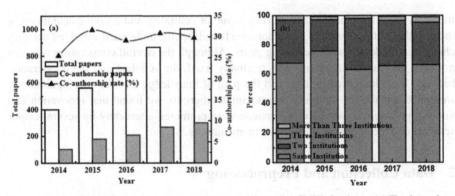

Fig. 1. The published journal papers were extracted from the CNKI database. (a) Total number of papers, coauthorship papers and coauthorship rate in 2014 − 2018. (b) Cumulative proportion of cooperative institutions in 2014 − 2018.

is the same as the rate of coauthorship. Although the overall situation is not regular, the cooperation of more than one institution is slowly increasing, especially the cooperation of more than two institutions, which has been significantly strengthened. This shows that inclusive finance is gradually evolving from the cooperation of a single institution or research group to broader cooperation.

3 Results

The coauthor network was constructed by 1053 papers with more than one author, where the author is the vertex (node). If two authors have collaborated on an article, then they were connected, and the edge weight is assigned to1. For each additional cooperated paper, the edge weight is increased by1. This method of constructing the network of coauthors is the simplest, which is also consistent with Ref [16]. Finally, there are a total of 2154 authors (denoted by N) (vertex) with 2329 edges (denoted by M) of cooperation. Here, we use the number instead of the author's real name. The weighted network was defined as $G = (V, E, W)$, where V was the vertex and $V = \{v_1, v_2, \ldots, v_N\}$. E was the edge,$E = \{e_1, e_2, \ldots, e_M\}$. W is the edge weight, $W = \{w(e_1), w(e_2), \ldots, w(e_M)\}$.

3.1 General Characteristics of the Coauthor Network

To present the change in the coauthor network more intuitively, the network dataset was divided by the year of publication. The coauthor network graphs in 2014 and 2018 are shown in Fig. 2. It is clear that the vertex in the network is increasing significantly. That is, an increasing number of authors have joined the academic community of inclusive finance. The changes in the vertex are shown in Fig. 3. There were only 226 nodes in the network in 2014. Although some authors left the network, the ID of nodes in the network increased to 660 in 2018. However, it is also clear that this coauthor network is a very sparse network, and the collaboration between authors is still not high. The degree distributions of these five coauthor networks were calculated, which follows the power-law distribution $\left(p(d) \sim d^{-k}\right)$. They passed the K-S test with the threshold value of 0.05.

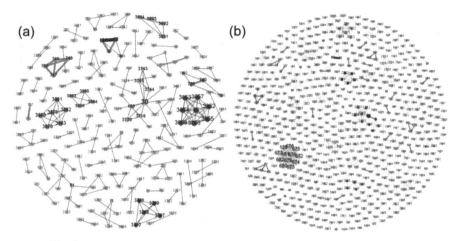

Fig. 2. The network topology of coauthor networks in 2014(a) and 2018(b).

To measure the strength of cooperation between authors, the average weighted degree of the coauthor network, which is the average mean of the total weights of the incident edges of all the vertices in the network, was calculated to represent the influence of authors. The average weighted degree represents the overall cooperation strength of the entire network. As shown in Fig. 3(a), there has been some increase in the average weighting over the five years. It is very interesting that although the cooperation rate was not the highest in 2016, its average weighting was the largest (up to 2.283), and there was a very significant decrease thereafter. This also means that the cooperation strength was not strong in the inclusive finance coauthor network in the five years from 2014 − 2018. However, the average weighted degree of the coauthor network in the field of inclusive finance is still greater than that of the English teaching research field (the average weighted degree is 1.914) [17].

The average clustering coefficient and average path length were also calculated to describe the general characteristics of the network. The higher the average clustering coefficient, that is, the higher the degree of coauthor network node aggregation, the smaller the average path length is. There is usually an inverse relationship between the average path length and the clustering coefficient. The average clustering coefficients of the coauthor networks per year are shown in Fig. 3(b). These values are basically above 0.9, and there is a slight upward trend. That is, scholars in the field of inclusive finance have obvious clustering characteristics, and the possibility of cooperation between different scholars who have cooperated with the same scholar is greater than 90%. Combined with the lower average weighted degree, it can be concluded that stable subgroups existed in the field of inclusive finance. In this field, the cooperation between scholars was relatively fixed. This is not conducive to innovation and the development of inclusive finance. The average path length has a slight downward trend, which shows

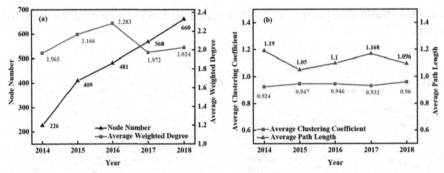

Fig. 3. The node number and the average weighted degree (a), the average cluster coefficient and the average path length (b) of coauthor networks in 2014 − 2018.

that the speed of innovation in this field can be gradually accelerated. Different authors can gradually and faster access the evolution of inclusive finance and transfer it to other authors.

3.2 Ego Characteristics of the Coauthor Network

To further evaluate the importance of the author and its role in the process of innovation in inclusive finance, we studied the centrality of the authors in the coauthor network. Centrality is closely related to the author's position, the role she/he plays, and social prestige in the coauthor network. Additionally, centrality has been proposed for studying the social learning properties of the eBay Green Team Facebook network [19]. In this article, the author's degree centrality, betweenness centrality and eigenvector centrality have been calculated. Degree centrality is used to answer authors who know more about other authors, which can be regarded as the link between the author and others. Degree centrality D_v is defined as the sum of all the edges connected to vertex v_i in coauthor network [18]. Authors with a higher degree of centrality represented that they were more central in the network of coauthors. The top 10 authors with the highest degree centrality are listed in Table 1. It should be noted that the top 10 authors in year 2018 were all from the same paper. The author's degree centrality in the field of inclusive finance is much lower than that in the field of computer science [15] and slightly similar to that in the field of English teaching research [14]. Of course, this had some relationship with the little publications by their authors. Obviously, the top 10 authors with the highest degree centrality are very different every year, which shows that scholars engaged in scientific research and innovation in inclusive finance have changed rapidly in the years 2014–2018, and few of them can maintain strong and long-lasting academic influence in the academic community of inclusive finance.

Eigenvector centrality is a natural extension of degree centrality. It is not only related to the number of coauthors but also reflects whether the coauthors of a certain author are important. If a scholar works with an author who is more important than himself, the higher the network centrality is. In a coauthor network, the eigenvector centrality of an author is proportional to the sum of the centrality of its coauthors [20]. In short, scholars who have had exchanges and cooperation with more important scholars are also more

Table 1. The top 10 authors with the highest degree centrality D_{vi} in the coauthor network in inclusive finance.

2014		2015		2016		2017		2018	
ID	D_{vi}	ID	D_{vi}	ID	D_{vi}	ID	D_{vi}	ID	D_{vi}
30	10	110	7	1	7	481	9	672	11
3853	7	846	7	1326	5	13	6	673	11
3852	7	261	7	25	5	2427	6	674	11
3854	7	847	7	362	5	2426	6	675	11
3916	7	262	7	1327	5	596	6	676	11
3855	7	848	7	1328	5	2428	6	677	11
3856	7	53	7	1329	5	2429	6	678	11
3857	7	849	7	1398	5	2430	6	679	11
3858	7	1030	7	1399	5	538	5	680	11
3879	5	1029	7	1400	5	261	5	128	11

important in the network. The eigenvector centrality reflects the tendency of nodes to be important nodes. Figure 4 presents the eigenvector centrality distribution of all authors in the coauthor network from 2014 to 2018. Most of the author's eigenvector centrality was concentrated around 0.2, while even around 0.02 in 2018. Only a few authors' eigenvector centrality was located at approximately 1 each year, which indicated that only a few authors were closely connected to the most important authors.

Due to the low frequency of cooperation in the field of inclusive finance, another kind of author may play a more important role, that is, the author who can exchange information in this coauthor network, that is, the author who plays the role of controlling the knowledge flows. Betweenness centrality is a measure of a person's ability as an intermediary. It is defined as the shortest number of paths that all authors have through a given author [21]. A higher betweenness centrality means that this author has an important pivotal role in the information exchange in this coauthor network. The top 10 authors with the highest betweenness centrality and their affiliation unit (AU)[1] are listed in Table 2. There is little difference between the betweenness centrality and degree centrality of scholars in inclusive finance. However, it is different that there are a few

---

[1] HU, Hubei University; PBC, People's Bank of China; CRF, China Rural Finance; HAU, Hunan Agricultural University; CUFE, Central University of Finance and Economics; EYECC, Ernst & Young (China) Enterprise Consulting Co., Ltd.; LNU, Liaoning Normal University; CBRC, China Banking Regulatory Commission; JU, Jinan University; AUFE, Anhui University of Finance and Economics; JMSU, Jiamusi University; BJ, Bank of Jiangsu; CPGC, China Post Group Corporation; SC, State Council; HUF, Hebei University of Finance; BO, Business Observer; CF, Chinese Financier; NUFE, Nanjing University of Finance and Economics; FE, Financial Economy; XVTC, Xi'an Vocational and Technical College; XAU, Xinjiang Agricultural University; RUC, Renmin University of China; CSCP, China Soft Capital Group; NAC, Nanjing Agricultural College; WBS, Wenzhou Business School.

authors who have high betweenness centrality in more than one year. This shows that only a few people in inclusive finance can master the core resources of information exchange, and they are also highly prolific authors. They can control the direction and speed of knowledge diffusion within the coauthor network. It should be noted that most of the authors with greater betweenness centrality come from financial universities or financial companies in China (Table 2).

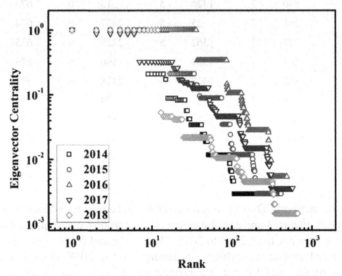

Fig. 4. The eigenvector centrality (EC) distribution of coauthor networks during 2014 − 2018. The X-axis is reordered according to the author's EC. Each point represents an author.

3.3   The Evolution of Cohesive Subgroups in the Coauthor Network

Based on the above analysis, although scientific research cooperation was not very frequent compared with other fields, the growth trend of the cooperation and the publication of papers was very obvious in the academic community of inclusive finance. Although only a few authors could occupy a more resourceful position and maintain a long-lasting influence in this coauthor network, there were also very few authors who could cooperate and exchange information with these high-quantity authors. Therefore, it is necessary to divide the social groups of the coauthor network of inclusive finance. Cohesive subgroups are a very effective quantity for the formal division of social groups [22]. There are relatively strong, direct and frequent connections within the cohesive subgroups. Due to its linear time complexity and intuitiveness, $k-$ core decomposition is one of the most widely accepted algorithms for clustering cohesive subgroups [23, 24]. In general, the $k-$ core of a network is the largest subgraph. Here, we use the $k-$ core decomposition method to study the cohesive subgroup of the coauthor network in the field of inclusive finance.

Table 2. The top 10 authors with the highest betweenness centrality and their affiliation unit (AU) in the coauthor network in inclusive finance.

2014		2015		2016		2017		2018	
ID	AU	ID	AU	ID	AU	ID	AU	ID	AU
30	HU	10	CUFE	1	BJ	481	PBC	88	CF
51	PBC	23	LNU	95	CPGC	12	BO	3	CF
69	CRF	247	PBC	15	SC	29	RUC	175	NAC
752	CRF	248	CBRC	392	HUF	13	HAU	633	PBC
9	HAU	125	JU	12	BO	3	CF	236	PBC
136	PBC	57	JU	65	NUFE	538	RUC	79	HU
739	PBC	9	HAU	99	NUFE	1	BJ	240	JU
740	PBC	7	CUFE	152	FE	221	CSCP	627	WBS
10	CUFE	283	AUFE	165	XVTC	579	PBC	628	HU
22	EYECC	98	JMSU	368	XAU	16	LU	230	PBC

Table 3. The sizes of the 2− core subgroup from 2014 to 2018 in the field of inclusive finance.

Year	Number of 2-core	Nodes of 2-core	Ratio $r_2(\%)$	Average clustering coefficient
2014	23	90	38.20	0.977
2015	46	179	43.77	0.997
2016	74	276	57.38	0.978
2017	73	263	46.30	0.991
2018	92	318	48.18	0.990

The large 2− core subgroups are the core for knowledge exchange and innovation. As listed in Table 3, the change in the sizes of 2− core subgroups was studied first. There were 232− core subgroups with 90 authors in 2014, which have a ratio (r_2) of 3 8.20% in total authors of the coauthor network, where r_2 =(the authors of 2− core subgroups)/(total authors in coauthor network). Although the r_2 reached the highest level (57.38%) in 2016, which is confirmed by the average weighted degree shown in Fig. 3(a). Due to the rapid increase in the number of authors in the coauthor network, the size of the 2− core subgroup in 2018 exceeded that of 2016, and the ratio r_2 in 2018 did not exceed that of 2016. Compared with 2014, the size of the 2− core subgroup in 2018 has reached more than 3.5 times that of 2014, and the scale of coauthorship in the coauthor network has been significant. Combined with the change of the average clustering coefficient in Fig. 3(b), it shows that the cooperation between the authors has gradually changed from the cooperation relying on the core nodes to the stable cooperation between most of them. An increasing number of authors play the role of "bridges" in the network so that an increasing number of authors can join the inclusive finance community of coauthors and form larger subgroups, which is very beneficial for information exchange and innovation in inclusive finance.

The maximal subgroup refers to the group with the largest proportion of authors in the coauthor network structure and the largest group. It is also the most active and knowledge spreading group in the coauthor network in a certain period. To some extent, the maximal subgroup controls the speed and direction of scientific research cooperation. The maximal subgroups of each year in the field of inclusive finance are shown in Fig. 5. In 2004, the most active and influential group in the coauthor network was centered on authors with ID 30, who were from Hubei University. Her/His collaborators came from not only the same affiliation but also Shanghai University of Finance and Economics and a bank. Scholar with ID 30 has played a very important role in promoting knowledge exchange and innovation in the research of inclusive finance in 2014. Different from 2014, although there was a highly prolific author in the maximal subgroup in 2015, there was no particularly obvious central author. In 2016 and 2017, there were also center authors with ID 1 and 481. Because the center author with ID 1 moved from a university to a bank, the diffusion and knowledge exchange of inclusive finance was brought. Similar to 2015, the maximal subgroup in 2018 also resulted from a large-scale cooperation with out obvious center authors. In the period of 2014 − 2018 we studied, the development of inclusive finance was promoted alternately by the promotion of several different central authors and the cooperation of many scholars.

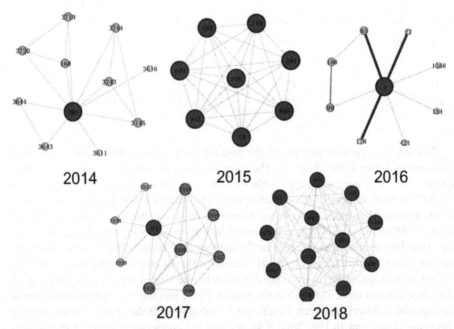

Fig. 5. The maximal subgroup in the coauthor network from 2014 to 2018.

4 Conclusions

In this paper, 3540 papers in inclusive finance from 2014 to 2018 have been extracted from CNKI database. Based on author name, affiliation unit, and publication year, the coauthorship dataset was prepared to form a coauthor network in the field of inclusive finance. Through social network analysis, we constructed an undirected weighted coauthor network with 2154 authors and studied its characteristics with evolution over time. The authors of the coauthor network increased very significantly. We found that the cooperation between them is mainly limited in the same affiliation unit, but the cooperation across affiliation units increased over time. By studying the evolution of the general characteristics of the coauthor network, such as the number of authors, the average weighted degree, the average clustering coefficient, and the average path length, we found that although the cooperation between authors in the field of inclusive finance was larger than that in the field of English education research, it was still not closer compared with many other fields, while the cooperation mainly occurred in the relatively fixed authors. This is not beneficial for the development of inclusive finance. With the increase in years, this problem has been greatly alleviated, and the possibility and speed of the knowledge flow has accelerated. Through the analysis of author centrality, we found that only a few authors can have high influence, but they cannot always be at the center position in the coauthor network. Financial universities and institutions were the main innovation promoters of inclusive finance. After calculating the $2-$ core subgroups and the maximal subgroups, we found that the number of $2-$ core subgroups in the coauthor network grew exponentially every year. The knowledge development and innovation of inclusive finance in $2014 - 2018$ were mainly promoted alternately by several center authors and cooperation by many scholars. More stable cooperation gradually formed from the promotion of several center authors. We believe that these studies are of great benefit for promoting the innovation and development of inclusive finance.

Acknowledgments. This work is supported by the Natural Science and Technology Foundation of Guizhou Province, China (QKHJC [2019]1045), the Scientific Research Fund of GUFE (2019XJC02), and the Plan Project for Guizhou Provincial Science and Technology ($No.QKH - PTRC$[2018]5803).

References

1. Corrado, G., Corrado, L.: Inclusive finance for inclusive growth and development. Curr. Opin. Env. Sust. **24**, 19-23 (2017)
2. Liu, T., He, G., Turvey, C.G.: Inclusive finance, farm households entrepreneurship, and inclusive rural transformation in rural poverty-stricken areas in China. Emerg. Mark. Financ. Tr. (2019). https://doi.org/10.1080/1540496X.2019.1694506.
3. Su, T., Yu, Y., Chen, Y., Zhang, J.: The experience, dilemma, and solutions of sustainable development of inclusive finance in rural china: based on the perspective of synergy. Sust. **11**, 5984 (2019)
4. Tian, K., Xiong, D., Yan, W., Yuan, G.X.: The study of dynamics for credit default risk by backward stochastic differential equation method. Int. J. Financ. Eng. **5**, 1850038 (2018)

5. Yu, X., Yang, Q., Wang, R., Fang, R., Deng, M.: Data cleaning for personal credit scoring by utilizing social media data: an empirical study. IEEE Intell. Syst. **35**, 7–15 (2020)
6. Guo, G., Zhu, F., Chen, E., Liu, Q., Wu, L., Guan, G.: From footprint to evidence: an exploratory study of mining social data for credit scoring. ACM Trans. Web **10**(Art. 22) (2016)
7. Zhou, N., Wu, Q., Hu, X.: Research on the policy evolution of China's new energy vehicles industry. Sustainability **12**, 3629 (2020)
8. Nunes, M., Abreu, A.: Managing open innovation project risks based on a social network analysis perspective. Sustainability **12**, 3132 (2020)
9. Newman, M.E.J.: Coauthorship networks and patterns of scientific collaboration. Proc. Natl. Acad. Sci. USA **101**, 5200-5205 (2004)
10. Alireza, A., Liaquat, H., Loet L.: Betweenness centrality as a driver of preferential attachment in the evolution of research collaboration networks. J. Informetr. **6**, 403–412 (2012)
11. Rong, X., Tao, Q.: Social network analysis in LIS field corelation of empirical research. Libr. World 2010 (2010)
12. Sun, N., Zhu, J., Cheng, H., Wu, Y.: Study coauthor relationship based on social network analysis papers (in Chinese). J. Suzhou Univ. **29**(09) (2014)
13. Daud, A., et al.: Finding rising stars in co-author networks via weighted mutual influence. In: WWW '17 Companion: Proceedings of the 26th International Conference on World Wide Web Companion, pp. 33-41 (2017)
14. Fujita, M., Inoue, H., Terano, T.: Searching promising researchers through network centrality measures of co-author networks of technical papers. In: 2017 IEEE 41st Annual Computer Software and Applications Conference, pp. 615-618 (2017)
15. Fu, C., Zeng, W., Ding, R., Mao, C., He, C., Chen, G.: Social Network Analysis of China Computer Federation Coauthor Network. HCC 2017, LNCS 10745, pp. 422–432 (2018)
16. Newman, M.E.J.: Scientific collaboration networks. I. Network construction and fundamental results. Phys. Rev. E **64**, 016131 (2001)
17. Zhang, X., Li, X., Jiang, S., Li, X., Xie, B.: Evaluation of scholar's contribution to team based on weighted coauthor network. In: ICPCSEE 2019, CCIS 1058, pp. 52–61 (2019)
18. Wasseman, S., Faust, K.: Social Network Analysis: Methods and Applications, vol.. Cambridge University Press, Cambridge (1994)
19. Kim, H.M., Oh, K.W., Jung, H.J.: Socialization on Sustainable Networks: The Case of eBay Green's Facebook. Sustainability 12, 3476 (2020)
20. Bonacich, P.: Some unique properties of eigenvector centrality. Soc. Netw. **29**, 555-564 (2007)
21. Borgatti, S.P.: Centrality and network flow. Soc. Netw. **27**, 55-71 (2005)
22. Frank, K.A.: Identifying cohesive subgroups. Soc. Netw. **17**, 27-56 (1995)
23. Batagelj, V., Zaversnik, M.: An O(m) algorithm for cores decomposition of networks. arXiv preprint 2003 arXiv:cs/0310049
24. Kong, Y.-X., Shi, G.-Y., Wu, R.-J., Zhang, Y.-C.: core: Theories and applications. Phys. Rep. **832**, 1-32 (2019)

Multirelationship Aware Personalized Recommendation Model

Hongtao Song, Feng Wang, Zhiqiang Ma$^{(\boxtimes)}$, and Qilong Han

College of Computer Science and Technology, Harbin Engineering University, Harbin,
HLJ 150001, People's Republic of China
mazhiqiang@hrbeu.edu.cn

Abstract. The existing methods using social information can alleviate the data sparsity issue in collaborative filtering recommendation, but they do not fully tap the complex and diverse user relationships, so it is difficult to obtain an accurate modeling representation of the user. To solve this, we propose a multirelationship aware personalized recommendation(MrAPR) model, which aggregates the various relationships between social users from two aspects of the user's personal information and interaction sequence. Based on the comprehensive and accurate relationship graphs established, the graph neural network and attention network are used to adaptively distinguish the importance of different relationships and improve the aggregation reliability of multiple relationships. The MrAPR model better describes the characteristics of user interest and can be compatible with the existing sequence recommendation methods. The experimental results on two real-world datasets clearly show the effectiveness of the MrAPR model.

Keywords: Graph neural networks · Attention · Relationship aware · Personalized recommendation

1 Introduction

Users' interaction sequence information can reflect the evolution of their interests, and an increasing number of researchers have focused on using such information to capture users' interests for personalized recommendations [1]. In real-world application scenarios, the target user probably interact with very limited items, and it is difficult to explore the true intentions using only his own interaction sequence information. Because the user's social relationship information implies its potential interest, the use of the user's explicit social relationship can enhance the target user representation, such as friend relationships and trust relationships [2]. With the rapid development of social networks, social recommendation has gradually become a hot direction of research [3].

At this stage, the main social information-based methods with the purpose improving recommendation performance include graph-based recommendation, deep learning-based recommendation and matrix-based decomposition recommendation [4]. Because of its simple operation, high recommendation accuracy and easy expansion, matrix decomposition becomes the preferred basic model for a large number of researchers

Y. Wang et al. (Eds.): ICPCSEE 2022, CCIS 1628, pp. 123–136, 2022.
https://doi.org/10.1007/978-981-19-5194-7_10

to build a social recommendation system. Early in 2008, Ma et al. proposed a social recommendation model (SoRec), which is groundbreaking for the first time to integrate social information into the personalized recommendation system [5]. Following that, the social matrix factorization (SocialMF) proposed by Jamali et al. introduced a trust propagation mechanism into a matrix decomposition model, which ignored the impact of the user's own interaction information while utilizing the user's social information [6]. Ma et al. also used user trust relationships as a regularization constraint and proposed a social regularization model (SR2) based on the regularization of social information [7]. Through analyzing four real datasets in detail, Guo et al. believed that both the explicit and implicit effects of scoring interaction should be considered in social recommendation, and then proposed a recommendation model based on a trust mechanism (TrustSVD) [8]. Xiong et al. proposed a TopK sort recommendation algorithm based on trust relationships based on user ratings and user trust relationships [9]. Recently, a social recommendation model using graph neural network technology called Diffnet has been proposed, which considers both the users' social information and their own historical interactive information [10]. In Diffnet, the portrayal of user characteristics is more abundant, but the user social relationship is not distinguished in detail. In fact, there are various relationships between users in social networks. Users not having direct social relationships may have common interests, such as fans of a blogger, comment users of a certain blog, users who make similar remarks, and so on. However, the implicit relationships between users are diverse and complex, and the type of relationships and the degree of mutual influence are both different. Accurately express and utilizing such relationships is still a challenging research topic.

In this paper, a novel multirelationship aware personalized recommendation (MrAPR) model is proposed. Starting with the construction and exploration of social user relationships, the model aggregates a variety of relationship information between social users from the two aspects of user personal information and user interaction sequence, which achieves more accurate modeling for users using graph neural networks (GNNs) [11] and attention networks [12]. Compared with the existing method, explicit relationships and implicit relationships between users are more comprehensively expressed and utilized, which alleviates the sparse problem of single-user interaction and better portrays the characteristics of user interest. Furthermore, the proposed model can be compatible with existing sequence recommendation methods. Our main contributions are listed as follows:

- We fully excavate a variety of relationships between users, including user social relationships, user interaction relationships and user potential relationships. Based on these relationships and the user-item heterogeneous graph, the user social relationship graph, user interaction relationship graph, user potential relationship graph based on personal information and user potential relationship graph based on the interaction sequence are constructed.
- We make more comprehensive and detailed use of the various relationships. Based on the four kinds of user relationship graphs, the MrAPR model is designed and implemented, in which the GNN and bilayer attention network are used to accurately model the target users.

- The effectiveness of the MrAPR model is experimentally verified on two commonly used and publicly available datasets, where multiple parameter combinations are used to explore the parameter settings that maximize the model performance.

2 Preliminary Preparation

2.1 Problem Definition

Table 1. Symbol definition and description.

Symbols	Definitions and Descriptions
U	User set, $|U| = N$
V	Item set, $|V| = M$
D	Embedding dimension
e_a	Embedding of user a, $e_a \in \mathbb{R}^{1 \times d}$
e_{aU}	Embedding of user a based on personal information, $e_{aU} \in \mathbb{R}^{1 \times d}$
e_{aI}	Embedding of user a based on interaction sequence, $e_{aI} \in \mathbb{R}^{1 \times d}$
$e_{I(m)}$	Embedding of item m, $e_{I(m)} \in \mathbb{R}^{1 \times d}$
P	User embedding matrix, $P \in \mathbb{R}^{N \times d}$
Q	Item embedding matrix, $Q \in \mathbb{R}^{M \times d}$
S	User-user social matrix, $S \in \mathbb{R}^{N \times M}$
R	User-item rating matrix, $R \in \mathbb{R}^{N \times M}$
$\hat{y}_{am}$	Predicted preference of user a to item m
$e_{aU}^{(l)}$	Embedding of personal information at the l-level of user a
$e_{aI}^{(k)}$	Embedding of interactive sequences at the k-level of user a
P_U	Initial embedding matrix of the user's personal information
P_I	Initial embedding matrix of the user's interaction sequence
G	User-item heterogeneous graph
G_s	User social relationship graph
G_h	User interaction relationship graph
G_{qU}	User potential relationship graph based on personal information
G_{qI}	User potential relationship graph based on interaction sequence

To express this clearly, the symbol definitions and descriptions are given in Table 1. The social recommendation problem can be defined as follows.

Definition: Given the user set U, item set V, user embedding matrix **P**, item embedding matrix **Q**, user-user social matrix **S**, and user-item rating matrix **R**, our goal is to predict the users' unknown preference for items, which can be abstracted into $\hat{R} = f(U, V, S, R, P, Q)$, where $\hat{R} \in \mathbb{R}^{N \times M}$ represents the users' predicted preference for items.

2.2 Data Preprocessing

Generally, in social recommendation, a user's data may include information such as ID, gender, age remarks and friend list, and an item's data may contain the information such as the item ID, price, type and reviews. These features can be divided into numeric type and text type, and they should be encoded in different ways. Numeric information, such as ID, age and price, can be encoded by one hot. Multiplying the one-hot encoding by the randomly initialized transformation matrix, we can obtain the initial embedding. For text-like information such as remarks, Word2Vec [13] can be used for the corresponding embedding.

2.3 User Relationship Graphs

We believe that the users affect each other through not only friendship but also their behaviors. The friendships are always direct, but the relationship between the two users established by behaviors will be indirect, both of which cannot be ignored. Furthermore, there may be some potential relationships between two users, such as those who have common hobbies or similar behaviors, and they should also be identified and considered.

By constructing the user-item heterogeneous graph G, we can find the direct and indirect relationships between users. In addition, we define two types of potential relationships according to users' personal information and interaction sequence in this paper. Finally, we obtain four kinds of user relationship graphs, which are the user social relationship graph G_s, user interaction relationship graph G_h, user potential relationship graph based on personal information G_{qU} and user potential relationship graph based on interaction sequence G_{qI}.

Graph G consists of user nodes, item nodes, user-user connections and user-item connections. The connections of nodes in graph G_s only contain the user-user connections in graph G, which indicate the users' direct relationship. If two user nodes are connected to the same item node in graph G, there is an edge between the two user nodes in graph G_h.

The graph G_{qU} is established by the following steps:

(1) Establish a full connection graph G^*_{qU}, where the nodes are the same as in G_s and G_h.
(2) Delete the edge in G^*_{qU}, if there is correspondingly an edge in G_s or G_h.
(3) For any two nodes, calculate the similarity by their embeddings based on personal information.
(4) For any node, sort the similarity values with the others in descending order. Taking the first K_u nodes, the edges between the target node and those are not in the first K_u nodes are deleted.

The graph G_{ql} is established similarly by the following steps:

(1) Establish a full connection graph G^*_{ql}, where the nodes are the same as in G_s and G_h.
(2) Delete the edge in G^*_{ql}, if there is correspondingly an edge in G_s or G_h.
(3) For any two nodes, calculate the similarity by their embeddings based on the interaction sequence.
(4) For any node, sort the similarity values with the others in descending order. Taking the first K_i nodes, the edges between the target node and those that are not in the first K_i nodes are deleted.

3 Modeling and Training

3.1 MrAPR Model

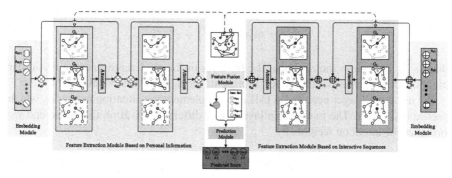

Fig. 1. The overall architecture of the MrAPR model.

Figure 1 exhibits the overall architecture of the MrAPR model, which includes five modules named the embedding module, the feature extraction module based on personal information, the feature extraction module based on the interactive sequence, the feature fusion module and the prediction module. Next, we analyze these five modules in detail.

(1) **Embedding Module.** Taking user a as an example, the initial embedding based on personal information is denoted as $\{e_{aU1}...e_{aUn}\}$, where e_{aUk} is the kth embedding obtained by its numeric class information or text class information. The initial fused embedding $e_{aU}^{(0)}$ based on the personal information of user a can be obtained by

$$e_{aU}^{(0)} = f_U(\mathbf{W}_U \times \mathbf{P}_U) + b_U \tag{1}$$

where $f_U()$ is a nonlinear function, $\mathbf{P}_U$ is the initial embedding matrix of the user's personal information, $\mathbf{W}_U$ is the transformation matrix, and b_U is the bias.

Similarly, the initial embedding based on the interactive sequence is denoted as $\{e_{aI1}...e_{aIn}\}$. The initial fused embedding $e_{aI}^{(0)}$ based on the interactive sequence of user a can be obtained by

$$e_{aI}^{(0)} = f_I(\mathbf{W}_I \times \mathbf{P}_I) + b_I \tag{2}$$

where $f_I()$ is a nonlinear function, $\mathbf{P}_I$ is the initial embedding matrix of the user's personal information, $\mathbf{W}_I$ is the transformation matrix, and b_I is the bias.

From the overall architecture of the MrAPR model, it is obvious that the existing sequence recommendation algorithms can be compatible with our model to fuse the user's interaction sequence information to obtain $e_{aI}^{(0)}$.

(2) Feature Extraction Module Based on Personal Information. This module aims to model the impact between users through different relationships based on personal information. To calculate the final representation vector of the target user, we use GNN to aggregate its multihop neighbors' information based on G_s, G_h and G_{qU}, and distinguish the degree based on the attention mechanism. As shown in Fig. 1, the process of the lth layer can be expressed by

$$e_{aUs}^{(l)} = AGG^s\left(N_a^s\right) + e_{aUs}^{(l-1)} \tag{3}$$

$$e_{aUh}^{(l)} = AGG^h\left(N_a^h\right) + e_{aUh}^{(l-1)} \tag{4}$$

$$e_{aUqu}^{(l)} = AGG^{qu}\left(N_a^{qu}\right) + e_{aUqu}^{(l-1)} \tag{5}$$

where $e_{aU*}^{(l)}$ represents the results of the lth-layer on G_*, N_a^* represents the first-order neighbors of user a on G_*, and AGG^* represents the aggregation calculation.

We use a multilayer perceptron (MLP) to implement the attention networks, which is set to two layers. The two hidden layers have different activation functions, and we normalize the attention weights by

$$\eta_{aUs}^l = MLP^l\left(e_{aUs}^{(l)}, e_{aU}^{(l-1)}\right) \tag{6}$$

$$\eta_{aUh}^l = MLP^l\left(e_{aUh}^{(l)}, e_{aU}^{(l-1)}\right) \tag{7}$$

$$\eta_{aUqu}^l = MLP^l\left(e_{aUqu}^{(l)}, e_{aU}^{(l-1)}\right) \tag{8}$$

$$\eta_{aU}^l = \eta_{aUs}^l + \eta_{aUh}^l + \eta_{aUqu}^l \tag{9}$$

where η_{aU*}^l corresponds to the weights of $e_{aU*}^{(l)}$, MLP^l represents the attention network at layer l, $e_{aU}^{(l-1)}$ represents the output at layer l-1 and η_{aU}^l ensures that the weights are normalized.

The final output $e_{aU}^{(l)}$ can be calculated by

$$e_{aUs}^{(l)'} = \frac{\eta_{aUs}^l}{\eta_{aU}^l} \times e_{aUs}^{(l)} \tag{10}$$

$$e_{aUh}^{(l)'} = \frac{\eta_{aUh}^l}{\eta_{aU}^l} \times e_{aUh}^{(l)} \tag{11}$$

$$e_{aUqu}^{(l)'} = \frac{\eta_{aUqu}^l}{\eta_{aU}^l} \times e_{aUqu}^{(l)} \tag{12}$$

$$e_{aU}^{(l)'} = e_{aUs}^{(l)'} + e_{aUh}^{(l)'} + e_{aUqu}^{(l)'} \tag{13}$$

$$e_{aU}^{(l)} = g_U^l \left(W_U'^l \times \left[e_{aU}^{(l)'}, e_{aU}^{(l-1)} \right] \right) \tag{14}$$

where $g_U^l()$ is a nonlinear function and $W_U'^l$ is a transformation matrix.

(3) Feature Extraction Module Based on Interactive Sequence. Similar to that above, this module aims to model the impact between users through different relationships based on the interactive sequence. Here, the three graphs G_s, G_h and G_{ql} are used.

As shown in Fig. 1, the process of the kth layer can be expressed by

$$e_{als}^{(k)} = AGG^s \left(N_a^s \right) + e_{als}^{(k-1)} \tag{15}$$

$$e_{alh}^{(k)} = AGG^h \left(N_a^h \right) + e_{alh}^{(k-1)} \tag{16}$$

$$e_{alqi}^{(k)} = AGG^{qi} \left(N_a^{qi} \right) + e_{alqi}^{(k-1)} \tag{17}$$

where $e_{al*}^{(k)}$ represents the results of the kth-layer on G_*, N_a^* represents the first-order neighbors of user a on G_*, and AGG^* represents the aggregation calculation.

The attention network calculation processes in this module are as follows:

$$\eta_{als}^k = MLP^k \left(e_{als}^{(k)}, e_{al}^{(k-1)} \right) \tag{18}$$

$$\eta_{alh}^k = MLP^k \left(e_{alh}^{(k)}, e_{al}^{(k-1)} \right) \tag{19}$$

$$\eta_{alqi}^k = MLP^k \left(e_{alqi}^{(k)}, e_{al}^{(k-1)} \right) \tag{20}$$

$$\eta_{al}^k = \eta_{als}^k + \eta_{alh}^k + \eta_{alqi}^k \tag{21}$$

where η_{al*}^k corresponds to the weights of $e_{al*}^{(k)}$, MLP^k represents the attention network at layer k, $e_{al}^{(k-1)}$ represents the output at layer k-1 and η_{al}^k ensures that the weights are normalized.

The final output $e_{al}^{(k)}$ can be calculated by

$$e_{als}^{(k)'} = \frac{\eta_{als}^k}{\eta_{al}^k} \times e_{als}^{(k)} \tag{22}$$

$$e_{alh}^{(k)'} = \frac{\eta_{alh}^k}{\eta_{al}^k} \times e_{alh}^{(k)} \tag{23}$$

$$e_{alqi}^{(k)'} = \frac{\eta_{alqi}^k}{\eta_{al}^k} \times e_{alqi}^{(k)} \tag{24}$$

$$e_{al}^{(k)'} = e_{als}^{(k)'} + e_{alh}^{(k)'} + e_{alqi}^{(k)'} \tag{25}$$

$$e_{al}^{(k)} = g_I^k \left(W_I'^k \times \left[e_{al}^{(k)'}, e_{al}^{(k-1)} \right] \right) \tag{26}$$

where $g_I^k()$ is a nonlinear function and $W_I'^k$ is a transformation matrix.

(4) Feature Fusion Module. To obtain the final user embedding e_a, the two results of $e_{aU}^{(l)}$ and $e_{al}^{(k)}$ are fused in this module utilizing the gated network [14], as shown by

$$e_a = \beta \cdot e_{aU}^{(l)} + (1 - \beta) \cdot e_{al}^{(k)} \tag{27}$$

$$\beta = h \left(W_{gate} \left[e_{aU}^{(l)}, e_{al}^{(k)} \right] \right) \tag{28}$$

where β is a weight coefficient, $h()$ denotes a nonlinear function, and W_{gate} is a trainable transformation matrix.

(5) Prediction Module. Finally, taking user a and candidate item m as examples, the predicted score $\hat{y}_{am}$ can be calculated by

$$\hat{y}_{am} = SoftMax \left(e_a \cdot e_m^T \right) \tag{29}$$

3.2 Model Training

At the model training step, we use a pairwise ranking-based loss function for optimization [15], which is widely used for implicit feedback, as shown as

$$LOSS = \min_{\Theta} \sum_{(a,i) \in R^+ \cup (a,j) \in R^-} -ln\sigma \left(\hat{y}_{ai} - \hat{y}_{aj} \right) + \lambda \Theta^2 \tag{30}$$

where R^+ represents the set of positive samples (the set of items that the user has actually interacted with), R^- represents the set of negative samples (the set of items that the user has not interacted with by random samples), $\hat{y}_{ai}$ represents the probability of user a predicting item i in the positive sample set, $\hat{y}_{aj}$ represents the probability of user a predicting item j in the negative sample set, $\sigma()$ is the Sigmiod function, λ is a regularization coefficient that controls the complexity of the user and the item embedding, and Θ is the trainable parameter used. In this paper, the Adam [16] optimizer is utilized to optimize the loss function.

4   Experiment

4.1   Dataset

In this paper, two representative datasets Ciao and Yelp are chosen for our experiment, which are available at the popular social networking websites Ciao (http://www.ciao.co.uk) and Yelp (https://www.yelp.com/dataset). Each social networking service allows

Table 2. Statistics of the datasets.

Dataset	Ciao	Yelp
Users	7360	17237
Items	110222	38342
Ratings	278696	204448
Total Links	118271	143765

users to rate items, browse/write reviews, and add friends to their 'Circle of Trust'. Hence, they provide a large amount of rating information and social information.

For data preprocessing, users with fewer than two items or social links and items with fewer than two interactive records are all deleted [17]. The statistics of these two datasets are presented in Table 2. For training, testing and validation set division, we select the commonly used 8:1:1 ratio as the basis.

4.2 Baselines and Evaluation Metrics

For the top-N ranking evaluation, we use two widely used metrics, the hit ratio (HR) and normalized discounted cumulative gain (NDCG), and the larger the values of HR and NDCG are, the higher the recommended accuracy of the model. We compared our MrAPR model with a variety of recommendation models, including several classic and newly published related recommendation models, which are listed as follows.

BPR [18]: a classical pairwise-based recommendation model.
FM [19]: a feature-enhanced latent factor model.
SocialMF: a social-based recommendation model.
TrustSVD: a social recommendation model based on a trust mechanism.
ContextMF [20]: a context-aware social recommendation model.
PinSage [21]: a recommended model based on graph convolution.
Diffnet: a deep graph neural network recommendation system.

The performance comparison of different recommendation systems is shown in Table 3. It is indicated that our MrAPR model improves the recommendation effect on both datasets compared with other competitive models. Furthermore, it is indicated that (1) social network information helps to improve the accuracy of the recommendation system. (2) The GNN can be a good way to capture the user's unknown preferences and accurately model the user. (3) The setting of the MrAPR model is reasonable.

4.3 Parameter Settings

In this paper, we conduct relevant parameter experiments to explore the combination of the hyperparameters that can achieve the best MrAPR model performance. The MrAPR model hyperparameters include learning rate r, training batch size match − size, dimension D of feature embedding vector, regularization parameter λ, propagation depth of l

Table 3. Performance comparison of different recommendation systems.

Model	Ciao		Yelp	
	HR@10	NDCG@10	HR@10	NDCG@10
BPR	0.2632	0.0711	0.2616	0.1573
FM	0.2768	0.1103	0.2835	0.1720
SocailMF	0.2785	0.1233	0.2709	0.1695
TrustSVD	0.2826	0.1922	0.2854	0.1710
ContestMF	0.2985	0.2022	0.3011	0.1808
Pinsage	0.2944	0.2503	0.2966	0.1786
Diffnet	0.3523	0.2121	0.3437	0.2095
MrAPR	0.3700	0.3112	0.3624	0.2871

in the feature extraction module based on personal information and propagation depth k in the feature extraction module based on interactive sequences.

Table 4. The influence of the number of propagation layers l and k.

Depth	Ciao		Yelp		
	HR@10	NDCG@10	HR@10	NDCG@10	Depth
$l = 1$	0.2599	0.2001	0.2529	0.1789	$k = 1$
$l = 1$	0.2651	0.1894	0.2602	0.1841	$k = 2$
$l = 1$	0.3378	0.2421	0.2933	0.2791	$k = 3$
$l = 2$	0.3528	0.2811	0.3189	0.2902	$k = 1$
$l = 2$	0.3700	**0.3112**	**0.3624**	0.2871	$k = 2$
$l = 2$	**0.3721**	0.3089	0.3577	0.2796	$k = 3$
$l = 3$	0.3644	0.2991	0.3588	**0.3011**	$k = 1$
$l = 3$	0.3699	0.2986	0.3468	0.2849	$k = 2$
$l = 3$	0.3291	0.3021	0.3489	0.2391	$k = 3$

After a series of tests, we chose to set the learning rate r to 0.001 and the training batch size match − size to 512. We try the regularization parameter λ in the set of 0.0001, 0.001, 0.01, and 0.1 and find that the value 0.001 reaches the best performance. For the dimension D of the feature embedding vector, we set it to 16,32 and 64 for comparative experiments, and finally, the value of 32 is selected.

For a GNN model, the number of propagation layers is an important parameter affecting the effect of the model. For the propagation depth l in the feature extraction module based on personal information and the propagation depth k in the feature extraction module based on the interaction sequence, we experiment by setting different layers in detail to explore the best model effect, as shown in Table 4.

In this experiment, the dimension of the feature embedding vector is set to a moderate 32-dimensional value, which can ensure the accuracy of the modeling effect while appropriately reducing the spatial complexity of the model. In general, the results in Table 4 show that the model works best when the values of propagation depth l and k are both 2. In addition, the results show the characteristics of a Gaussian distribution. The model performance rises with the increasing number of propagation layers, reaches the best when l and k are both 2, and then slowly declines as the number of propagation layers increases, which is also in line with most researchers' experience in using GNN.

4.4 Ablation Experiments

The MrAPR model includes a feature extraction module based on personal information, a feature extraction module based on an interactive sequence and a feature fusion module, To test and verify the effectiveness, we take the experimental data of Ciao as an example to analyze the rationality of the three modules.

(1) Feature Extraction Module Based on Personal Information. First, the graph G_s for the aggregation of neighbor information is only used, the propagation depth l is set to 2, there is no feature extraction module based on the interaction sequence and the feature fusion module is set to the method of adding two vectors. Second, add the graph G_h to the module. Third, add the graph G_{qU} to the module. Finally, the attention network ($ATTENTION^l$) is added to the module. With all the other settings that are the same as in section 4.3, the experimental results are shown in Table 5. Here, $MrAPR^*$ represents a basic model that does not include the feature extraction module based on personal information, the feature extraction module based on the interactive sequence and the feature fusion module, and it uses only the original features of the user and the item for prediction.

Table 5. Analysis of feature extraction modules based on personal information.

Model	HR@10	NDCG@10
$MrAPR^* + G_s$	0.3122	0.2873
$MrAPR^* + G_s + G_h$	0.3156	0.2921
$MrAPR^* + G_s + G_h + G_{qU}$	0.3188	0.2979
$MrAPR^* + G_s + G_h + G_{qU} + ATTENTION^l$	**0.3195**	**0.3011**

(2) Feature Extraction Module Based on Interactive Sequences. Similarly, we perform experiments analyze the feature extraction module based on interactive sequences. First, the graph G_s for the aggregation of neighbor information is only used, the propagation depth k is set to 2, there is no feature extraction module based on the personal information and the feature fusion module is set to the way of adding two vectors. Second, add the graph G_h to the module. Third, add the graph G_{ql} to the module. Finally, the attention network ($ATTENTION^k$) is added to the module. With all the other settings that are the same as in section 4.3, the experimental results are shown in Table 6. Here, $MrAPR^*$ also represents the basic model.

Table 6. Analysis of feature extraction modules based on interactive sequences.

Model	HR@10	NDCG@10
$MrAPR^* + G_s$	0.3089	0.2751
$MrAPR^* + G_s + G_h$	0.3153	0.2795
$MrAPR^* + G_s + G_h + G_{ql}$	0.3290	0.2835
$MrAPR^* + G_s + G_h + G_{ql} + ATTENTION^k$	**0.3351**	**0.2982**

(3) Feature Fusion Module. In this part, we discuss the feature fusion module. First, the feature fusion module is set to add two vectors. Second, the feature fusion module is set to the way of catenating two vectors. Finally, the feature fusion module is set to the gated network. The experimental results are shown in Table 7. where $MrAPR^\#$ represents the model after adding a feature extraction module based on personal information and a feature extraction module based on an interactive sequence to the basic model, $[A + B]$ indicates the vector addition method, $[A,B]$ indicates the vector catenating method, and $GATE$ represents the gated network mode.

Table 7. Analysis of the feature fusion module.

Model	HR@10	NDCG@10
MrAPR# + [A + B]	0.3588	0.3083
MrAPR# + [A,B]	0.3591	0.3083
MrAPR# + GATE	**0.3698**	**0.3105**

Summary, the three experimental results verify the rationality of the MrAPR model structure, and clearly show that the corresponding module is necessary.

5   Conclusion

This paper starts by exploring the various relationships between users, studies how to model users more comprehensively and accurately, and proposes a novel MrAPR model. In this paper, we mainly focus on the rationality of capturing various relationships between users, and a large number of experimental results verify the effectiveness of the model well. In addition, easily seen from the structure, the MrAPR model has strong flexibility, and the existing sequence recommendation methods can be easily applied to the model.

Acknowledgements. This research was supported by the National Key R&D Program of China under Grant No.2020YFB1710200, and the National Natural Science Foundation of China under Grant No.61872105 and No.62072136.

References

1. Adomavicius, G., Tuzhilin, A.: Toward the next generation of recommender systems: A survey of the state-of-the-art and possible extensions. IEEE Trans. Knowl. Data Eng. **17**(6), 734–749 (2005)
2. Du, S., Chen, Z., Wu, H., Tang, Y., Li, Y.: Image recommendation algorithm combined with deep neural network designed for social networks. Complex **9**, 1–5196190 (2021)
3. Wang, X., Yang, X., Lei Guo, Y., Han, F.L., Gao, B.: Exploiting social review-enhanced convolutional matrix factorization for social recommendation. IEEE Access **7**, 82826–82837 (2019)
4. Sarwar, B.M., Karypis, G., Konstan, J.A., Riedl, J.: Itembased collaborative filtering recommendation algorithms. In: Shen, V.Y., Saito, N., Lyu, M.R., Zurko, M.E. (eds.) Proceedings of the Tenth International WorldWide Web Conference, WWW 10, Hong Kong, China, May 1–5, 2001, 285–295. ACM (2001)
5. Ma, H., Yang, H., Lyu, M.R., King, I.: Sorec: social recommendation using probabilistic matrix factorization. In: Shanahan, J.G., Amer-Yahia, S., Manolescu, I., Zhang, Y., Evans, D.A., Kolcz, A., Choi, K-S., Chowdhury, A. (eds.) Proceedings of the 17th ACM Conference on Information and Knowledge Management, CIKM 2008, Napa Valley, California, USA, October 26–30, 2008, pp. 931–940. ACM (2008)
6. Jamali, M., Ester, M.: A matrix factorization technique with trust propagation for recommendation in social networks. In: Amatriain, X., Torrens, M., Resnick, P., Zanker, M. (eds.) Proceedings of the 2010 ACM Conference on Recommender Systems, RecSys 2010, Barcelona, Spain, September 26–30, 2010, pp. 135–142. ACM (2010)
7. Ma, H., Zhou, D., Liu, C., Lyu, M.R., King, I.: Recommender systems with social regularization. In: King, I., Nejdl, W., Li, H. (eds.) Proceedings of the Forth International Conference on Web Search and Web Data Mining, WSDM 2011, Hong Kong, China, February 9–12, 2011, pp. 287–296. ACM (2011)
8. Guo, G., Zhang, J., Yorke-Smith, N.: Trustsvd: collaborative filtering with both the explicit and implicit influence of user trust and of item ratings. In: Bonet, B., Koenig, S. (eds.) Proceedings of the Twenty-Ninth AAAI Conference on Artificial Intelligence, January 25–30, 2015, Austin, Texas, USA, pp. 123–129. AAAI Press (2015)
9. Xiong, L.-R., Wang, L.-Y., Huang, Y.-Z.: An approach for top-k recommendation based on trust information. In: 10th IEEE Conference on ServiceOriented Computing and Applications, SOCA 2017, Kanazawa, Japan, November 22–25, 2017, pp. 198–205. IEEE Computer Society (2017)
10. Wu, L., Sun, P., Fu, Y., Hong, R., Wang, X., Wang, M.: A neural influence diffusion model for social recommendation. In: Piwowarski, B., Chevalier, M., Gaussier, E., Maarek, Y., Nie, J-Y., Scholer, F. (eds.) Proceedings of the 42nd International ACM SIGIR Conference on Research and Development in Information Retrieval, SIGIR 2019, Paris, France, July 21–25, 2019, pp. 235–244. ACM (2019)

11. Jaume, G., Nguyen, A., Mart´ınez, M.R., Thiran, J.-P., Gabrani, M.: edgnn: a simple and powerful GNN for directed labeled graphs. In: CoRR, abs/1904.08745 (2019)

12. Luong, T., Pham, H., Manning, C.D.: Effective approaches to attention-based neural machine translation. In: Ma´rquez, L., CallisonBurch, C., Su, J., Pighin, D., Marton, Y. (eds.), Proceedings of the 2015 Conference on Empirical Methods in Natural Language Processing, EMNLP 2015, Lisbon, Portugal, September 17–21, 2015, pp. 1412–1421. The Association for Computational Linguistics (2015)

13. Qian, Y., Du, Y., Deng, X., Ma, B., Ye, Q., Yuan, H.: Detecting new chinese words from massive domain texts with word embedding. J. Inf. Sci. **45**(2) (2019)

14. Zhou, K., Zhao, W.X., Bian, S., Zhou, Y., Wen, J.-R., Yu, J.: Improving conversational recommender systems via knowledge graph based semantic fusion. In: Gupta, R., Liu, Y., Tang, J., Prakash, B.A. (eds.), KDD '20: The 26th ACM SIGKDD Conference on Knowledge Discovery and Data Mining, Virtual Event, CA, USA, August 23–27, 2020, pp. 1006–1014. ACM (2020)

15. Liu, S., Ounis, I., Macdonald, C.: Social regularisation in a bpr-based venue recommendation system. In: Frommholz, I., Liu, H., Moshfeghi, Y. (eds.) Proceedings of the 9th PhD Symposium on Future Directions in Information Access colocated with 12th European Summer School in Information Retrieval (ESSIR 2019), Milan, Italy, July 17th - to - 18th, 2019, volume 2537 of CEUR Workshop Proceedings, pp. 16–22. CEUR-WS.org (2019)

16. Wojna, Z., Gorban, A.N., Lee, D.-S., Murphy, K., Yu, Q., Li, Y., Ibarz, J.: Attention-based extraction of structured information from street view imagery. In: 14th IAPR International Conference on Document Analysis and Recognition, ICDAR 2017, Kyoto, Japan, November 9–15, 2017, pp. 844–850. IEEE (2017)

17. Phon-Amnuaisuk, S.: Exploring music21 and gensim for music data analysis and visualization. In: Tan, Y., Shi, Y. (eds.) Data Mining and Big Data - 4th International Conference, DMBD 2019, Chiang Mai, Thailand, July 26–30, 2019, Proceedings, volume 1071 of Communications in Computer and Information Science, pp. 3–12. Springer (2019)

18. Rendle, S., Freudenthaler, C., Gantner, Z., Schmidt-Thieme, L.: BPR: bayesian personalized ranking from implicit feedback. CoRR, abs/1205.2618 (2012)

19. Rendle, S.: Factorization machines. In: Webb, G.I., Liu, B., Zhang, C., Gunopulos, D., Wu, X. (eds.) ICDM 2010, The 10th IEEE International Conference on Data Mining, Sydney, Australia, 14–17 December 2010, pp. 995–1000. IEEE Computer Society (2010)

20. Jiang, M., Cui, P., Wang, F., Zhu, W., Yang, S.: Scalable recommendation with social contextual information. IEEE Trans. Knowl. Data Eng. **26**(11), 2789–2802 (2014)

21. Hell, F., Taha, Y., Hinz, G., Heibei, S., Mu¨ller, H., Knoll, A.C.: Graph convolutional neural network for a pharmacy cross-selling recommender system. Inf. **11**(11), 525 (2020)

Machine Learning for Data Science

Preliminary Study on Adapting ProtoPNet to Few-Shot Learning Using MAML

Yapu Zhao, Yue Wang$^{(\boxtimes)}$, and Xiangyang Zhai

School of Information, Central University of Finance and Economics, Beijing, China
yuelwang@163.com

Abstract. ProtoPNet proposed by Chen et al. is able to provide interpretability that conforms to human intuition, but it requires many iterations of training to learn class-specific prototypes and does not support few-shot learning. We propose the few-shot learning version of ProtoPNet by using MAML, enabling it to converge quickly on different classification tasks. We test our model on the Omniglot and MiniImagenet datasets and evaluate their prototype interpretability. Our experiments show that MAML-ProtoPNet is a transparent model that can achieve or even exceed the baseline accuracy, and its prototype can learn class-specific features, which are consistent with our human recognition.

Keywords: Prototypical features · Interpretability · Few-shot learning · MAML

1 Introduction

Taking advantage of large and high-quality datasets, deep learning models have shown satisfactory results in the field of image classification. However, when sufficient samples with supervised information are not easy to obtain, traditional machine learning methods have difficulty converging to good performance. Therefore, research on few-shot learning has become more popular and has made rapid progress recently. A large number of studies on few-shot learning construct models based on prior knowledge to narrow the hypothetical space to learn quickly for new tasks on smaller datasets [1], such as the Siamese Network [21] constructed by referring to human comparative thinking when recognizing images. Constructed in a similar way, the metric learning model restricts a model to learn a benchmark of classification first and then classify according to the distance from the benchmark [23]. Although the number of studies on few-shot learning

Yue Wang is the corresponding author of this paper (yuelwang@163.com). Yapu Zhao proposes the idea of combing MAML and ProtoPNet. And Yue Wang proposes the consistency of features (CoF) indicator in this paper. This work is supported by: National Defense Science and Technology Innovation Special Zone Project (No. 18-163-11-ZT-002-045-04); Engineering Research Center of State Financial Security, Ministry of Education, Central University of Finance and Economics, Beijing, 102206, China; Program for Innovation Research in Central University of Finance and Economics; National College Students' Innovation and Entrepreneurship Training Program "Research and development of interpretable algorithms and prototype system for small sample image recognition".

Y. Wang et al. (Eds.): ICPCSEE 2022, CCIS 1628, pp. 139–151, 2022.
https://doi.org/10.1007/978-981-19-5194-7_11

has increased, little attention has been given to the interpretability of few-shot learning [2]. The learning model of prototypical networks, which is a branch of metric learning, may provide perspectives for the interpretability of few-shot learning. It constructs distinctive prototypes and makes classification based on comparison with examples, which fits the way of human cognition of new things.

Prototype networks have been applied in few-shot learning models for a long time. Snell et al. [7] map the support set of a class to a higher dimension and take the mean as the prototype of the class. However, this well-known model did not provide interpretability. Here, Chen et al. [8] greatly inspired us. The model, called ProtoPNet, builds prototype layers upon convolutional base layers to scan the prototypical feature regions, and the model is constrained by a loss function to force it to learn prototypical features highly related to the corresponding class. However, this method requires many iterative updates, which make it difficult to obtain good classification results under limited datasets. Can there be any way to help ProtoPNet converge quickly on datasets of new classes?

The MAML proposed by Finn et al. [10] gives us a possible solution. The classifier in MAML acquires the ability to converge faster by training through a large number of different classification tasks. Therefore, by constantly constructing prototypical features in different tasks, ProtoPNet may also achieve good initialization, which means that a transparent image classification model with fast convergence can be obtained. We propose MAML-ProtoPNet, which combines MAML and ProtoPNet, and test our idea on the Omniglot and MiniImagenet datasets. The classification accuracies of our experiments approach or even exceed the baseline performance. We further test the interpretability of prototypical features and prove their interpretability potentials in preliminary experiments.

Contributions

- We propose a meta-learning version of ProtoPNet by using MAML so that it can work and interpret for new classes of small data.
- We test our model on the Omniglot and MiniImagenet datasets, and the results approach or even exceed the baseline performance. We further evaluate their interpretability and find that they have learned class-specific features that are consistent with human cognition.

2 Related Work

2.1 Few-Shot Learning

At present, there are three main perspectives to optimize the performance of few-shot learning [1]: data perspective, model perspective and algorithm perspective. These three ideas start from different stages of the machine learning process to achieve optimal results.

Data Perspective. First, the data dimension is a simple idea to deal with few-shot problems. Data augmentation can reduce the upper bound of the generalization error. Traditional data augmentation methods add training samples by simple transformations, such as rotation and scaling, to the existing datasets [1]. Recently, there have also been some methods used to generate data from similar and larger datasets. For example, Mehrotra et al. [3] applied GAN to few-shot learning and proposed a generative adversarial residual pairwise network. However, data augmentation methods are usually used as an auxiliary method in few-shot learning to expand available training samples.

Model Perspective. By using prior knowledge, which adds constraints to models, the assumption space can be narrowed. Therefore, fewer samples are needed to search for optima in the narrowed assumption space. For example, CNN networks that train similar but different tasks were constrained to jointly update parameters in Luo et al. [4]. Metric learning models are built on comparison by learned distance functions [5, 6]. This kind of classification based on comparison is more intuitive, suitable for direct reasoning, and has the potential for interpretability. In 2018, Chen et al. [8] proposed a model that decomposed its classification criterion into a weighted sum of similarity scores of meaningful feature regions, which had a higher level of interpretability. However, as we mentioned before, the training cost of this model was high. Gao et al. [9] applied an attention model on smaller datasets to identify prototypes and proposed feature-level attentions. However, their model only told where a query image was highly activated and did not tell where it was similar to the images learned before.

Algorithm Perspective. Finn et al. [10] proposed model-agnostic meta-learning (MAML), which is model agnostic and can be compatible with any model using gradient descent. To date, MAML has been applied to many models and has produced good performance. After one or several iterations, it can converge quickly [11–13]. Nevertheless, little attention is given to the interpretability of this kind of meta-learning. If we apply a transparent model on MAML, which in our work is by applying ProtoPNet [8], we may obtain a model that can not only converge quickly on smaller datasets but also give quantitative interpretability.

2.2 Interpretability

The mainstream interpretability methods can be divided into post hoc interpretability and endogenous interpretability models [19]. Common post hoc interpretability models include decision tree [14], deconvolution, separate representation [15], knowledge distillation [16], etc. Decision trees interpret black box models by fitting classification rules [17], and deconvolution can visualize learned features through convolutions of reverse filters [18]. These models imitate the classification process by fitting black-box models, but the actual working mechanism of the black-box models is still unknown, and most methods are qualitative [19]. The bag-of-words model maps an image to a collection of meaningful patches [20]. Although bag-of-words models can quantitatively analyze interpretability, they require a large number of manual annotations on meaningful areas in images, which are expensive to label and costly to train. Prototypical features, as an alternative to meaningful patches of pictures, can be captured by training prototype networks. When prototypical features are used as comparison benchmarks, quantitative classification criteria can be obtained, which are interpretable, real and intuitive.

3 Proposed Methods

Our model is called MAML-ProtoPNet. It adopts the meta-learning method in the interpretable neural network ProtoPNet proposed in [8]. By learning prototypical features in a meta-learning version, the model can obtain the ability to rapidly construct prototypical features and converge to new tasks.

The training process combines MAML and ProtoPNet. The outer loop is simply MAML, which consists of an across-task training, across-task validation and across-task testing process. Every process in an outer loop has several inner loops to learn the current best parameters θ of the classification model. As shown in Fig. 1, the across-task training process in an outer loop contains multiple inner loops starting from the same initial state θ. In an inner loop, we use ProtoPNet as the classification model containing the base convolutional layer, the prototype layer and the fully connected layer (without bias). Each inner loop is a typical classification task based on learned prototypes, which consists of within-task training, prototype projection and within-task testing process. Prototypical features are trained by gradient descent in the within-task training process. The prototype projection process chooses the closest feature value of prototypical features in the training set. The within-task testing process classifies images by using the most distinct feature regions (i.e., the chosen prototypical features) in the training set, and the loss value of a within-task testing process in an inner loop serves as the loss value of that inner loop. As shown in Fig. 1, the across-task training process takes the average of the loss value of multiple inner loops to update the initial parameters θ of the current ProtoPNet model.

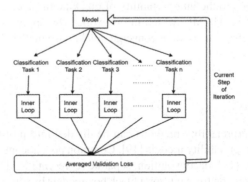

Fig. 1. Across-task training process

3.1 Adapting ProtoPNet to MAML

We set the inner-loop task to be the n-way k-shot image classification task. As shown in Fig. 2, let x be an image in the training set (support set). In the within-task training process, the base convolutional layers convert x to the feature map $f(x)$ with the shape of (h, w, d). The prototype layer consists of m prototypical features per class { $o_i | o_i \epsilon classi$}, and each prototypical feature has a shape of (h_1, w_1, d_1), where $h_1 < h$, $w_1 < w$ and

$d_1=d$. Each prototypical feature scans the feature map $f(x)$ to calculate the L^2 distances. The distance matrix is converted to the similarity matrix, and by global max pooling, we obtain the maximum similarity score between $f(x)$ and prototypical features. The maximum similarity scores lead to the classification scores through the fully connected layers. We refer to the loss function in [8], which encourages o_i to learn the most relevant features of class i and penalize irrelevant features. The best convergence condition of prototypical features is achieved in the form of the best classification performance.

Necessary Adaptations to MAML. Since MAML is trained on different classes during each loop, it is difficult to learn the stable prototypical features of each class. Therefore, we need to make necessary adaptations to MAML. First, the pretraining method is adopted in the inner loop: training base convolutional layers first and then adding prototype and fully connected layers after achieving good classification performance. Second, we should change prototypical features during each loop of classification to adapt to MAML. While prototypical features are projected to nearest feature maps, the prediction performance is not affected when the projection does not move much by the proof in [8]. Through these two adaptations, it will be easier for the model to learn prototype features in each inner loop. Hopefully, we can obtain the best initial parameters for few-shot learning in outer loops.

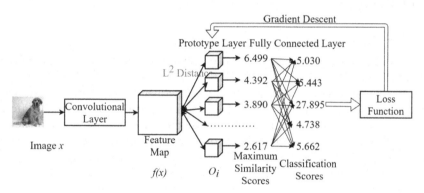

Fig. 2. Within-task training process

3.2 Evaluating Models

Interpretability can only be evaluated by human experience. First, we randomly sample an equal number of images from each class in the test set. For each image, we record its top k features given by similarity scores. After we obtain the top k features, the name of a feature is given by human experts by looking at the feature region in the original image, generated by the deconvolution or receptive field methods or simply direct mapping. Using the names of the top k features, we can analyze the interpretability performance of prototypical features.

In datasets that are composed of subjects that have physical meanings, we can evaluate the interpretability of a model through expert annotations and scoring from a preliminary perspective. However, for abstract datasets, expert annotations based on actual semantics might be difficult to carry out. However, we have found two metrics to measure a model's interpretability. First, machine learned features are annotated based on their positions in their original images by both machine and human expert. After obtaining the labels given by expert and our model, we calculate the matching degree between these two labels and define the score of the matching degree as *Consistency of Semantics (CoS)*. Second, the consistency of the top k features is identified by the machine. We can use the entropy function to measure the inconsistency for abstract recognition. Concretely, we record the percentage p_i of a type of feature where there are n types of labeled features. The *consistency of Features* (called *CoF*) is defined by Eq. (1).

$$CoF = 1 - \frac{\sum_{i=1}^{n} -p_i log p_i}{log n} \tag{1}$$

where $log n$ is the maximum entropy when each of the n types of features has a percentage of $1/n$. The more consistency a model identifies the same feature with, the more it is similar to the way humans learn abstract images.

4   Experiments

4.1   Datasets

For few-shot learning, we perform our experiments on Omniglot [22] and MiniImagenet [24].

Omniglot is a dataset composed of 1623 handwritten characters (i.e., classes) collected from 50 alphabets. Each character has 20 samples, which are drawn by 20 different human subjects. It is an abstract dataset. We use data augmentation, which rotates the training and validation datasets in multiples of 90 degrees to make more character classes. Therefore, we use 3200 character classes for training (800 original characters and 1600 characters generated by rotations), 656 character classes for validation (164 original characters and 492 characters generated by rotations), and 659 characters for testing. Both the height and width of an image are 28 pixels.

The MiniImagenet dataset is a small part extracted from the ImageNet dataset. The dataset contains 60,000 color images in 100 actual categories. A typical class contains 600 samples, and the height and width of an image is 84 pixels. There are 64 classes for training, 16 classes for validation and 20 classes for testing.

4.2   Experiment 1: Omniglot Few-Shot Classification

We perform our experiments in the 1-shot and 5-shot on 5-way classification. We construct a four-layer CNN to classify the Omniglot dataset. A convolutional layer is followed by a batch-norm layer, an activation layer (ReLU) and a max-pooling layer (kernel-size = 2, stride = 2) except for the last convolutional layer. The stride of the last max-pooling layer is set to 1. All the layers have 64 filters. The last layer's feature map is of size $4 \times 4 \times 64$ after calculation. If we further construct a ProtoPNet, we add prototype layers upon the CNN. We let each character class have four prototypical features. Each prototypical feature is of size $1 \times 1 \times 64$ for 64 filters. Since each training task takes 5 classes from the training classes and selects 4 prototypical features, there are 20 prototypical features in total. There is a similarity score between each prototypical feature and the target image. Finally, we have a fully connected layer upon the prototype layers, mapping the similarity scores to the classes. When training the model, we fix the weight from a prototypical feature to its corresponding class to 1 and fix the weights from it to the other classes to –0.5. In this way, we can accelerate the training speed without affecting the classification performance.

In Table 1, we compare the classification performance of MAML-CNN and MAML-ProtoPNet. MAML-CNN is the baseline model that combines MAML and the four-layer CNN. MAML-ProtoPNet is our model that combines MAML and Pro-toPNet. Each score is averaged over 5 classes of multiple tasks of the test set. In the 5-way 5-shot experiment, the scores of MAML-ProtoPNet are very close to those of MAML-CNN. In the 5-way 1-shot experiment, the scores of MAML-ProtoPNet are lower than those of MAML-CNN, but the accuracy and recall scores are still above 0.9. The possible reason why the scores of our model on 5-way 1-shot drops is as follows. While prototypical features of a class can be selected after comparing five images of any other class in the 5-shot case, they are selected only by one image in the 1-shot case, making it difficult to select class-specific features in this case. How-ever, comparing 1 shot and 5 shots, we find that slightly increasing the number of training sets can swiftly improve the performance of ProtoPNet (e.g., the precision score increases from 0.877 to 0.967).

Table 1. Omniglot experiment results on 5-way 1-shot and 5-shot learning

	Accuracy	Precision	Recall	F1
	5-way 1-shot			
MAML-CNN	0.963	0.946	0.963	0.952
MAML-ProtoPNet (ours)	0.912	0.877	0.912	0.889
	5-way 5-shot			
MAML-CNN	0.985	0.988	0.985	0.984
MAML-ProtoPNet (ours)	0.977	0.967	0.977	0.970

4.3 Experiment 2: MiniImagenet Few-Shot Classification

We perform our experiments in the 1-shot and 5-shot on 5-way classification. The height and width of the input images are set to 224 pixels to match the model's structure. We choose to take the convolutional layers of VGG-16 [25] as the convolutional base layer, whose structure is shown in Fig. 3. There are five convolutional blocks in total, and each convolution block contains 2 or 3 convolutional layers. Each convolution layer is followed by an activation layer (ReLU), and each convolution block is followed by a max-pooling layer (kernel-size = 2, stride = 2). The last layer's feature map is of size $7 \times 7 \times 512$ after calculation. We let each category have twenty prototypical features. Each prototypical feature is of size $1 \times 1 \times 512$ for 512 filters. Since each training task takes 5 classes from the training classes and selects 20 prototypical features, there are 100 prototypical features in total. The parameter setting of the fully connected layer is the same as that of Experiment 1.

In Table 2, we compare the classification performance of MAML-VGG and MAML-ProtoPNet. MAML-VGG is the baseline model that combines MAML and the convolutional layers of VGG-16, as shown in Fig. 3. MAML-ProtoPNet is our model. In both the 1-shot and 5-shot experiments, the performance of MAML-ProtoPNet is better than that of MAML-VGG (e.g., the accuracy increases almost 6% from MAML-VGG to MAML-ProtoPNet in the 5-way 5-shot experiment), which shows that the added prototype layer can improve the performance.

Table 2. MiniImagenet experiment results on 5-way 1-shot and 5-shot learning

	Accuracy	Precision	Recall	F1
	5-way 1-shot			
MAML-VGG	0.540	0.406	0.540	0.442
MAML-ProtoPNet (ours)	0.575	0.453	0.575	0.491
	5-way 5-shot			
MAML-VGG	0.615	0.632	0.615	0.567
MAML-ProtoPNet (ours)	0.675	0.743	0.675	0.665

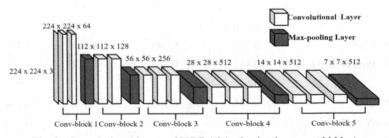

Fig. 3. Convolutional layers of VGG-16 (activation layers are hidden)

4.4 Experiment 3: Interpretability Analysis on Omngilot

We conduct interpretability verification analysis on 100 classes. We examine 5 test samples with their top 4 prototypical features for each class, therefore 2000 prototypical features in total. Due to our four convolutional layer structure, every image outputs an activation map of size $2 \times 2 \times 64$ while the size of our prototypical feature is set to $1 \times 1 \times 64$. Therefore, prototypical features roughly represent four parts of each image.

We visualize prototypical features and use this position of the four parts (upper-left, upper-right, lower-left, lower-right) as the labels of prototypical features. We also obtain feature region labels given by human experts, and then we examine whether the two regions are matched to examine the interpretability of the top 4 prototypical features.

Table 3 shows 5 query samples from different categories and their top 4 prototypical features and images. Prototypical features are built from prototypical images. We can see that almost all prototypical features are from the same class of each query image, which shows prototypical features do learn the typical features that are class-specific. Table 4 examines the consistency of the scores of prototypical features in two dimensions. *Consistency of semantics (CoS)* defined in 3.2 is the score between labels given by expert and labels given by our model. The *CoS* score is 0.904, which shows that the meanings of the prototypical features are highly consistent with our human recognition. The *CoF* defined in 3.2 examines the label consistency of the top 4 features of each query in the same class. The *CoF* score is 0.811, which shows that the top 4 prototypical features of each query in the same class identify the same areas. Combining the above two discussions on prototype consistency, we may say that the prototypical features in our model learn class-specific features that are consistent with our human cognition on abstract datasets.

Table 3. Visualization of five query images and their top 4 prototypical features

	query	prototype 1		prototype 2		prototype 3		prototype 4	
		feature	image	Feature	image	feature	image	feature	image
1	ਓ	ਂ	ਓ	ਓ	ਓ	ਂ	ਓ	ਰ	ਓ
2	ਸ਼	ਂ	ਸ਼	੍	ਚ	ਸ਼	ਸ਼	ਂ	ਸ਼
3	ੜ	੩	ੜ	੍	ੜ	੩	ੜ	੩	ੜ
4	ੜ	੭	ੜ	੭	ੜ	੭	ੜ	੨	ੜ
5	ਹੀ	ਹੀ	ਹੀ	ਹਾ	ਹੀ	ੁ	ਹੀ	ੲ	ਹੀ

Table 4. Consistency of prototypical features in *CoS* and *CoF* views

CoS	CoF
0.904	0.822

4.5 Experiment 4: Preliminary Interpretability Analysis on MiniImagenet

MiniImagenet is composed of images that have practical meanings, so we can further discuss learned prototypical features with expert annotations. We randomly select a 5-way task from the test dataset, which is made up of 5 images for training (support set) and 60 images for testing (query set). Then, we put the task into the MAML-ProtoPNet model to obtain prototypical features for each query and select the top 4 features of the same class according to its corresponding similarity scores. We artificially annotate the top 4 features for all images in the query set referring to the expert annotation rules defined in Table 5 and score the coincidence degree between the extracted prototype feature area

Table 5. Expert annotation rules

Photo					
Class	Golden Retriever	Malamute	Dalmatian	African Hunting Dog	Lion
Labels	·eye_black ·ear_gold ·mouth_black ·nose_black ·face_gold ·head_gold ·body_gold ·leg_gold ·tail_gold ·other_none	·eye_black ·ear_white ·mouth_black ·nose_black ·face_white ·head_black ·body_black andwhite ·leg_white ·tail_white ·other_none	·eye_black ·ear_balck andwhite ·mouth_black ·nose_black ·face_black andwhite ·head_black andwhite ·body_black andwhite ·leg_black andwhite ·tail_white ·other_none	·eye_black ·ear_balck ·mouth_black ·nose_black ·face_black andyellow ·head_black andyellow ·body_black andyellow ·leg_black andyellow ·tail_white ·other_none	eye_black ·ear_balck ·mouth_black ·nose_black ·face_black andyellow ·head_black andyellow ·body_black andyellow ·leg_black andyellow ·tail_white ·other_none

and the annotation semantic area for each query. The score of the coincidence degree ranges from 1 to 10.

Table 6 shows the score of the coincidence degree for every label in each class. We can see from Table 6 that class-specific features, such as the ear, face, head, and body, have relevant higher scores than the other features. This condition shows that MAML-ProtoPNet has the potential to learn class-specific features. In addition, most scores of coincidence degree are above 7. Therefore, we might consider that the interpretability shown by MAML-ProtoPNet is consistent with human cognition from a preliminary perspective.

Table 6. Expert annotation results

Golden Retriever	Score	Malamute	Score	Dalmatian	Score	African Hunting Dog	Score	Lion	Score
eye_black	6.41	eye_black	7.70	eye_black	7.71	eye_black	7.45	eye_yellow	7.06
ear_gold	7.65	ear_white	8.46	ear_blackandwhite	8.24	ear_black	7.70	ear_yellow	7.82
mouth_black	7.47	mouth_black	7.17	mouth_black	7.47	Mouth_black	8.37	mouth_white	7.76
nose_black	6.53	nose_black	7.34	nose_black	7.56	nose_black	7.17	nose_black	6.40
face_gold	9.83	face_white	8.63	face_blackandwhite	9.28	face_blackandyellow	8.72	face_yellow	7.98
head_gold	8.32	head_black	8.99	head_blackandwhite	8.41	head_blackandyellow	8.61	head_yellow	8.04
body_gold	7.99	body_blackandwhite	9.19	body_blackandwhite	9.21	body_blackandyellow	8.95	body_yellow	7.77
leg_gold	7.28	leg_white	7.47	leg_blackandwhite	8.52	leg_blackandyellow	8.29	leg_yellow	6.28
tail_gold	5.89	tail_white	6.52	tail_white	6.88	tail_white	6.90	tail_yellow	7.71
other_none	8.21	other_none	9.02	other_none	8.99	other_none	7.50	hair_yellow	8.14
								other_none	7.82

5   Conclusion and Future Work

In this work, we have proposed a meta-learning version of ProtoPNet by using MAML, which is a transparent model and has the ability to quickly converge in different classification tasks. We trained this model on the Omniglot and MiniImagenet datasets for 5-way 1- and 5-shot via a meta-learning method. We have achieved near or above baseline accuracy and obtained a quantitative interpretability of the classification process. We have also discussed the interpretability of prototypical features and proved that they can learn class-specific features in our preliminary experiments.

Our future work will try to compare more advanced and stable models to further verify the feasibility and stability of our model. We will try to apply more model structures to improve the performance of our model and further explore the interpret-ability of learned prototypical features.

Supplementary Material and Code: The supplementary material and code are available at https://github.com/Luoluopista/MAML-ProtoPNet.

References

1. Wang, Y., Yao, Q., Kwok, J.T., Ni, L.M.: Generalizing from a few examples: a survey on few-shot learning. arXiv: Learning (2019)
2. Xue, Z., Duan, L., Li, W., Chen, L., Luo, J.: Region comparison network for interpretable few-shot image classification. arXiv preprint arXiv:2009.03558 (2020)
3. Mehrotra, A., Dukkipati, A.: Generative adversarial residual pairwise networks for one shot learning. ArXiv, abs/1703.08033 (2017)
4. Luo, Z., Zou, Y., Hoffman, J., Fei-Fei, L.F.: Label efficient learning of transferable representations across domains and tasks. In: Advances in Neural Information Processing Systems, 30 (2017)
5. Gidaris, S., Komodakis, N. Dynamic few-shot visual learning without forgetting. In: 2018 IEEE/CVF Conference on Computer Vision and Pattern Recognition, pp. 4367–4375 (2018)
6. Suárez, J.L., García, S., Herrera, F.: A tutorial on distance metric learning: mathematical foundations, algorithms, experimental analysis, prospects and challenges. Neurocomputing **425**, 300–322 (2021)
7. Snell, J., Swersky, K., Zemel, R.: Prototypical networks for few-shot learning. In: Advances in Neural Information Processing Systems, 30 (2017)
8. Chen, C., Li, O., Barnett, A., Su, J., Rudin, C. This looks like that: deep learning for interpretable image recognition. NeurIPS (2019)
9. Gao, T., Han, X., Liu, Z., Sun, M. Hybrid attention-based prototypical networks for noisy few-shot relation classification. In: Proceedings of the AAAI Conference on Artificial Intelligence, vol. 33, no. 1, pp. 6407–6414 (2019)
10. Finn, C., Abbeel, P., Levine, S.: Model-agnostic meta-learning for fast adaptation of deep networks. In: ICML 2017, pp. 1126–1135 (2017)
11. Jiang, X., et al.: On the importance of attention in meta-learning for few-shot text classification. ArXiv, abs/1806.00852 (2018)
12. Jamal, M., Qi, G., Shah, M.: Task agnostic meta-learning for few-shot learning. In: 2019 IEEE/CVF Conference on Computer Vision and Pattern Recognition (CVPR), pp. 11711–11719 (2019)
13. Phillips, P.J., Hahn, C.A., Fontana, P.C., Broniatowski, D.A., Przybocki, M.A.: Four Principles of Explainable Artificial Intelligence (2020)
14. Bastani, O., Kim, C., Bastani, H.: Interpreting blackbox models via model extraction. ArXiv, abs/1705.08504 (2017)
15. Zhou, B., Bau, D., Oliva, A., Torralba, A.: Interpreting deep visual representations via network dissection. IEEE Trans. Pattern Anal. Mach. Intell. **41**, 2131–2145 (2019)
16. Cheng, X., Rao, Z., Chen, Y., Zhang, Q.: Explaining knowledge distillation by quantifying the knowledge. In: 2020 IEEE/CVF Conference on Computer Vision and Pattern Recognition (CVPR), pp. 12922–12932 (2020)
17. Zhang, Q., Yang, Y., Wu, Y.N., Zhu, S.: Interpreting CNNs via decision trees. In: 2019 IEEE/CVF Conference on Computer Vision and Pattern Recognition (CVPR), pp. 6254–6263 (2019)
18. Brahimi, M., Mahmoudi, S., Boukhalfa, K., Moussaoui, A. (2019, September). Deep interpretable architecture for plant diseases classification. In: 2019 Signal Processing: Algorithms, Architectures, Arrangements, and Applications (SPA), pp. 111–116. IEEE (2019)
19. Rudin, C.: Stop explaining black box machine learning models for high stakes decisions and use interpretable models instead. Nat. Mach. Intell. **1**(5), 206–215 (2019)
20. Qader, W.A., Ameen, M.M., Ahmed, B.I.: An overview of bag of words; importance, implementation, applications, and challenges. In: 2019 International Engineering Conference (IEC), pp. 200–204. IEEE (2019)

21. Melekhov, I., Kannala, J., Rahtu, E.: Siamese network features for image matching. In: 2016 23rd International Conference on Pattern Recognition (ICPR), pp. 378–383. IEEE (2016)
22. Lake, B.M., Salakhutdinov, R., Tenenbaum, J.B.: The Omniglot challenge: a 3-year progress report. Curr. Opin. Behav. Sci. **29**, 97–104 (2019)
23. Kulis, B.: Metric learning: a survey. Found. Trends Mach. Learn. **5**(4), 287–364 (2013)
24. Vinyals, O., Blundell, C., Lillicrap, T., Wierstra, D.: Matching networks for one shot learning. In: Advances in Neural Information Processing Systems, 29 (2016)
25. Simonyan, K., Zisserman, A.: Very deep convolutional networks for large-scale image recognition. arXiv preprint arXiv:1409.1556 (2014)

A Preliminary Study of Interpreting CNNs Using Soft Decision Trees

Qilong Zhao and Yue Wang[✉]

School of Information, Central University of Finance and Economics, Beijing, China
yuelwang@163.com

Abstract. Decision trees can be used to enhance the interpretability of neural networks. In this work, we compare the classification and interpretability performance of the normal decision tree and a type of soft decision tree when they are used to interpret the decision paths of CNN networks. With the help of feature visualization and human-labeled features, we demonstrate that the soft decision trees identify more consistent features while maintaining much higher classification performance than the normal decision tree.

Keywords: CNNs · Interpretability · Decision trees

1 Introduction

CNNs perform well but lack interpretability. Although we can interpret them by highlighting key image regions, it can be difficult to find the connection between a feature and the semantics of the relevant image region. The first and most important step in this process is to find the features that have the most influence on the prediction. Some try to mine the correspondence between semantics and features inside the neural network [9, 14], while others choose to use interpretable models to decipher the decision basis of the neural network, e.g., using proxy decision trees. Among the latter, the application of tree models is very common and natural [7, 20].

Decision trees have good interpretability, and they can clearly show how a prediction is made. However, when faced with raw pixels in images, decision trees are poorly suited. With the hierarchical representations of data by CNNs, we can extract the high-level features of the images to train a decision tree. On the other hand, we can also use the interpretability of decision trees to obtain reliable decision rules.

Yue Wang is the corresponding author (yuelwang@163.com), who proposes the consistency of interpretability (CoI) indicator in this paper. Qilong Zhao and Yue Wang contributed equally to this work. This work is supported by: National Defense Science and Technology Innovation Special Zone Project (No. 18-163-11-ZT-002-045-04); Engineering Research Center of State Financial Security, Ministry of Education, Central University of Finance and Economics, Beijing, 102206, China; Program for Innovation Research in Central University of Finance and Economics; National College Students' Innovation and Entrepreneurship Training Program "Research and development of interpretable algorithms and prototype system for small sample image recognition".

We compare the interpretability and classification performance of two types of decision trees (normal/soft decision tree) and explore the decision rules of both. The results show that a type of soft decision tree [1] can achieve better performance than a normal decision tree obtained under the same conditions. This advantage is even more pronounced in multiclassification tasks. The soft decision trees can achieve an accuracy of 96.7%, while VGG16 has an accuracy of only 92.9% on the same multiclassification task in our experiments. Furthermore, we show that since a soft decision tree uses the entire feature set to decide at every internal node, it can identify more consistent features than a normal decision tree.

2 Related Work

We briefly list the related works on CNN interpretability and especially tree models.

CNN Interpretability. Khan et al. [4] reviewed the progress on CNN architectures and presented a classification of recent CNN architectures. They mentioned some of the challenges faced by deep CNNs, including the lack of interpretability. Zhang et al. [21] conducted a survey of studies in understanding neural-network representations and learning neural networks with interpretable/disentangled representations. Zeiler et al. [16] used a multilayered deconvolutional network to visualize the intermediate feature layers. Bau et al. [17] proposed a general framework called network dissection for quantifying the interpretability of CNNs. Qi et al. [18] selected CNN filters and used feature map visualization to semantically annotate the important filters. Hong et al. [19] proposed a method to reveal semantic representations at hidden layers of CNNs by occluding. On the other hand, some works have tried to combine tree models and neural networks to obtain the respective advantages of both [3]. A very relevant work was done by Zhang et al. [7], who used decision trees to interpret CNNs at the semantic level. They developed a method to modify CNNs and establish a tight coupling of CNNs and decision trees. In addition, Wang et al. proposed using normal decision trees to interpret the last layers of CNNs [20].

Soft Decision Trees. Olaru et al. [15] proposed a type of decision tree that can make soft decisions. Irsoy and others [5] also proposed a soft decision tree that makes soft decisions, which uses a soft gating function to merge the decision results of subtrees. Both obtained better performance than the normal decision tree. Hinton et al. [2] proposed a training method called distillation to transfer the knowledge learned from a large model to a smaller model, which is more suitable for deployment. Subsequently, Hinton and others [1] proposed distilling a soft decision tree out of deep neural networks. We use it to compare with a normal decision tree in this paper. Luo et al. [6] proposed a hierarchical differentiable neural regression model called the soft decision tree regressor, which can achieve good performance in regression tasks with low cost.

Other Tree-Like Models. Tanno et al. [3] proposed adaptive neural trees (ANTs) combining the operational mechanisms of deep neural networks and decision trees, reflecting the characteristics of each and performing well in both classification and regression tasks. Kontschieder et al. [8] proposed deep neural decision forests - a new model trained in an end-to-end manner that unifies the divide-and-conquer principle in decision trees with the representation learning functionality in deep convolutional networks. Zhang et al. [13] recently proposed a new decision tree-based meta-learning framework for interpretable few-shot learning (FSL), called MetaDT. They replaced the black-box FSL classifier with an interpretable decision tree. Wu et al. [9] explicitly regularized the deep model so that human users can quickly step through the prediction process of the deep model.

Local Interpretable Methods. There are also ways to partially explain the black box model. Ribeiro et al. [10] proposed LIME, a new interpretation technique that explains the predictions of any classifier in an interpretable and trustworthy manner by learning a locally interpretable model around the prediction. After that, Ribeiro and others [11] introduced a new set of rule-based, model-agnostic interpretation methods called anchors. Lundberg et al. [12] proposed a unified framework for explaining predictions: SHAP. For a given prediction, SHAP assigns an importance value to each feature.

Our Contributions. Although there are many interpretable models for CNNs, the performance comparisons of different models are not clear. In this paper, we focus on comparing the performance of using normal and soft decision trees to interpret the last layers of CNNs.

3   Proposed Methods

We examine the performance of interpreting CNNs using the normal decision tree and Frosst & Hinton's soft decision tree. In this section, we first discuss the differences and relations of the two models and then illustrate how to design experiments to fairly compare their interpretability performance.

3.1   Model Foundations

Decision Tree. A decision tree is an old and well-known machine learning algorithm used to organize decision paths in a tree structure. The CART (classification and regression tree) algorithm is currently used to train a decision tree. The idea of the algorithm is to greedily divide the training set into two subsets according to some feature k and its threshold tk. The algorithm attempts to find k and tk by minimizing the cost function defined in Eq. (1).

$$J(k, t_k) = \frac{m_{left}}{m}G_{left} + \frac{m_{right}}{m}G_{right} \tag{1}$$

where G_{left} (G_{right}) measures the impurity of the left (right) subset, m_{left} (m_{right}) is the size of the left (right) subset, and m is the size of the training set. The impurity measure can be a function of gini or entropy. The process is repeated to apply to the subsets until meeting stop conditions, e.g., the maximum depth of a decision tree (max_depth) or the minimum number of samples required to split an internal node (min_samples_split).

Frosst & Hinton's Soft Decision Tree. One type of soft decision tree proposed by Frosst and Hinton [1] is similar to a hierarchy of logistic regressions. One node uses a logistic function σ to differentiate the data into the two subsets. At an internal node, if $\sigma(\beta(wx + b)) > 0.5$, it goes to the right branch; otherwise, it goes to the left branch, where x is the feature vector of a sample point, w is the weights of the features and β is called an inverse temperature to adjust the probability of the σ function. $w_i x + b$ can be seen as a filter at the internal node i. The soft decision tree is trained using a loss function that minimizes the cross entropy between each leaf weighted by its path probability and the target distribution, and the path probability is calculated by the probabilities of each node on the path.

Similarities and Differences Between Normal and Soft Decision Trees. Although both types of trees classify samples by dividing them into two branches, they have many differences. First, the decision tree is constructed using a greedy algorithm, and the soft decision tree is trained by optimizing a cost function. Second, at each internal node, the decision is made by a single feature in the decision tree, while it is made by the weighted sum of all features in the soft decision tree. In addition, although the success of classifying the dataset of the MINST handwritten digits was shown for the soft decision tree in [1], it handles raw pixels of images and thus cannot be applied to realistic pictures (see Sect. 4). However, the two trees can be trained using processed features such as the high-layer CNN features to interpret the decision path of the CNN classification.

3.2 Using Normal/Soft Decision Trees to Interpret CNNs

The process of interpreting CNNs by normal/soft decision tree is shown in Fig. 1.

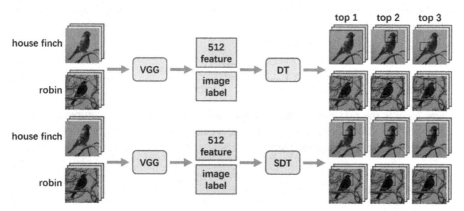

Fig. 1. The process of interpreting CNNs by normal/soft decision tree (DT: normal decision tree, SDT: soft decision tree)

Step 1. We train a CNN to obtain the features (i.e., the activation of each CNN filter) at the highest CNN layer.

Step 2. We use the features and the labels of images to train a normal/soft decision tree.
Step 3. We compare the interpretability between the normal and soft decision trees according to human experts.

3.3 Evaluating Interpretability

Interpretability can only be evaluated by human experience. First, we randomly sample an equal number of images from each class in the test set. For each image, we record its true and predicted classes, as well as the names of the top k features selected according to the order of the internal nodes in the tree hierarchy.

For the decision tree, it is natural to obtain the features from the judging conditions of the internal nodes in the decision path of an image. However, for the soft decision tree, it is somewhat obscure how to do so since the soft decision tree uses all features at each internal node in the decision path of an image. We note that the internal node goes to the right branch when $wx + b > 0$ and to the left branch when $wx + b \leq 0$. Therefore, it is natural to obtain the largest positive wixi component when $wx+b > 0$ and obtain the smallest negative wixi component when $wx+b \leq 0$ because the corresponding component contributes most to the decision. In this way, we obtain the most contributing features in the decision path of an image.

After we obtain features, the name of a feature is given by human experts by looking at the feature region of the original image, which is generated by a deconvolution or receptive field method. Using the names of the top k features, we can then analyze the interpretability performance of the tree models to CNNs.

There are two metrics to measure a model's interpretability. First, the percentage of features learned by the machine compared to those annotated by humans. Second, the consistency of the top features identified by the machine. We can use the entropy function to measure the inconsistency. Concretely, we record the percentage pi of a type of feature where there are m types of labeled features. The consistency of interpretability (called CoI) is defined by Eq. (2).

$$\text{CoI} = 1 - \frac{\sum_{i=1}^{m} -p_i log p_i}{log m} \qquad (2)$$

where *logm* is the maximum entropy when each of the m types of features has a percentage of 1/m. The more consistency a model can identify the same features, the more similar it is to the way humans learn.

4 Experiments

4.1 Dataset and Experimental Setup

We carry out experiments according to Steps 1–3 in Sect. 3.2 to explore the classification and interpretability performance of the normal decision tree (abbr. DT) and the soft decision tree (abbr. SDT). For binary classification and interpretability, images of two species of birds from the Mini-ImageNet dataset are used. In the multiple classification experiment, images of four bird species from the Mini-ImageNet dataset are used as the dataset. Mini-ImageNet is actually a subset of ImageNet [22].

There are two ways to train a soft decision tree, either using ground truth (i.e., true labels) or soft targets (the probabilities of predicted classes made by the CNN). For a pretrained CNN model, the outputs of the softmax layer can be used as soft targets [1]. We use the VGG16 model pretrained on ImageNet to provide features and soft targets in the experiments. Features are provided by layer 28 of VGG16, with 512 dimensions. Later, we use "512 features" to refer to them.

In Experiment 2, Experiment 3 and Experiment 4, the maximum depth of the two decision trees was 3.

4.2 Experiment 1: Classification Performance

For 600 images in each class, 60% are used as the training set, 20% as the validation set, and the remaining 20% as the test set. Tables 1 and 2 show the results of binary classification and multiclass classification, respectively.

In Table 1, VGG16 test results are first displayed as the benchmark. Then, we list the results of different interpretable models for different max_depths (maximum tree depth). VGG-DT (or VGG-SDT) refers to the normal (or soft) decision tree trained using 512 features of the top CNN layer of VGG16. As mentioned before, a soft decision tree can be trained using true labels or soft targets. We also use original images to train the soft decision tree.

We can see from the two tables that the performance of VGG-SDT (512 features) is much better than that of VGG-DT (512 features), showing the advantage of using soft decision trees to interpret CNNs. However, the performance of VGG-SDT (original images) is very low because it uses the raw pixels of images as features. Therefore, we need high-level features extracted by CNNs for complicated images (note that [1] used MINST, whose images are much simpler than ImageNet). In addition, VGG-SDT (512 features, true labels) is very close to or even surpasses VGG-SDT (512 features, soft labels); thus, it is unnecessary to use soft targets.

Table 1. Binary classification (house finch, robin)

Model	max_depth	Accuracy	Precision	Recall	F1
VGG16	n/a	**0.942**	**0.934**	**0.950**	**0.942**
VGG16-DT (512 features, true labels)	3	0.823	0.825	0.823	0.822
	4	**0.852**	**0.893**	**0.801**	**0.844**
	5	0.832	0.862	0.793	0.825
VGG16-SDT (512 features, true labels)	3	0.958	0.959	0.958	0.958
	4	**0.967**	**0.967**	**0.967**	**0.967**

(*continued*)

Table 1. (*continued*)

Model	max_depth	Accuracy	Precision	Recall	F1
	5	0.946	0.947	0.946	0.946
VGG16-SDT (512 features, soft targets)	3	0.946	0.946	0.946	0.946
	4	0.946	0.948	0.946	0.946
	5	**0.954**	**0.954**	**0.954**	**0.954**
VGG16-SDT (original images, soft targets)	4	0.479	0.356	0.479	0.337
	8	0.538	0.540	0.538	0.531

Table 2. Multiclass classification (house finch, robin, goose, toucan)

Model	max_depth	Accuracy	Precision	Recall	F1
VGG16	n/a	**0.929**	**0.932**	**0.929**	**0.929**
VGG16-DT (512 features, true labels)	3	0.619	0.768	0.620	0.683
	4	0.679	0.751	0.681	0.705
	5	**0.688**	**0.800**	**0.690**	**0.734**
VGG16-SDT (512 features, true labels)	**3**	**0.960**	**0.963**	**0.960**	**0.961**
	4	0.958	0.960	0.958	0.958
	5	0.960	0.961	0.960	0.960
VGG16-SDT (512 features, soft targets)	3	0.958	0.958	0.958	0.958
	4	**0.967**	**0.967**	**0.967**	**0.967**
	5	0.956	0.957	0.956	0.956
VGG16-SDT (original images, soft targets)	4	0.256	0.273	0.256	0.228
	8	0.265	0.273	0.265	0.241

4.3   Experiment 2: Visualization of Normal/Soft Decision Trees' Top Features

We randomly select four images from each of the two classes (house finch and robin) for visualization, and the results are shown in Fig. 2. The left half of the figure shows the results for the house finch, and the right half shows the results for the robin.

For each image, we follow the steps described in Sect. 3.3 to visualize the top 3 features of the normal decision tree and the soft decision tree. The visualization results for each feature are circled in red. The different sizes and positions of the red boxes are due to the different corresponding features. We can see that DT and SDT can often select useful features of the two kinds of birds.

Note that the soft decision tree prefers to choose the same feature. The reason is as described in Sect. 3.3; the mechanism of selecting the top k features is different between the two models. When a sample passes an internal node of a tree, only one feature is considered in the case of DT, but all features are considered in the case of SDT; thus,

SDT tends to select the same important features at every internal node along a decision path.

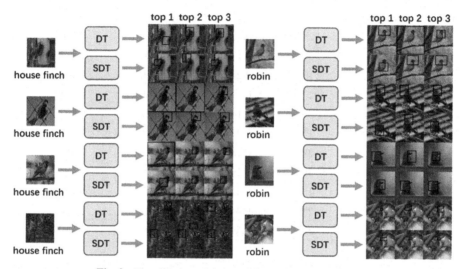

Fig. 2. Visualization of the top 3 features for DT and SDT

4.4 Experiment 3: Interpretability Performance

We calculate CoI (Eq. (2)) to measure the interpretability performance. The larger the CoI is, the more consistent features that are identified. In the experiment, the semantic annotation of features is performed manually, and we randomly select 10% of the samples from the dataset as annotated data. For the features visualized, we label them according to part and color (i.e., "(body, brown)").

Table 3 shows the CoI of the kth top features identified by DT and SDT. It also shows the most common feature tag and its proportion in all top k features. We can see that SDT has a much higher CoI than DT (because SDT tends to identify the same features illustrated in 4.2), and they both identify the meaningful most common feature tags.

4.5 Experiment 4: Scores of Human Experts on Tag Clarity

We score the tag clarity of the first three important features of DT and SDT. Scores range from 0 to 10; the higher the score is, the higher the clarity of the tag. For each tag of the two classes of targets (house finches and robins), we calculate the average scores of DT and SDT, respectively. Table 4 shows the results.

The second column of Table 4 shows all the tags of the corresponding class. In each row of the table, the two right columns show the average scores of DT and SDT. It can be seen that for most tags, the average score of SDT is higher than that of DT. This proves that the semantic meaning of the features identified by SDT is clearer and more consistent with human experience.

Table 3. CoIs of DT and SDT

Model	Class	Feature	CoI	Most common tag	Proportion
DT	House finch	1st	0.158	(body, brown)	0.233
		2nd	0.232	(eye, black)	0.400
		3rd	0.133	(beak, brown)	0.217
	Robin	1st	0.247	(head, black)	0.383
		2nd	0.191	(chest, gold)	0.333
		3rd	0.141	(eye, white)	0.267
SDT	House finch	1st	0.278	(eye, black)	0.500
		2nd	0.285	(eye, black)	0.467
		3rd	0.299	(eye, black)	0.483
	Robin	1st	0.447	(eye, white)	0.667
		2nd	0.444	(eye, white)	0.667
		3rd	0.418	(eye, white)	0.650

Table 4. Tag clarity averages of DT and SDT

Class	Tag	DT average	SDT average
House finch	(body, brown)	4.042	6.960
	(wing, brown)	6.286	7.846
	(chest, red)	6.714	7.000
	(head, red)	6.541	7.227
	(eye, black)	6.857	5.545
	(beak, brown)	7.400	7.000
	(other)	4.800	6.000
Robin	(body, black)	3.710	6.471
	(wing, black)	7.400	8.286
	(chest, gold)	7.333	7.368
	(head, black)	7.300	7.444
	(eye, white)	7.233	7.500
	(beak, yellow)	8.500	8.200
	(other)	3.667	6.200

5 Conclusion and Future Work

In this work, we compare the interpretability of a normal decision tree and a soft decision tree for the VGG CNN. By visualizing the top k features on the decision paths, we find that the soft decision tree tends to use the same features, and the normal decision tree often uses different features. Combined with manual semantic annotation, we compare the feature recognition consistency of the two kinds of decision trees. The results indicate that the soft decision trees are more capable of identifying consistent features. Combined with the results of scoring the clarity of tags for the first three important features, we demonstrate that the soft decision trees identify features that are clearer and more consistent.

We also test the classification performance of the two decision trees in a binary classification task and a multiclass classification task. The soft decision tree is significantly better than the normal decision tree, showing that the soft decision tree is more reliable for approximating CNNs.

In the future, we hope to test on additional datasets to evaluate the performance of DT and SDT in explaining the models and to compare them with other models.

References

1. Frosst, N., Hinton, G.: Distilling a neural network into a soft decision tree. arXiv preprint arXiv:1711.09784 (2017)
2. Hinton, G., Vinyals, O., Dean, J.: Distilling the knowledge in a neural network. arXiv preprint arXiv:1503.02531 (2015)
3. Tanno, R., Arulkumaran, K., Alexander, D., Criminisi, A., Nori, A.: Adaptive neural trees. In: International Conference on Machine Learning. PMLR, pp. 6166–6175, May 2019
4. Khan, A., Sohail, A., Zahoora, U., Qureshi, A.S.: A survey of the recent architectures of deep convolutional neural networks. Artif. Intell. Rev. **53**(8), 5455–5516 (2020). https://doi.org/10.1007/s10462-020-09825-6
5. Irsoy, O., Yıldız, O.T., Alpaydın, E.: Soft decision trees. In Proceedings of the 21st International Conference on Pattern Recognition (ICPR 2012), pp. 1819–1822, November 2012
6. Luo, H., Cheng, F., Yu, H., Yi, Y.: SDTR: soft decision tree regressor for tabular data. IEEE Access **9**, 55999–56011 (2021)
7. Zhang, Q., Yang, Y., Ma, H., Wu, Y.N.: Interpreting CNNs via decision trees. In: Proceedings of the IEEE/CVF Conference on Computer Vision and Pattern Recognition, pp. 6261–6270 (2019)
8. Kontschieder, P., Fiterau, M., Criminisi, A., Bulo, S.R.: Deep neural decision forests. In Proceedings of the IEEE International Conference on Computer Vision, pp. 1467–1475 (2015)
9. Wu, M., Hughes, M., Parbhoo, S., Zazzi, M., Roth, V., Doshi-Velez, F.: Beyond sparsity: tree regularization of deep models for interpretability. In: Proceedings of the AAAI Conference on Artificial Intelligence, vol. 32, no. 1, April 2018
10. Ribeiro, M.T., Singh, S., Guestrin, C.: "Why should i trust you?" explaining the predictions of any classifier. In Proceedings of the 22nd ACM SIGKDD International Conference on Knowledge Discovery and Data Mining, pp. 1135–1144, August 2016
11. Ribeiro, M.T., Singh, S., Guestrin, C.: Anchors: high-precision model-agnostic explanations. In: Proceedings of the AAAI Conference on Artificial Intelligence, vol. 32, no. 1, April 2018
12. Lundberg, S.M., Lee, S.I.: A unified approach to interpreting model predictions. In: Advances in Neural Information Processing Systems 30 (2017)
13. Zhang, B., Jiang, H., Li, X., Feng, S., Ye, Y., Ye, R.: MetaDT: meta decision tree for interpretable few-shot learning. arXiv preprint arXiv:2203.01482 (2022)

14. Sabour, S., Frosst, N., Hinton, G.E.: Dynamic routing between capsules. In: Advances in Neural Information Processing Systems 30 (2017)
15. Olaru, C., Wehenkel, L.: A complete fuzzy decision tree technique. Fuzzy Sets Syst. **138**(2), 221–254 (2003)
16. Zeiler, M.D., Fergus, R.: Visualizing and understanding convolutional networks. In: Fleet, D., Pajdla, T., Schiele, B., Tuytelaars, T. (eds.) ECCV 2014. LNCS, vol. 8689, pp. 818–833. Springer, Cham (2014). https://doi.org/10.1007/978-3-319-10590-1_53
17. Bau, D., Zhou, B., Khosla, A., Oliva, A., Torralba, A.: Network dissection: quantifying interpretability of deep visual representations. In Proceedings of the IEEE Conference on Computer Vision and Pattern Recognition, pp. 6541–6549 (2017)
18. Qi, C., et al.: Analyzing interpretability semantically via CNN visualization. In: Zeng, J., Qin, P., Jing, W., Song, X., Lu, Z. (eds.) ICPCSEE 2021. CCIS, vol. 1452, pp. 88–102. Springer, Singapore (2021). https://doi.org/10.1007/978-981-16-5943-0_8
19. Hong, Q., Wang, Y., Li, H., Zhao, Y., Guo, W., Wang, X.: Probing filters to interpret CNN semantic configurations by occlusion. In: Zeng, J., Qin, P., Jing, W., Song, X., Lu, Z. (eds.) ICPCSEE 2021. CCIS, vol. 1452. Springer, Singapore (2021). https://doi.org/10.1007/978-981-16-5943-0_9
20. Wang, S., Wang, Y., Zhao, Q., Yang, Z., Guo, W., Wang, X.: Exploring Classification Capability of CNN Features. In: Zeng, J., Qin, P., Jing, W., Song, X., Lu, Z. (eds.) ICPCSEE 2021. CCIS, vol. 1451, pp. 268–282. Springer, Singapore (2021). https://doi.org/10.1007/978-981-16-5940-9_21
21. Zhang, Q.-S., Zhu, S.-C.: Visual interpretability for deep learning: a survey. Front. Inf. Technol. Electron. Eng. **19**(1), 27–39 (2018). https://doi.org/10.1631/FITEE.1700808
22. Vinyals, O., Blundell, C., Lillicrap, T., Wierstra, D.: Matching networks for one shot learning. In: Advances in Neural Information Processing Systems 29 (2016)

Deep Reinforcement Learning with Fuse Adaptive Weighted Demonstration Data

Baofu Fang and Taifeng Guo[✉]

School of Computer Science and Information Engineering, Hefei University of Technology, Hefei, China
1136062482@qq.com

Abstract. Traditional multi-agent deep reinforcement learning has difficulty obtaining rewards, slow convergence, and effective cooperation among agents in the pretraining period due to the large joint state space and sparse rewards for action. Therefore, this paper discusses the role of demonstration data in multiagent systems and proposes a multi-agent deep reinforcement learning algorithm from fuse adaptive weight fusion demonstration data. The algorithm sets the weights according to the performance and uses the importance sampling method to bridge the deviation in the mixed sampled data to combine the expert data obtained in the simulation environment with the distributed multi-agent reinforcement learning algorithm to solve the difficult problem. The problem of global exploration improves the convergence speed of the algorithm. The results in the RoboCup2D soccer simulation environment show that the algorithm improves the ability of the agent to hold and shoot the ball, enabling the agent to achieve a higher goal scoring rate and convergence speed relative to demonstration policies and mainstream multi-agent reinforcement learning algorithms.

Keywords: Multiagent deep reinforcement learning · Exploration · Offline reinforcement learning · Importance sampling

1 Introduction

Reinforcement learning improves by interacting with the environment through rewards and punishments to solve sequential decisions for optimal returns to produce a general-purpose agent that solves complex problems [1]. Reinforcement learning has been combined with deep learning in recent years and, combined with the representational power of deep learning, is used to solve many challenging problems, such as IoT security [2], traffic control [3], and resource scheduling [4], reaching and exceeding human capabilities on these tasks.

Multi-Agent Reinforcement Learning (MARL) [5] is a new field formed by combining reinforcement learning with multi-agent systems, which divide the system into several self-consistent and agent subsystems, each of which can perform tasks independently while communicating and coordinating with each other to accomplish tasks together. Multiagent reinforcement learning has already solved many problems that

Y. Wang et al. (Eds.): ICPCSEE 2022, CCIS 1628, pp. 163–177, 2022.
https://doi.org/10.1007/978-981-19-5194-7_13

cannot be solved by single agents in areas such as fleet scheduling management problems [6] and massively multiplayer online role-playing games [7]. However, multiagent reinforcement learning explores potential policies through iterative trials, requiring the agents to continuously interact with the simulation environment and learn from their collective traces, leading to low sample efficiency of these multi-agent deep reinforcement learning methods. Due to the huge state space and action space, this random initialization algorithm makes it difficult to obtain rewards in the early stages of training due to the high cost of global exploration, and its slow boosting policy is not suitable for complex environments such as RoboCup2D.

Offline reinforcement learning is inspired by deep learning in large datasets to learn policies from fixed datasets [8], which can use large amounts of existing recorded interaction data to solve real-world decision problems. In offline reinforcement learning, the agent does not receive feedback from the online environment, improving the sample utilization. However, because the policy generating the sample is different from the optimized policy (off-policy), there is a bias in the sampling of the two policies, which makes it difficult for ordinary reinforcement learning methods to cope with the new state space and can even harm the performance of the reinforcement learning algorithm. The policies in the collected demonstration dataset are often suboptimal. It also limits the upper limit of offline reinforcement learning.

The main contributions of this paper are as follows:

1. We propose the distributed combining of demands actor-critic (DCDAC) algorithm. using a scalable distributed actor-critic architecture, which can accelerate the data throughput and improve the learning speed while using policy gradients for learning on demonstration data to give a good initial guide to the policy network.
2. We propose adaptive weighting parameters generated based on the evaluation of performance in the training environment to adjust the weights of the demonstration data to eliminate the detrimental effects of the demonstration data on the policies of the agents.

2   Related Work

2.1   Deep Reinforcement Learning

Reinforcement learning means learning by interacting with the environment and receiving feedback. The balance between exploration and exploitation is a major issue with reinforcement learning [1]. Exploration means that the agent tries some new actions that may result in higher rewards or nothing; exploitation means that the agent repeatedly adopts actions that are known to maximize rewards because the agent already knows that a certain number of rewards can be obtained. Some of the commonly used methods for exploration include ε − greedy methods with decay (1–0) values to return random actions (exploration if the random number is less than ε) or greedy actions (exploitation if the random number is greater than ε). Stochastic sampling models output random samples of action distributions (e.g., actions and variances) for exploration based on Gaussian distributions and OU (Ornstein-Uhlenbeck) noise to add noise to the model output actions. In real environments, the above methods of exploration are difficult to exploit because in real environments, exploration can damage expensive objects such as robotic arms and cars, while offline reinforcement learning only uses demonstration data for learning, improving the applicability of deep reinforcement learning in many real-world tasks. Its representative method is imitation learning, where successful demonstration trajectories are provided to the agent and the agent learns accordingly [9], and a simple approach to imitation learning is behavior cloning (BC) [10], which uses data from demonstrations as labels and trains the agent using supervised learning. Thus, pure imitation learning methods cannot go beyond the capabilities of demonstration data.

2.2 Multiagent Reinforcement Learning

Multiagent reinforcement learning is an emerging field where deep reinforcement learning (DRL) and multiagent systems are combined. There are problems with nonstationary environments, global exploration, and relative generalization caused by the increase in the number of agents [11]. Learning in a multiagent environment is fundamentally more difficult than in a single-agent environment.

The current methods commonly used for multiagent can be divided into three categories: decentralized methods, centralized methods, and value decomposition methods. The decentralized approach, also called independent learning (IL) [13], is an early method of multiagent reinforcement learning in which each agent learns only the individual action-value function and considers other agents as part of the environment, avoiding the curse of dimensionality. However, because the policies of other agents are constantly changing, the policies learned by the agents will also constantly change, resulting in nonstationary problems. The centralized approach learns the joint action-value function directly by considering all the agents' information, which can mitigate the adverse effects of nonstationarity. However, as the number of agents increases, the parameter space grows exponentially, and the joint action-value function is difficult to use in an environment with a large number of agents due to the problem of scalability. The value decomposition method [12] first "decentralizes" the learning of the individual action-value function of each agent and then "centralizes" the fitting of the joint action-value function using the individual action-values. In this approach, the computational complexity of the joint action-value function grows linearly with the number of agents, and the information of all agents is taken into account. However, the value decomposition approach is prone to the problem of relative overgeneralization [14], which means that the policy network incorrectly converges to suboptimal joint actions. Relative overgeneralization occurs when multiple agents must coordinate their actions but receive negative rewards if only some of them adopt the wrong behavior. The academic community then turned its attention back to decentralized learning approaches. Recently, there has also been independent proximal policy optimization (IPPO) [15] combined with proximal policy optimization (PPO) [16], which states the use of gradient clipping to control the magnitude of differences between old and new policies, thus mitigating nonstationarity in a multiagent environment, and experimental results also show that the use of independent learning in multiple agents can also achieve better results than value decomposition methods. However, the high cost of global exploration due to the huge state and action space leads to its slow convergence in complex environments.

Several approaches have combined reinforcement learning with demonstrative data to achieve faster learning, hopefully with good results in terms of obtaining policies that outperform demonstrators. For example, Deep Q-learning from Demonstrations (DQfD) [17], which combines behavioral cloning from imitation learning with deep q-learning, learns a policy for Atari games by using a loss function that combines a large number of supervised learning loss functions with an n-step Q-learning loss function, which helps ensure that the network satisfies the Bellman equation. However, its training speed is too slow, and in practice, behavioral cloning does not guarantee that the cloned policy will be effective due to the distribution bias between the states in the demonstration data and the policy's states [9]. It is also mentioned in some recent studies that using behavioral cloning on demonstration data does not show better results than using the policy gradient directly [18].

3   Methods

The framework of a distributed actor-critic algorithm with joint demonstration data is shown in Fig. 1, which includes a learner and multiple workers for a given agent i. The workers synchronize their parameters with the learner at regular intervals and interact with the environment to generate trajectories. Multiple workers can greatly increase

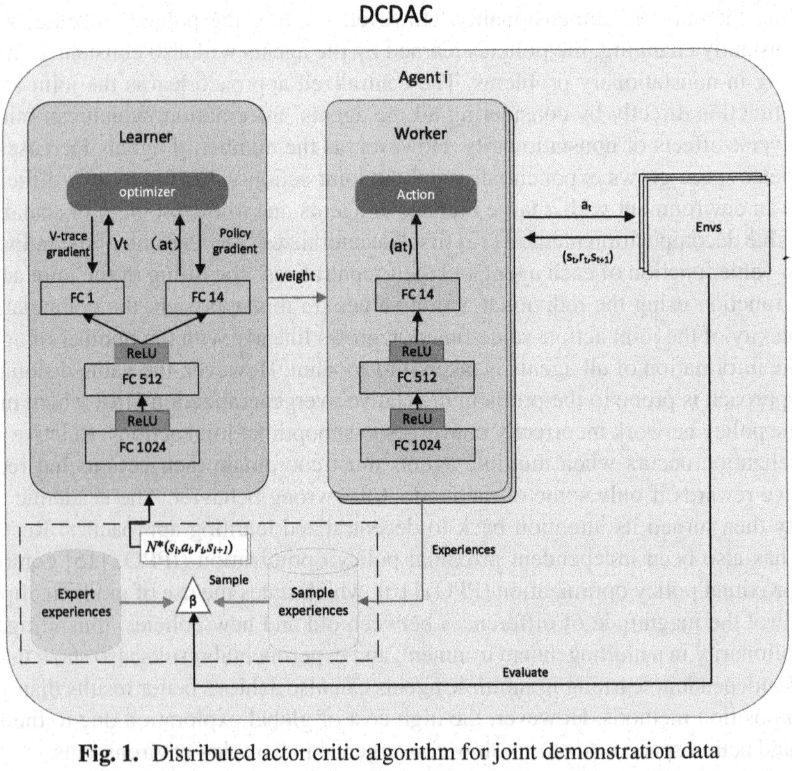

Fig. 1. Distributed actor critic algorithm for joint demonstration data

the sample generation speed and improve learning efficiency. The learner decides to use interaction-generated trajectories or demonstration data for learning based on the parameter β, which is generated by the performance metric of the training environment on the current worker.

The learner uses a fully connected (FC) layer to map the raw data to the hidden feature space, as shown in Fig. 1, and uses the ReLU activation function to improve the model representation. Both actors and critics use states as input, so we make actors and critics share the representation of states, and one head outputs action probabilities (n action types output n-dimensional vectors), and the other head outputs the value of the state. Whereas in workers, the neural network is similar to the learner but only uses the actor function, the trajectories for each learner's training are collected from many workers. In the case of a worker, for example, it updates its local policy μ to the latest learning policy π at the beginning of the trajectory generated by interacting with the environment and runs n steps in its environment. After running n-steps, the worker puts the trajectories $trace_i = (s_0, a_0, r_0, s_1, a_1, r_1, , \ldots, s_T, a_T, r_T) \in$ TRACE   of state, action, and reward. After running n-steps, the trajectory generator saves the state, action, and reward trajectories along with the corresponding policy distribution $\mu(a_t|s_t)$ and puts them in a queue, while the demonstration data need to be collected in advance and sent to the learner with the trajectories in the trajectory generator according to the weight in learning. Then, the learner continuously updates the trajectory of its policy batches. However, when parameters are updated, the learner's policy π may be updated several times earlier than the worker's policy μ, so there is a policy lag between the worker and the learner. For this reason, we use V-trace to relieve this lag.

We consider the infinite discount reward model optimization problem in a Markov decision process (MDP) [1], where the objective is to find a policy that maximizes the expected sum of future discount rewards: $\text{argmax}_\pi V^\pi(s) = E_\pi \left[\sum_{t\geq 0} \gamma^t r_t \right]$, where $\gamma \in [0, 1)$ is the discount factor, $r_t = r(s_t, a_t)$ is the reward at time t, s is the state at time t (initialized $s = s_0$), and $a_t \sim \pi(s_t)$ is the action generated by following some policy.

The goal of the off-policy reinforcement learning algorithm is to use the trajectory generated by a certain policy μ, called the behavioral policy, to learn the value function V_π of another policy π (which may be different from μ), called the target policy. In the trajectory $(s_t, a_t, r_t)_{t=s}^{t=s+n}$ generated by the agent according to the given policy μ, the n-step V-trace goal is defined as follows:

$$v_t = V(s) + \sum_{t=t_0}^{t_0+n-1} \gamma^{t-t_0} \left(\prod_{i=t_0}^{t-1} c_i \right) \delta_t V \tag{1}$$

where $\delta_t V = \min\left(\overline{\rho}, \frac{\pi(a_t|s_t)}{\mu(a_t|s_t)}\right) * (r_t + \gamma V(s_{t+1}) - V(s_t))$ is the TD-target of the value function, $c_i = \min\left(\overline{c}, \frac{\pi(a_i|s_i)}{\mu(a_i|s_i)}\right)$ is the truncated importance sampling, and $\overline{\rho}$ and $\overline{c}$ are the truncation parameters. Using this approach can effectively reduce the variance in the evaluation of the value function.

There is a policy lag between worker and learner. We update the gradient parameters using importance sampling for the behavioral policy μ and the target policy π:

$$l_{pg} = \rho_t \log \pi_\theta(a_t, s_t)(r_t + \gamma v_{t+1} - V_\theta(s_t)) \tag{2}$$

Similar to PPO, where $\rho_t = \min\left(\bar{\rho}, \frac{\pi(a_t|s_t)}{\mu(a_t|s_t)}\right)$ is truncated importance sampling, limiting the difference between the old and new policies both safeguards the results of importance sampling and reduces the influence of other agents on the optimal policy of the current agent because the agent has fewer differences between the new and original policies. v_{s+1} is the value of the next state.

Define the set of demonstration data D for successive states s^D and actions a^D and returns r^D generated over time steps during task execution according to a fixed rule of demonstration policy:

$$D = \left\{s_0^D, a_0^D, r_0^D, s_1^D, a_1^D, r_1^D, , \ldots, s_T^D, a_T^D, r_T^D\right\} \tag{3}$$

Using the policy gradient to train on the demonstration data, the loss function of the demonstration data can be defined as:

$$l_D = -\left[log\pi_\theta(a_t|s_t)A(s_t, a_t)\right] \tag{4}$$

The actor network is initialized by θ and fed with a given state s_t. The output action distribution $log\pi_\theta(a_t|s_t)$ represents the direction of the policy gradient. $A(s_t, a_t)$ is the advantage function, which represents the difference between the performance of the current action a_t and the mean of the performance of all possible actions and is used to determine the direction of the policy update—a positive value increases the probability of this action, while a negative value decreases the probability of this action. The advantage of using the policy gradient for training the demonstration data is that it is similar to the way the collected trajectories are trained. This makes it possible to make the best use of the demonstration data.

The V-trace objective is defined by (1). Gradient descent is used to reduce the value parameter θ to the target value of $V(s)$ during training.

$$l_{V(s)} = \frac{1}{2}(v_t - V_\theta(s_t))^2 \tag{5}$$

To prevent premature convergence, add an entropy reward to the loss function:

$$l_c = -\pi_\theta(a|s_t) \log \pi_\theta(a|s_t) \tag{6}$$

However, the demonstration data are not valid for all training phases because the demonstration data are not perfect, and there is room for improvement and a negative impact on reinforcement learning in the second stage of training. Therefore, we introduce β to reduce the impact of the demo data when the performance reaches a stable level. Connect the above loss function as follows:

$$l_{DCD} = \beta(l_D) + (1 - \beta)(l_{pg} + l_{vs} + l_c) \tag{7}$$

In the selection of the weight parameter, we found that it is not the higher weight (0.5) of demonstration data that is more beneficial to the performance. Instead, higher weighted demonstration data can negatively affect reinforcement learning due to large policy differences. We experimentally observe that using a β value of 0.1 minimizes the negative impact of large policy differences on the offline data. When the performance of the demonstration data with 0.1 weight is unimproved, the β value is corrected to 0, and the next stage of training is performed entirely using reinforcement learning.

Algorithm 1 Distributed Actor Critic Algorithm for Combined Demonstration

---

Input: training count T, demonstration data D ,maximum trajectory length L_{tarce} , paramter β, batch size N.

Output: Neural network of actors and critics after training.

Randomly initialize the parameters of the actors and critics network.

1: **def** worker:

2: **for** episode = 0 to T **do**

3:   Initialize the environment and get the initial state s_0

4:   Synchronize neural network parameters from the learner

5:   **for** t = 1 to L_{tarce} **do**

6:   select action according to current policy $a_t = \pi\left(s_t \mid \theta_\pi\right)$

7:    Execution of action a_t yields a return r_t and the next state s_{t+1}

8:    Storage Status, action, distribution of action, next action

9:    **if** end of current trajectory **then**

10:    Quit the current for loop and proceed to the next epsiode.

11:    **end if**

12:   **end for**

13:   Send data to trainer

14: **end for**

1: **def** learner:

2: **for** episode = 1 to T **do**

3:   Randomly select sampling from the demonstration data and the worker's data according β Calculate the loss function based on the sampled data

4:   Generate the gradient update model according to Equation (7).

5: **end** for

---

4 Experimental Results and Analysis

4.1 Experimental Environment and Data

In this paper, the RoboCup2D multiagent soccer environment Fig. 2 is used as the simulation experiment environment, and the opponent is an agent2d standard soccer player, which has been used as the base for the RoboCup2D World Cup championship for the last 5 years. The task scenario is described in detail as follows: The abstract concept of soccer is used on a two-dimensional plane, where the players, the ball, and the field are all two-dimensional objects. In this, each agent receives its state perception and must independently choose its actions. At the beginning of each episode, the agents and the ball are randomly placed in the attacking half of the field. The episode ends when a goal is scored, when the ball leaves the field, at 500 steps, or when the ball is intercepted by the goalkeeper. We use the state space relative to the agent itself. This includes the agent's position, position relative to the ball, goal, penalty area, opponents, teammates, and angle, and whether one can kick the ball or not. Constitute a vector of $58 + 9 * n$ dimensions.

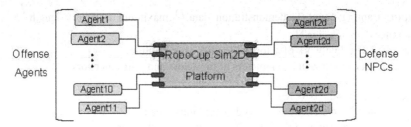

Fig. 2. RoboCup2D standard opponent platform

As opposed to HFO [19], a half-field offensive soccer environment, we added some mid-level action to make it more similar to gamepad controls. The game is a very successful and mature approach in RL. The action space used in the experiment is shown in Table 1. Action spaces.

Table 1. Action spaces

TORIGHT	TOPLEFT
BOTTOMRIGHT	TOP
BOTTOM	TOPRIGHT
BOTTOMLEFT	TOLEFT
AUTOPASS	AUTOTACKLE
MOVE	SHOOT
GO_TO_BALL	NOOP

The movement space is shown in Table 1. Action spaces, which first includes eight movement actions: left, top left, top, top right, right, bottom right, bottom, and bottom left. When holding the ball, the movement becomes a ball-carrying action. AUTOPASS is a passing movement that automatically selects the target and passes the ball based on the angle and distance between teammates and yourself, as well as whether an opponent player is on the passing route; MOVE is moving to the formation point (a preset catch point according to the ball's position); GO_TO_BALL is running to the ball; SHOOT is a shot action, which selects the optimal shot angle based on the goalkeeper and the enemy player. For the current reinforcement learning task, it is easier to use the above action space for effective control of the agent. It is also possible to obtain good results after experiments. The reward uses the HFO setting.

In this paper, we collect consecutive trajectories from the demonstration policy according to fixed rules. The goal sequence of the left demonstration policy is shown in Fig. 3, and the state of the current moment, including the relative position coordinates, velocity, body angle, and head facing angle of each agent; the position and velocity of the ball; and the position coordinates, velocity, body angle, current action, and next action of all teammates and opponents, sorted according to the distance from the current agent state at the next step, and combined with the rewards generated by our reward function. As demonstrated by the different agents listed in the dictionary. At the beginning of the sampling process, adding higher weight to the demonstration, the data gives some guidance to the policy of the agents and stabilizes the transition from demonstration learning to the policy of reinforcement learning. After the training reaches stability (no further improvement for a long time), the adaptive weights β are adjusted, and the demonstration data are stopped to prevent the negative impact of the demonstration data on reinforcement learning.

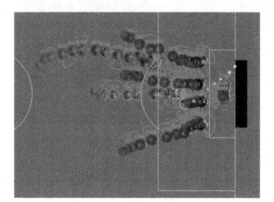

Fig. 3. Collection of demonstration policy trajectories.

To verify the performance of the method in this paper, the convergence speed is reflected in a 2 vs. 2 offensive environment where the adversary is the agent2d base [20], all World Cup teams currently use agent2d for their base, and the agent of agent2d is compared in a 4 vs. 5 scenario to further verify the ability of the algorithm to converge in a large-scale scenario, and the impact of different β is compared in a 2 vs. 5 scenario.

For comparison, the following algorithms are used: IMPALA [21], which uses distributed data collectors to explore in parallel in the environment, improving exploration capabilities also used to solve multiagent problems; Behavioral Cloning [9], which uses supervised learning to exactly imitate demonstration policies; and IPPO [15], which extends PPO to multiagent tasks where each agent runs a set of PPO [16] algorithms independently, with good results in the StarCraft environment. The deep reinforcement learning network framework in the experimental scenario of this paper is implemented by torch with a Core i7-10700 processor, 16 GB RAM, GeForce RTX 1060 GPU, and Adam as the optimizer with a learning rate of 0.0006.

4.2 Results and Analysis

4.2.1 2 vs. 2 Experiment

In the 2 vs. 2 experiment, the number of offensive players is set to 2, the number of defensive players to 1, and a goalkeeper is added. A 2 vs. 2 offensive scenario is set. At each time step, the agent observes their own and their teammates' positions in the environment, as well as the positions of the opposing defensive players. Each iteration is 3,000 steps, and the number of rounds in each iteration is different because the length of each training trajectory is different. The number of goals scored in the current iteration divided by all the rounds is the goal rate.

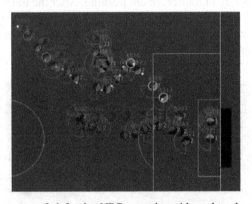

Fig. 4. 2 offensive agents vs. 2 defensive NPC scenarios with goal tracks. (Color figure online)

To illustrate the training effect of successful DCDAC, we take one of the goal trajectory graphs as an example. As shown in Fig. 4, yellow is the trained agent, red is the agent2d standard soccer agent, and white is the ball. Where the light color is the initial position and the brighter color is the later position, the yellow No. 7 player agent learns to move upward first after getting the ball, to lure the goalkeeper upward to make a defensive hole, and then shoot with the ball afterward.

The goal rate comparison of the algorithm is shown in Fig. 5. Comparison of goal rate under 2 offensive agents vs 2 defensive player scenarios. As the number of iterations increases, the behavioral clone remains stable after 300 iterations, and the goal rate of the agents in the other methods also increases, meaning that the agents have gradually learned the policies for successful goals. The addition of demonstration data will make the agents obtain the reward of scoring goals and the demonstration of behavior earlier, thus helping the agent learn the optimal policy faster and better. The IMPALA algorithm, on the other hand, shows better results relative to other algorithms because of its use of parallel workers, which can explore different states in parallel. When the number of iterations reaches 1500, the success rate of the IMPALA algorithm and DCDAC

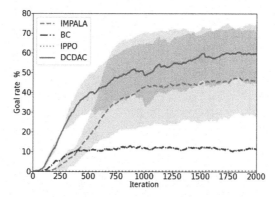

Fig. 5. Comparison of goal rate under 2 offensive agents vs 2 defensive player scenarios.

algorithm stabilizes. When the number of iterations reaches 2000, the goal rate of the DCDAC algorithm stabilizes at approximately 0.6, which can achieve better results than IMPALA but is still inferior to the algorithm proposed in this paper. It can be seen that in the complex multiagent environment, the method proposed in this paper makes a great contribution to the training of the agents.

4.2.2 4 vs. 5 Large-Scale Experiments

In this experiment, the difficulty is increased in terms of the number of agents. Set the number of offensive agents to 4 and the number of defensive players to 5, a scenario with 4 defensive players and one goalkeeper. Compared with the abovementioned 2 vs. 2 half-court offensive experiment, at this time, the agents need to cooperate to break through the defense of the opposing defender to score a goal. The increase in the number of agents also increases the amount of information that the neural network needs to process, which increases the burden on the deep neural network. The perfect NPC that the enemy has won the World Cup also makes it more difficult for the agent to score goals, and the agent's training becomes more difficult.

The algorithm success rate comparison is shown in Fig. 6. Due to the increased difficulty of the experiment, the success rate of all four groups of algorithms has been decreased due to the lack of effective rewards in the huge action space, and state-space IMPALA and PPO algorithms have failed with a goal rate of only 1% after 5000 iterations. The BC algorithm directly through the demonstration data on the imitation in 500 iterations after the goal success rate was maintained at approximately 2%, due to the demonstration data containing only part of the policy, does not contain all cases. In the environment, the agent will encounter states that never occur, which leads to the BC approach being far below the example policy. Since the starting β was set to 0.1 in this experiment, the DCDAC algorithm did not converge as fast as the behavioral clone in the early stage, but because it was able to obtain many goal trajectories in the early stage, it was able to use them to learn and have a good initialization of the policy, which effectively reduced the cost of global exploration. Then, getting rid of the demonstration data to explore on its own after 2000 iterations also had some degree of performance decrease in the early stage but eventually far outperformed the other algorithms in the

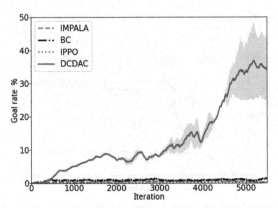

Fig. 6. Comparison of goal rate under 4 offensive agents vs 5 defensive NPC scenarios.

later stages of training. It can be seen that in the 4 vs. 5 attack environment, compared to 2 vs. 2, although the state space is larger, the task is more difficult and convergence is more difficult, the method proposed in this paper is still effective and adaptable.

The policy with a fixed rule that provides demonstration data versus the goal rate after the completion of DCDAC training is shown in Table 2.

Table 2. Comparison of goal rates between the demonstration policy and the DCDAC method

Scenes	Demonstration policy	DCDAC
2 vs. 2	52%	63%
4 vs. 5	20.5%	35%

The goal rate comparison between the DCDAC algorithm and the demonstration policy is shown in Table 2, where our method can produce better policies than the demonstration policy after training stabilization. Since policies generated based on complex rules (decision trees, evaluation of states under complex conditions, dynamic formation changes) are difficult to characterize directly with neural networks, we also do not obtain demonstration policies using BC. For methods such as IPPO, it is difficult to generate goal-scoring behavior in a huge space of states and actions, indicating that reinforcement learning without good demonstration has difficulty surpassing policies for which humans already have complex rules through pure exploration.

4.2.3   Impact of β Weighting on Performance

Finally, we study the effects of different β on the performance of the algorithm in an experimental scenario of 2 vs. 5, with two offensive agents and five defensive players. The goal-scoring rate is shown in Fig. 7. During the training process, when β is 0, there is no demonstration guideline, which leads to no good policy emerging from the agents. When β is 0.1, giving a little demonstration to the agents will make the agents' policies improve. When β is 0.2, it produces some improvement in the early training period when the performance of the agent is weak, but the negative effect brought by too many off-policy samples in the second half of the training period is more obvious, and the negative effect in the second half of the training period is more obvious when β is 0.5. Therefore, demonstration data that are too high will not be better.

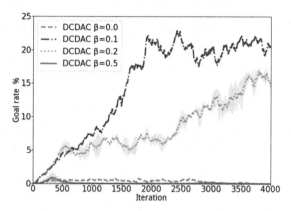

Fig. 7. Comparison of different parameters in 2 offensive agents vs 5 defensive NPC scenarios.

5 Discussion

In this paper, the role of demonstration data in multiagent systems is studied to solve the problem of difficult convergence in large-scale multiagent reinforcement learning due to the difficulty of emerging effective rewards from initial policies. A multiagent deep reinforcement learning algorithm with adaptive weight fusion demonstration data is proposed. The algorithm combines demonstration data with a distributed multiagent reinforcement learning algorithm and experiments in a RoboCup2D simulation environment for comparison, which can make full use of the respective advantages of demonstration data and reinforcement learning and improve the convergence speed and robustness of the algorithm. The experimental results also verify the effectiveness of the algorithm in this paper. The algorithm in this paper is a discrete action space, which leads to the inability to produce more accurate and reasonable actions, and further research on the hybrid action space will also be conducted. In addition, it is found that the cooperation level is still low in the case of multiagents in the experiments. Future research will also be conducted on these aspects of planning to improve them through course learning and self-play.

References

1. Sutton, R.S., Barto, A.G.: Reinforcement Learning: An Introduction. MIT Press, Cambridge (2018)
2. Li, Y., Liu, T., Zhu, J., Wang, X.: IoT security situational awareness based on Q-learning and Bayesian game. In: Zeng, J., Qin, P., Jing, W., Song, X., Lu, Z. (eds.) ICPCSEE 2021. CCIS, vol. 1452, pp. 190–203. Springer, Singapore (2021). https://doi.org/10.1007/978-981-16-5943-0_16
3. Chu, T., Wang, J., Codeca, L., Li, Z.: Multi-agent deep reinforcement learning for large-scale traffic signal control. IEEE Trans. Intell. Transp. Syst. **21**(3), 1086–1095 (2019)
4. Hausknecht, M., Mupparaju, P., Subramanian, S., Kalyanakrishnan, S., Stone, P.: Half field offense: an environment for multiagent learning and ad hoc teamwork. In: AAMAS Adaptive Learning Agents (ALA) Workshop (2016)
5. Chang-Yin, S., Chao-Xu, M.: Important scientific problems of multi-agent deep reinforcement learning. Acta Automatica Sinica **46**(7), 71–79 (2020)
6. Nguyen, D.T., Kumar, A., Lau, H.C.: Policy gradient with value function approximation for collective multiagent planning. In: Advances in Neural Information Processing Systems: Proceedings of NIPS, pp. 4–9 (2017)
7. Peng, P., et al.: Multiagent bidirectionally-coordinated nets: emergence of human-level coordination in learning to play starcraft combat games. arXiv preprint arXiv:1703.10069 (2017)
8. Fujimoto, S., Meger, D., Precup, D.: Off-policy deep reinforcement learning without exploration. In: International Conference on Machine Learning, pp. 2052–2062. PMLR (2019)
9. Levine, S., Kumar, A., Tucker, G., Fu, J.: Offline reinforcement learning: tutorial, review, and perspectives on open problems. arXiv preprint arXiv:2005.01643 (2020)
10. Zhan, E., Zheng, S., Yue, Y., Sha, L., Lucey, P.: Generative multi-agent behavioral cloning. arXiv (2018)
11. Hernandez-Leal, P., Kartal, B., Taylor, M.E.: A very condensed survey and critique of multi-agent deep reinforcement learning. In: Proceedings of the 19th International Conference on Autonomous Agents and MultiAgent Systems, pp. 2146–2148 (2020)
12. Sunehag, P., et al.: Value-decomposition networks for cooperative multi-agent learning. arXiv preprint arXiv:1706.05296 (2017)
13. Matignon, L., Laurent, G.J., Le Fort-Piat, N.: Independent reinforcement learners in cooperative Markov games: a survey regarding coordination problems. Knowl. Eng. Rev. **27**(1), 1–31 (2012)
14. Yu, C., Velu, A., Vinitsky, E., Wang, Y., Bayen, A., Wu, Y.: The surprising effectiveness of MAPPO in cooperative, multi-agent games. arXiv preprint arXiv:2103.01955 (2021)
15. de Witt, C.S., et al.: Is independent learning all you need in the StarCraft multi-agent challenge? arXiv preprint arXiv:2011.09533 (2020)
16. Schulman, J., Wolski, F., Dhariwal, P., Radford, A., Klimov, O.: Proximal policy optimization algorithms. arXiv preprint arXiv:1707.06347 (2017)
17. Hester, T., et al.: Deep Q-learning from demonstrations. In: Thirty-Second AAAI Conference on Artificial Intelligence (2018)
18. Wang, Q., Xiong, J., Han, L., Sun, P., Liu, H., Zhang, T.: Exponentially weighted imitation learning for batched historical data. In: NeurIPS, pp. 6291–6300 (2018)
19. Hausknecht, M., Mupparaju, P., Subramanian, S., et al.: Half field offense: an environment for multiagent learning and ad hoc teamwork. In: AAMAS Adaptive Learning Agents (ALA) Workshop (2016)

20. Akiyama, H., Nakashima, T.: Helios base: an open source package for the RoboCup soccer 2D simulation. In: Behnke, S., Veloso, M., Visser, A., Xiong, R. (eds.) RoboCup 2013. LNCS (LNAI), vol. 8371, pp. 528–535. Springer, Heidelberg (2014). https://doi.org/10.1007/978-3-662-44468-9_46

21. Espeholt, L., et al.: IMPALA: scalable distributed Deep-RL with importance weighted actor-learner architectures. In International Conference on Machine Learning, pp. 1407–1416. PMLR (2018)

DRIB: Interpreting DNN with Dynamic Reasoning and Information Bottleneck

Yu Si, Keyang Cheng$^{(\boxtimes)}$, Zhou Jiang, Hao Zhou, and Rabia Tahir

School of Computer Science and Communication Engineering, Jiangsu University, Zhenjiang 212013, China

{siyu,2222008027,2212108025}@stmail.ujs.edu.cn, kycheng@ujs.edu.cn

Abstract. The interpretability of deep neural networks has aroused widespread concern in the academic and industrial fields. This paper proposes a new method named the dynamic reasoning and information bottleneck (DRIB) to improve human interpretability and understandability. In the method, a novel dynamic reasoning decision algorithm was proposed to reduce multiply accumulate operations and improve the interpretability of the calculation. The information bottleneck was introduced to the DRIB model to verify the attribution correctness of the dynamic reasoning module. The DRIB reduces the burden approximately 50% by decreasing the amount of computation. In addition, DRIB keeps the correct rate at approximately 93%. The information bottleneck theory verifies the effectiveness of this method, and the credibility is approximately 85%. In addition, through visual verification of this method, the highlighted area can reach 50% of the predicted area, which can be explained more obviously. Some experiments prove that the dynamic reasoning decision algorithm and information bottleneck theory can be combined with each other. Otherwise, the method provides users with good interpretability and understandability, making deep neural networks trustworthy.

Keywords: Dynamic reasoning · Information bottleneck · Interpreting DNNs

1 Introduction

Deep neural networks (DNNs) have achieved superb performance and widespread acceptance in many application areas, for example, in image classification [1], object detection [2], sentiment analysis [3], and other applications [4, 5]. It is well known that the success of deep learning models comes from efficient mathematical learning algorithms and a large parameter weights. This huge parameter space makes DNNs a complex black-box model, as the parameter space contains hundreds of network layers and millions of parameters. We use many data and GPUs to train the deep learning model. As illustrated in Fig. 1, deep learning models are called black-boxes and are also referred to as end-to-end systems. The input is the original data, and then the output is directly the final goal. The intermediate process is unknown and difficult to know. It is also be

Supported by National Natural Science Foundation of China [61972183].

expressed as uninterpretable and untrustworthy. However, why do we need interpretability? Although the performance of DNNs is solid and even outperforms humans on many specific tasks, we cannot understand these models. We do not apply them, especially in high-risk domains such as medical diagnosis [6], financial risk forecasting [7], and automatic driving [8]. Therefore, it is essential to open the black box and raise trust and transparency to DNNs.

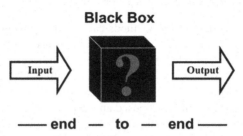

Fig. 1. The DNN is a black box trained from end to end. We cannot understand the mechanism of the inner process in it.

This paper proposes the dynamic reasoning and information bottleneck (DRIB) technique to construct an attribution interpretable convolution model. The contributions of our model are as described below:

1. A novel dynamic reasoning decision algorithm was proposed to reduce multiply accumulate operations and improve the interpretability of calculations.
2. The information bottleneck was introduced to the DRIB model to verify the attribution correctness of the dynamic reasoning module.
3. Some experiments prove that the dynamic reasoning decision algorithm and information bottleneck theory can be combined with each other. Otherwise, the method provides users with good interpretability and understandability, making deep neural networks trustworthy.

2   Related Works

2.1   Explain the Existing Deep Learning Models

Some of the post-hoc explainable methods work after building the deep neural networks. The sensitive features of DNNs can be identified by the feature analysis method. This method analyzes the explainability of the model through a specific example by modifying the local input and observing the impact on the prediction [9–11]. Network dissection [12] depends on the emergence of disentangled or human-explainable units during training [10, 13, 14]. However, when this method is applied to large and deep networks, its analysis is ineffective. In addition, the class activation mapping method shines brilliantly in the explainability field. They obtain the visualization graph by multiplying the feature graph by the weight. The difference between them is that CAM [15] obtains the weight from the fully connected layer; Grad-CAM [16] flows the gradient of a specific class to each feature graph, and then uses the average gradient as the weight; Score-CAM [17] does not need the gradient, but generates the weight for each feature graph through its forwarding score. All the above methods generate class activation features from the convolution layer, but our DRIB uses the information bottleneck theory to compress information and does not need to perform additional calculations for visualization.

2.2 Construction of Interpretable Deep Learning Models

Another part of ad hoc interpretable methods can avoid bias in post hoc explainability analysis. An interpretable representation technique is used to make the constructed neural network more interpretable [18, 19]. These methods replace parts of DNNs with interpretable machine learning models, such as decision trees [20, 21] and rule systems [22, 23]. Other approaches [24, 25] fuse deep learning into each decision tree node. Zhu Songchun and Zhang Quanshi proposed an interpretable CNN [26], which allows filters at a high level to represent specific local objects, which assists us in understanding the internal logic of CNNs. However, the method of building interpretable models is not universal. Therefore, the DRIB is attached to each convolution layer to make the convolution layer interpretable, and layer by layer to achieve model interpretability.

3 Method

An overview of the dynamic reasoning and information bottleneck (DRIB) is illustrated in Fig. 2. It shows the l-th convolution layer. Two modules are attached to each convolution layer. One is the Dynamic Reasoning Decision; the other is Information Bottleneck Verification. When the neural network performs a feedforward operation on a single input, the dynamic reasoning decision module first outputs the weight channel masks. It then determines the weight used to perform the actual calculation. Then the information bottleneck verification module verifies whether the feature selected by the dynamic reasoning decision module is within the result of the information bottleneck. If so, the dynamic reasoning decision module result is adopted and sent to the next round of calculation.

This section describes in detail the dynamic reasoning decision module and the information bottleneck verification module in the following sections. This paper will first introduce the dynamic reasoning decision module. We introduce the principle of information bottleneck theory for deep neural network validation on the basis of dynamic inference decision module.

3.1 Dynamic Reasoning Decision Module

In the feedforward calculation of a convolutional neural network with an L layer, the l-th convolutional layer computes the output features $x_l \in \mathbb{R}^{N \times C_o \times H_o \times W_o}$ with $x_{l-1} \in \mathbb{R}^{N \times C_i \times H_i \times W_i}$ and weights $\theta_l \in \mathbb{R}^{C_o \times C_i \times k \times k}$ by Formula 1.

$$\mathbf{x}_l = conv(\mathbf{x}_{l-1}, \theta_l) \tag{1}$$

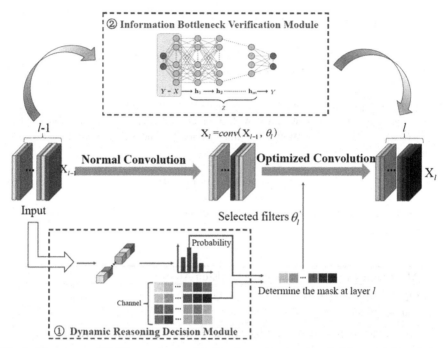

Fig. 2. The proposed DRIB overview model. The dynamic reasoning decision module optimizes the features with the best contribution in each convolution layer to reduce the amount of computation. The information bottleneck verification module verifies the accuracy of the whole network attribution and provides explanation and analysis.

where (N, C, H, W) are the corresponding batch, channel, height and width dimensions, respectively. The subscripts i, o denote variables for the input and output, respectively. Besides, the variable k is kernel size, and the symbol $conv$ is the convolution operation.

Suppose that x_{l-1} is the input features of the l-th layer. As indicated in Formula 2. Through the *ReLU* layer and *AvgPool* layer, we obtain an m-dimensional vector, $A(x_{l-1}) \in \mathbb{R}^m$.

$$A(\mathbf{x}_{l-1}) = Linear\ (GlobalAvgPool(Relu(\mathbf{x}_{l-1}))) = [p_1, p_2, \ldots, p_m] \qquad (2)$$

In addition, m kinds of masks are saved as $G_l \in \mathbb{R}^{m \times C_l}$ in layer l. $\pi(\mathbf{x}, \phi)$ is associated with each convolution layer. We take the largest $G_l[i]$ as the mask of layer l to run, where $i = argmax(A(x))$. How can the value of $G_l[i]$ be obtained? Given an m-dimensional output channel vectors with probability $P = [p_1, p_2, \ldots, p_m]$. The selection function I of the channel mask is represented formally as a categorical random variable. Gumbel random variables sampled from the Gumbel distribution $G = -log(-log(X))$ with $X \sim U[0, 1]$. Nevertheless, the problem of nondifferentiability to the underlying probability p is very troublesome. To solve this problem, the index I of $G_l[i]$ is substituted with a softmax form [27], as shown in Formula 3.

$$I_i = \frac{exp\left((\log p_i + G_i)/\tau\right)}{\sum_{j=1}^{m} exp\left((\log p_j + G_j)/\tau\right)}, \quad \forall i = 1, \cdots, m \tag{3}$$

where τ is a temperature parameter that governs the probability.

Finally, we can select weights with nonzero channel selection values $\theta_l' = \{\theta_l[j] \mid G[i][j] \neq 0\}$, and compute the final output for the l-th convolution layer by Formula 4.

$$x_l = conv\left(x_{l-1}, \theta_l'\right) * G_l[i] \tag{4}$$

Therefore, the dynamic reasoning decision can be summarized as Algorithm 1 from the above steps.

Algorithm 1 Dynamic Reasoning Decision Algorithm

Input: feature $x_{l-1} \in \mathbb{R}^{N \times C_i \times H_i \times W_i}$, weight $\theta_l \in \mathbb{R}^{C_o \times C_i \times k \times k}$

Output: feature $x_l \in \mathbb{R}^{N \times C_o \times H_o \times W_o}$

1: Get an m-dimensional vector from $Linear\left(GlobalAvgPool(Relu(\mathbf{x}_{l-1}))\right)$;

2: **for** k in $[0, ..., C\text{-}1]$ **do**

3:    Get m kinds of masks $G_l \in \mathbb{R}^{m \times C_l}$ in layer k;

4:    $I_i = \frac{exp\left((\log p_i + G_i)/\tau\right)}{\sum_{j=1}^{m} exp\left((\log p_j + G_j)/\tau\right)}$;

5:    $i = argmax(A(x))$;

6:    $\theta_k' = \{\theta_k[j] \mid G[i][j] \neq 0\}$;

7:    $conv\left(x_{l-1}, \theta_l'\right) * G_l[i]$;

8:    $k \leftarrow k + 1$;

9: **end for**

10: **return**

3.2 Information Bottleneck Verification Module

The theory of an information bottleneck is based on mutual information. Given the variable X and label Y, the mutual information $p(x, y)$ is defined as Formula 5. For the distributions p and q, $D_{KL}[p\|q]$ is their Kullback-Liebler divergence. $H(X)$ is the entropy of X and Y. $H(X \mid Y)$ is the conditional entropy of X and Y. The mutual information $I(X; Y)$ is an analytical representation of the degree of correlation between two variables. This method quantifies the total amount of information represented by another variable through one variable.

$$I(X; Y) = D_{KL}[p(x, y)\|p(x)p(y)]$$
$$= \sum_{x \in X, y \in Y} p(x, y)\log\left(\frac{p(x, y)}{p(x)p(y)}\right) \tag{5}$$
$$= H(X) - H(X \mid Y)$$

The information bottleneck verification module is used to take a proficient and compacted representation of the unique information X without losing the ability to predict the label Y. This means that we want to obtain the minimum sufficient statistic. However, the detailed minimal sufficient statistics only exist for extraordinary distributions. The information bottleneck can be taken as a measure of the solution to rate-distortion theory. In addition, it provides an idea to obtain an approximate minimum sufficient statistic. Minimal sufficient statistics $Z(X)$ are maps or partitions that compress all the information that X has on Y. We can formulate minimal sufficient statistics through the Markov chain: $Y \rightarrow X \rightarrow Z(X)$. Z indicates the key information between the data and the predicted label.

Suppose $x \in X$ is the input and $z \in Z$ is the compressed and effective representation output of x. $p(z \mid x)$ denotes the probabilistic mapping of the compressed representation. The information bottleneck can be formulated by Formula 6.

$$\min_{p(z|x), Y \rightarrow X \rightarrow Z} \{I(X; Z) - \beta I(Z; Y)\} \tag{6}$$

where $I(X; Z)$ and $I(Z; Y)$ denote the mutual information, and β determines those trade-offs in the process of anticipating these labels, guaranteeing that the minimum sufficient statistic is obtained.

4 Experiments

The experimental settings will be introduced first in this section. The first part includes datasets, pretraining models and evaluation metrics. Then, comparisons of the amount of calculation displayed by the dynamic reasoning decision of the DRIB are made. Furthermore, a series of attribution results and pixel disturbances shows the faithfulness of the information bottleneck in the DRIB. Finally, some SOTA visualization methods are used to verify the interpretability of the DRIB model.

4.1 Experimental Settings

Two public and popular datasets were applied in our experiments: (1) **CIFAR-10** is a sub-group of the Tiny Images dataset. It comprises 60000 32 × 32 color images. Those pictures were named with a standout amongst 10 fundamentally unrelated classes. There are 6000 pictures for every class for 5000 preparations and 1000 testing pictures. (2) **MNIST** is an expansive gathering about written by hand digits. It has a training set of approximately 60,000 samples, and a test set of 10,000 samples. The size of these images is 20 × 20 pixels. (3) ILSVRC2012 is a subset of the abundant hand-labeled ImageNet dataset arranged by the WordNet hierarchy, at which point each bud of the hierarchy is described by large group and thousands of images. The dataset includes 1000 leaf node categories, all of which consist of 1860 nodes. The pretraining models used in the experiments in this paper include VGG16 and ResNet-50. We set the learning rate to 0.01, the training batch to 128, and the training epoch to 100 for both datasets and models. The computing equipment used for the work was four TITAN-RTXs at 3.60 GHz, 32 GB of RAM and 500 GB of hard disk.

The evaluation has two aspects. One is for dynamic reasoning decisions, and the other is for information bottleneck assessment. (1) **Multiple Accumulate Operations (MACs)**. In network computing, multiply-accumulation is a common step to calculate the product of two numbers and add them to the accumulator. One MAC is calculated twice, assuming that both the input channel and the output channel are 1, the kernel size is $k \times k$, the input map size is (H_{in}, W_{in}), and the output map size is (H_{out}, W_{out}). The input channel C_{in} and the output channel C_{out} are considered; then $MACs = C_{in} * K * K * H_{out} * W_{out} * C_{out}$. (2) **The reduction ratio** measures the percentage of reduction in MACs. Let M_{before} be the MAC value before reduction, and M_{after} be the MAC value after reduction. Then Reduction Ratio $= ((M_{before} - M_{after})/M_{after}) * 100\%$. (3) **Accuracy** indicates the number of correct samples divided by the number of all samples. Its value represents the model effect. (4) **A bounding box (Bbox)** is used to quantify how sound attribution methods identify and localize the object of interest. Assuming that the bounding box contains n pixels, we assess how many of the top k pixels in the predicted mask P, where $k \leq n$. We can take the Bbox score by $Bbox = (k/n) \times 100$.

Table 1. Comparisons of MACs, reduction ratio and accuracy for pruning methods with dynamic reasoning decisions of the VGG16 model on the CIFAR10 dataset.

Method	MACs (M)	Reduction ratio (%)	Accuracy (%)
Baseline	313.2	–	93.50
Luo and Wu [28]	156.5	50	93.36
Li and Kadav [29]	206.6	34	93.00
He and Zhang [30]	156.5	50	93.18
Liu and Li [31]	153.4	51	93.31
Lin and Rao [32]	156.5	50	92.65
Gao and Zhao [33]	156.5	50	93.03
Wang and Zhang [27]	155.2	50.4	93.45
DRIB	**156.5**	**50.2**	**93.30**

4.2 Interpretability of Calculation in Dynamic Reasoning Decision

The excessive and complex parameters are the reasons for the inexplicability of deep neural networks. We reduce the calculation of parameters through the dynamic reasoning decision module in this experiment. By comparing some of the related methods mentioned in the SOTA work [27], we find that we can improve 156.7 M in MACs, with the reduction ratio increasing 0.2% and up to 93.30% accuracy compared with the baseline in Table 1. It also has obvious improvement compared with other methods, especially in the logical calculation. Especially in the ResNet50 model as shown in Table 2, on the basis of the reduction ratio reaching 51.4%, the accuracy can reach 93.27%. This is enough to prove that the dynamic reasoning decision is highly understandable and explainable in the calculation.

Table 2. Comparisons of MACs, reduction ratio and accuracy for pruning methods with dynamic reasoning decisions of the ResNet-50 model on the CIFAR10 dataset.

Method	MACs (M)	Reduction ratio (%)	Accuracy (%)
Baseline	125.8	–	92.80
Luo and Wu [28]	62.9	50	91.98
He and Zhang [30]	62.9	50	91.80
He and Kang [34]	65.4	48	92.56
He and Lin [35]	62.9	50	90.20
Wang and Zhang [27]	59.6	52.6	92.57
DRIB	**60.2**	**51.4**	**93.27**

4.3 Explainability of Attribution in the Information Bottleneck

The theory of information bottlenecks measures the effectiveness of the residual feature attribution of the above dynamic reasoning decision module after reducing the amount of computation. $I[Z, Y]$ denotes the amount of information that Z learns about label Y. $I[X, Z]$ indicates the degree of simplification of information in the learning process X. The transition point refers to the point at which information is formed during the transition, and the convergence point is a parameter included after excessive stability. The data in Table 3 for VGG16 and Table 4 show that the degree of information attribution is 82.15% and 88.34%, respectively, and fewer epochs are needed. This result proves that the DRIB model has an excellent effect on the attribution function.

Table 3. The key parameters of the VGG16 model.

	$I[X, Z]$	$I[Z, Y]$	Epochs	Accuracy
Transition point	8.753	2.285	7	45.56
Convergence point	4.673	2.732	69	82.15

Table 4. The key parameters of the ResNet-50 model.

	$I[X, Z]$	$I[Z, Y]$	Epochs	Accuracy
Transition point	9.375	2.376	2	55.47
Convergence point	4.895	2.983	55	88.34

In addition, we quantify the influence of input samples on classification accuracy in the attribution process. As depicted in Fig. 3, the input sample is compressed by the information bottleneck to obtain the mask after attribution. Finally, the mask was covered on the original image to obtain the attributed picture. To further verify the effect of attribution, we view the classification confidence by deleting and inserting pixels in the attribution weight area. We perform 113 significant pixel deletion or insertion operations on each sample, each of which performs 0.9% of the pixel range in the graph. The figure contains three parts: the dog, the cat, and the bird. The horizontal axis of the coordinates represents the number of pixels inserted or deleted in each table. The number of steps is counted a total of 113 times. The vertical axis represents the classified confidence of each step. The blue value is the confidence of the deleted pixels, and the orange value is the confidence of the inserted pixels.

In Fig. 3(A), because the picture of the dog is much larger than that of the bird, we can see from the table that the change of in confidence of the classified dog is more evident than that of the bird. The proportion of cats (B) in the image is similar to that of dogs in (A), so there is little change in the table data. The middle bird (C) environment is more complex than other samples. The table data show that the environment significantly influences the confidence of classification. The results of this experiment can help us to more intuitively understand the application of information bottleneck theory in interpretability.

4.4 Visualization of Understandability

To better validate the effectiveness of our method in feature attribution, we further identify and localize targets through the ILSVRC2012 dataset. Because this dataset possesses borderline annotation. In the process of attribution verification using information bottleneck theory, the feature graphs in the process are randomly selected. In each iteration of the experimental step, the highest ranked value is replaced with a constant. The modified inputs are used for network prediction, and then the degree of decline in the target category scores is measured based on the prediction results. The results are exhibited in Table 5. The DRIB model achieves better performance in VGG-16 by 8 * 8 pixels and ResNet-50 by 8 * 8 pixels. In addition, VGG Bbox and ResNet Bbox are better than the other models. In addition, as demonstrated in Fig. 4, the visualization obtains the feature map attributed by the information bottleneck theory.

Table 5. The decline of the target category score and bounding box score.

Model	Vgg-16 8 * 8	Vgg-16 14 * 14	ResNet-50 8 * 8	ResNet-50 14 * 14	Vgg Bbox	ResNet Bbox
Smooth Grad	43.8%	45.5%	48.5%	50.2%	39.9%	43.9%
Grad-CAM	51.0%	51.7%	53.6%	54.1%	39.9%	46.5%
Guided Grad-CAM	55.5%	57.6%	56.5%	57.7%	41.9%	46.8%
DRIB	**56.2%**	**54.3%**	**57.1%**	**53.1%**	**53.7%**	**55.1%**

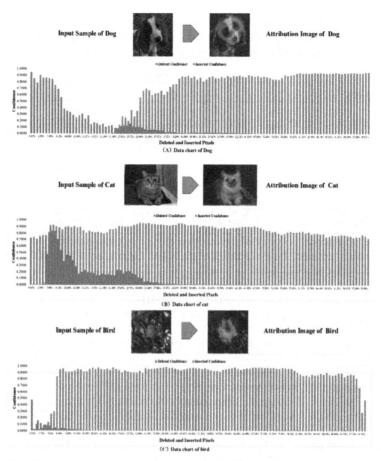

Fig. 3. The influence of some samples' classification confidence after deleting or inserting pixels.

Fig. 4. Visualization of the feature maps attributed by the information bottleneck theory.

5 Conclusion

This paper proposes a new method named the dynamic reasoning and information bottleneck (DRIB). In the method, a novel dynamic reasoning decision algorithm was proposed to reduce multiply accumulate operations and improve the interpretability of the calculation. The information bottleneck was introduced to the DRIB model to verify the attribution correctness of the dynamic reasoning module. Some experiments prove that the dynamic reasoning decision algorithm and information bottleneck theory can be combined with each other. Otherwise, the method provides users with good interpretability and understandability, making deep neural networks trustworthy.

References

1. Kim, Y.J., Bae, J.P., Chung, J.W., et al.: New polyp image classification technique using transfer learning of network-in-network structure in endoscopic images. Sci. Rep. **11**(1), 1–8 (2021)
2. Fan, Q., Zhuo, W., Tang, C.K., et al.: Few-shot object detection with attention-RPN and multi-relation detector. In: Proceedings of the IEEE/CVF Conference on Computer Vision and Pattern Recognition, pp. 4013–4022 (2020)
3. Yadav, A., Vishwakarma, D.K.: Sentiment analysis using deep learning architectures: a review. Artif. Intell. Rev. **53**(6), 4335–4385 (2019). https://doi.org/10.1007/s10462-019-09794-5
4. Wu, M., Parbhoo, S., Hughes, M., et al.: Regional tree regularization for interpretability in deep neural networks. In: Proceedings of the AAAI Conference on Artificial Intelligence, vol. 34, no. 04, pp. 6413–6421
5. Kubara, K.J., Manczak, B., Dolicki, B., et al.: Towards transparent and explainable attention models. In: ML Reproducibility Challenge 2020 (2021)
6. Misheva, B.H., Osterrieder, J., Hirsa, A., et al.: Explainable AI in credit risk management. arXiv preprint arXiv:2103.00949 (2021)
7. Torrent, N.L., Visani, G., Bagli, E.: PSD2 explainable AI model for credit scoring. arXiv preprint arXiv:2011.10367 (2020)
8. Loquercio, A., Segu, M., Scaramuzza, D.: A general framework for uncertainty estimation in deep learning. IEEE Robot. Autom. Lett. **5**(2), 3153–3160 (2020)
9. Zhang, Q., Cao, R., Shi, F., et al.: Interpreting CNN knowledge via an explanatory graph. In: Proceedings of the AAAI Conference on Artificial Intelligence, vol. 32, no. 1 (2018)
10. Bau, D., Zhou, B., Khosla, A., et al.: Network dissection: quantifying interpretability of deep visual representations. In: Proceedings of the IEEE Conference on Computer Vision and Pattern Recognition, pp. 6541–6549 (2017)
11. Nguyen, A., Clune, J., Bengio, Y., et al.: Plug & play generative networks: conditional iterative generation of images in latent space. In: Proceedings of the IEEE Conference on Computer Vision and Pattern Recognition, pp. 4467–4477 (2017)
12. Bau, D., Zhu, J.Y., Strobelt, H., et al.: Understanding the role of individual units in a deep neural network. Proc. Natl. Acad. Sci. **117**(48), 30071–30078 (2020)
13. Zhou, B., Bau, D., Oliva, A., et al.: Interpreting deep visual representations via network dissection. IEEE Trans. Pattern Anal. Mach. Intell. **41**(9), 2131–2145 (2018)
14. Fong, R., Vedaldi, A.: Net2Vec: quantifying and explaining how concepts are encoded by filters in deep neural networks. In: Proceedings of the IEEE Conference on Computer Vision and Pattern Recognition, pp. 8730–8738 (2018)
15. Zhou, B., Khosla, A., Lapedriza, A., et al.: Learning deep features for discriminative localization. In: Proceedings of the IEEE Conference on Computer Vision and Pattern Recognition, pp. 2921–2929 (2016)
16. Selvaraju, R.R., Cogswell, M., Das, A., et al.: Grad-CAM: visual explanations from deep networks via gradient-based localization. In: Proceedings of the IEEE International Conference on Computer Vision, pp. 618–626 (2017)
17. Wang, H., Wang, Z., Du, M., et al.: Score-CAM: score-weighted visual explanations for convolutional neural networks. In: Proceedings of the IEEE/CVF Conference on Computer Vision and Pattern Recognition Workshops, pp. 24–25 (2020)

18. Lage, I., Ross, A., Gershman, S.J., et al.: Human-in-the-loop interpretability prior. In: Advances in Neural Information Processing Systems, p. 31 (2018)
19. Subramanian, A., Pruthi, D., Jhamtani, H., et al.: SPINE: Sparse interpretable neural embeddings. In: Proceedings of the AAAI Conference on Artificial Intelligence, vol. 32, no. 1 (2018)
20. Ho, D.: NBDT: neural-backed decision trees. Master's thesis, EECS Department, University of California, Berkeley (2020)
21. Nauta, M., van Bree, R., Seifert, C.: Neural prototype trees for interpretable fine-grained image recognition. In: Proceedings of the IEEE/CVF Conference on Computer Vision and Pattern Recognition, pp. 14933–14943 (2021)
22. Fan, F., Wang, G.: Fuzzy logic interpretation of quadratic networks. Neurocomputing **374**, 10–21 (2020)
23. Ribeiro, M.T., Singh, S., Guestrin, C.: Anchors: High-precision model-agnostic explanations. In: Proceedings of the AAAI Conference on Artificial Intelligence, vol. 32, no. 1 (2018)
24. Zhang, Q., Yang, Y., Ma, H., et al.: Interpreting CNNs via decision trees. In: Proceedings of the IEEE/CVF Conference on Computer Vision and Pattern Recognition, pp. 6261–6270 (2019)
25. Shen, W., Guo, Y., Wang, Y., et al.: Deep differentiable random forests for age estimation. IEEE Trans. Pattern Anal. Mach. Intell. **43**(2), 404–419 (2019)
26. Zhang, Q., Wu, Y.N., Zhu, S.C.: Interpretable convolutional neural networks. In: Proceedings of the IEEE Conference on Computer Vision and Pattern Recognition, pp. 8827–8836 (2018)
27. Wang, Y., Zhang, X., Hu, X., et al.: Dynamic network pruning with interpretable layerwise channel selection. In: Proceedings of the AAAI Conference on Artificial Intelligence, vol. 34, no. 04, pp. 6299–6306 (2020)
28. Luo, J.H., Wu, J., Lin, W.: ThiNet: a filter level pruning method for deep neural network compression. In: Proceedings of the IEEE International Conference on Computer Vision, pp. 5058–5066 (2017)
29. Li, H., Kadav, A., Durdanovic, I., et al.: Pruning filters for efficient ConvNets. arXiv preprint arXiv:1608.08710 (2016)
30. He, Y., Zhang, X., Sun, J.: Channel pruning for accelerating very deep neural networks. In: Proceedings of the IEEE International Conference on Computer Vision, pp. 1389–1397 (2017)
31. Liu, Z., Li, J., Shen, Z., et al.: Learning efficient convolutional networks through network slimming. In: Proceedings of the IEEE International Conference on Computer Vision, pp. 2736–2744 (2017)
32. Lin, J., Rao, Y., Lu, J., et al.: Runtime neural pruning. In: Advances in Neural Information Processing Systems 30 (2017)
33. Gao, X., Zhao, Y., Dudziak, Ł., et al.: Dynamic channel pruning: feature boosting and suppression. arXiv preprint arXiv:1810.05331 (2018)
34. He, Y., Kang, G., Dong, X., et al.: Soft filter pruning for accelerating deep convolutional neural networks. arXiv preprint arXiv:1808.06866 (2018)
35. He, Y., Lin, J., Liu, Z., Wang, H., Li, L.-J., Han, S.: AMC: AutoML for model compression and acceleration on mobile devices. In: Ferrari, V., Hebert, M., Sminchisescu, C., Weiss, Y. (eds.) ECCV 2018. LNCS, vol. 11211, pp. 815–832. Springer, Cham (2018). https://doi.org/10.1007/978-3-030-01234-2_48

Multimedia Data Management
and Analysis

Advanced Generative Adversarial Network for Image Superresolution

Mei Jia[1], Mingde Lu[2], and Yu Sang[1(✉)]

[1] School of Electronic and Information Engineering, Liaoning Technical University,
Huludao 125105, China
sangyu2008bj@sina.com
[2] Institute of Computing Technology, Research Institute of Liaohe Oilfield Company,
Panjin 124010, China

Abstract. The superresolution (SR) method based on generative adversarial networks (GANs) cannot adequately capture enough diversity from training data, resulting in misalignment between input low resolution (LR) images and output high resolution (HR) images. GAN training has difficulty converging. Based on this, an advanced GAN-based image SR reconstruction method is presented. First, the dense connection residual block and attention mechanism are integrated into the GAN generator to improve high-frequency feature extraction. Meanwhile, an added discriminator is added into the GAN discriminant network, which forms a dual discriminator to ensure that the process of training is stable. Second, the more robust Charbonnier loss is used instead of the mean square error (MSE) loss to compare similarities between the obtained image and actual image, and the total variation (TV) loss is employed to smooth the training results. Finally, the experimental results indicate that global structures can be better reconstructed using the method of this paper and texture details of images compared with other SOTA methods. The peak signal-to-noise ratio (PSNR) values by the method of this paper are improved by an average of 2.24 dB, and the structural similarity index measure (SSIM) values are improved by an average of 0.07.

Keywords: Generative adversarial networks (GANs) · Superresolution (SR) · Residual dense block · Attention mechanism · Dual discriminator

1 Introduction

The image SR process converts LR images to HR images by using a specific algorithm [1]. The SR technique is an application of computer vision that is significant to several fields, including surveillance devices [2], remote sensing of satellite images [3], face recognition [4], and medical images [5]. HR images provide people with more information and allow them to locate the desired part of the image faster and more accurately.

Recently, deep learning (DL) has penetrated the field of image SR with good developments [6–10], which brings unique benefits to the SR technique. GANs are the most

© The Author(s), under exclusive license to Springer Nature Singapore Pte Ltd. 2022
Y. Wang et al. (Eds.): ICPCSEE 2022, CCIS 1628, pp. 193–208, 2022.
https://doi.org/10.1007/978-981-19-5194-7_15

widely used generator-discriminator deep network structure that works well in SR processing, where the generator falsifies the data by the initial input signals or noises, and the discriminator determines whether the input data are falsified by the generator or are the real data. Subsequently, some variants of GANs [11–16] have been proposed to improve network performance.

The GAN model by Ian Goodfellow et al. [17] provided a new approach for learning-based SR techniques. The GAN model was the first to be applied in the field of SR reconstruction with good results. SRGAN [11] is part of the more advanced learning-based SR reconstruction methods. In the SR reconstruction tasks, images generated by SRGAN recovered clear texture details for the first time, and the network performance was significantly better than other DL networks. Compared with convolutional neural network (CNN)-based methods, GAN-based SR methods can reconstruct HR images that are closer to the actual perception of human vision. However, in some GAN-based SR methods, the diversity of training data is overlooked, resulting in mismatched input LR images and output HR images. In addition, GAN training requires the generator and discriminator to work simultaneously and to be optimized alternately, so the learning speed is slow and the training is very difficult or even difficult to converge. Therefore, it is still possible to improve the SRGAN.

In the text, we propose an advanced GAN for SR. Based on the instability of the output image information of the generator, the network structure of the generator is improved to improve its feature expression capability. The structure of the discriminator is modified to improve the balance of the network training in response to the unbalanced training process; to improve the visual impact of the reconstructed image, a loss function suitable for SR reconstruction is introduced. In brief, our contributions in this paper are fourfold:

- We integrate a dense block with residuals-in-residuals (RRDB) and an attention mechanism into a GAN generator to improve high-frequency feature extraction and obtain richer information features in the form of dense connections.
- We extend an added discriminator into a GAN discriminant network, which forms a dual discriminator to enhance the stability of training.
- We employ the more robust Charbonnier loss instead of the MSE loss to evaluate the similarity between the obtained image and the actual image and use total variation (TV) loss to smooth the training results.
- The experimental results from three public datasets (Set5, Set14, and BSD100) demonstrate that our method enables sharper generated images and reconstruction of images with better visual quality compared with other methods.

2 Related Work

Because of its wide range of applications, SR plays a crucial role in specific applications, where it has become a hot topic. Currently, deep learning is undergoing rapid development; many scholars are combining deep learning and image SR methods in an effort to develop a more efficient approach to solve the problems arising from traditional methods. A convolution-based SR reconstruction network (SRCNN) was proposed by Dong et al. [6], which has high accuracy compared with traditional methods, but the network layer count is low, convergence is slow, the convolution kernel size is too small, and the extracted features are local, resulting in poor reconstruction. Based on SRCNN, ESPCN was proposed by Shi et al. [7] based on subpixel layers of a neural network. He et al. [18] proposed the residual network (ResNet), which can increase network deepness by a simple stacking of residual blocks and solves the problem of gradient disappearance caused by increasing the number. Based on ResNet, a convolutional neural network was proposed by Kim et al. [8], which can accurately transform LR images into HR images. The VDRS network model uses a deeper VGG network structure model and can control the training of deep models well. EDSR was proposed by Lim et al. [9], and the batch normalization (BN) layer was removed from the residual structure, saving more than 40% of memory and enabling the construction of models with better performance. Lai et al. [10] proposed the LapSRN, replacing the L2 loss function with the Charbonnier loss function and implementing deeply supervised training of the network to achieve higher quality reconstruction.

GAN-based methods include DCGAN [19], WGAN [20], and Patch GAN [21]. Many GAN-based SR image reconstruction methods [11–16] have also emerged, such as SRGAN [11], ESRGAN [12], and CGAN [13]. These methods have achieved good results. Compared with other methods, in the GAN model, image information can be used to enhance the accuracy of the model. Leding et al. [11] proposed the SRGAN algorithm for GANs. The SRGAN improves the perception of images by improving the realism of details. Wang et al. [12] proposed ESRGAN, which replaces the residual block with the dense block and removes the BN layer. The PNSR is not ideal for the generated image; however, the sensory effect of the naked eye is greatly improved. Gao et al. [13] proposed a CGAN-based image SR network. When GAN is directly applied to SR, a possible mismatch between input and output is solved, and its generator adopts a symmetric encoder-decoder structure and applies a jump connection to achieve cross-layered transfer of information at the low level between input and output. An SRGAN-based multiscale parallel learning generative network was proposed by Zun et al. [14], which consists of two blocks of residual networks at different scales. Yan et al. [15] presented a method of SR incorporating attention-GANs to achieve adaptive feature refinement in feature mapping. At large scale factors, it is able to improve the reconstruction of images when there is not enough high frequency information and texture details. Guo et al. [16] proposed an optimization model based on SRGAN. The BN layer is removed from the SRGAN generator, and channel attention is introduced so that each residual block can obtain corresponding weights to generate a feature map that handles more image details.

3 GAN and SRGAN

GAN [17] can be recognized as an effective structure. As shown in Fig. 1, GAN is a generative model with zero-sum game thinking, consisting of generator (G) and discriminator (D). G falsifies the data by the initial input signal or noise, while D determines whether the input data are forged by the generator or is the real data. The two play against each other repeatedly through such a process, continuously passing back information and updating and optimizing their respective network capabilities until finally D

can accurately determine the authenticity of the data while G generates data powerful enough to blur the judgment of D. The training process of D and G is represented by the function $V(G,D)$, and the objective function is expressed as follows:

$$\min_{G} \max_{D} V(G, D) = E_{x \sim Pdata(x)}[\log D(x)] + E_{z \sim P(z)}[\log(1 - D(G(x)))], \quad (1)$$

where $P_{data(x)}$ shows the true sample distribution and $P_{(z)}$ shows the generator distribution. The optimization goal of GAN is the loss of both G and D. The learning objective of D is to make $D(x)$ large and $D(G(x))$ small, maximizing the equation. The learning objective of G is to make $D(G(x))$ large, minimizing Eq. (1). The following transformation is performed in Eq. (2).

$$
\begin{aligned}
V(D, G) &= E_{x \sim Pdata(x)}[\log D(x)] + E_{z \sim P(z)}[\log(1 - D(G(x)))] \\
&= \int_x Pdata(x) \log(D(x))dx + \int_z P_z(z) \log(1 - D(g(z)))dz \\
&= \int_x Pdata(x) \log(D(x)) + P_g(x) \log(1 - D(x))dx.
\end{aligned}
\quad (2)
$$

The network works best if the discriminator is able to fully identify the differences between generated data and real data, which should then maximize Eq. (3).

$$D_G^*(x) = \frac{P_{data}(x)}{P_{data}(x) + P_g(x)}. \quad (3)$$

Equation (3) is the optimal discriminator for generating adversarial networks, $D(x)$ is maximized when the above equation is satisfied, $D(G(x))$ is small, and Eq. (1) is maximized. Network training results are best achieved when the generated data of G are in line with the real data distribution.

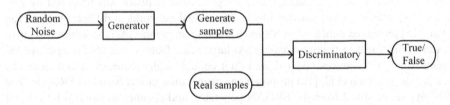

Fig. 1. The network structure of GAN

GANs generally deal with relatively small magnifications in SR reconstruction. When the magnification exceeds 4×, it is easy to make the reconstructed images not smooth, and the image details are not realistic enough. To improve the realism of reconstructed images, perceptual loss and adversarial loss are used in SRGAN and obtain a high PSNR after using MSE to optimize the generative network. The SRGAN consists of a generator and discriminator, and its network structure is shown in Fig. 2.

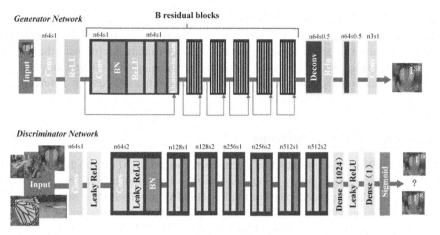

Fig. 2. SRGAN network structure

4 Proposed Method

This paper is based on the SRGAN with three improved parts: ① we integrate the RRDB into the generator G and fuse the attention mechanism in the RRDB to form the attention residual dense block (RRADB) and replace the leaky ReLU activation function in the original structure with an exponential linear unit (ELU) activation function; ② we extend an added discriminator into the GAN discriminant network, which forms a dual discriminator to ensure the stability of the training process; ③ we employ a more robust Charbonnier loss instead of the MSE loss to measure the similarity between the obtained and actual images and use TV loss to smooth the training results.

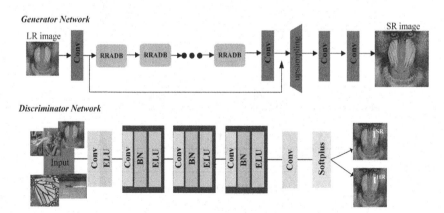

Fig. 3. Our network architecture

The overall structure is shown in Fig. 3. First, a preprocessing step is performed for an input image. The HR image is converted into an LR image by downsampling and sent to the generative network for learning. The result is to generate a "fake" HR sample image; then, the generated HR sample images and real primitive HR images are stitched together and sent to the discriminator network for learning. The discriminator network outputs a judgment for feedback. If it cannot correctly discriminate, the result is passed back to the generator network for further learning, while itself and the generator are updated and iterated until the discriminator network has difficulty in distinguishing between true and false input images.

4.1 Generator Network Structure

We integrate RRDB and the attention mechanism, formed RRADB, into the GAN generator to improve high-frequency feature extraction and obtain richer information features in the form of dense connections. For training stability, the BN layer is removed, which is able to improve performance and reduce complexity [22]. The attention mechanism in RRADB is the convolutional block attention module (CBAM) [23], consisting of channel attention (CA) and spatial attention (SA). Taking into account the interdependence of the feature channels, CA can increase the expressiveness of the network by adjusting its features appropriately. In SA, high-frequency details can be enhanced by focusing more on high-frequency information regions. Therefore, these two modules can effectively extract salient features. Meanwhile, all leaky ReLU activation functions in RRADB are replaced by ELU activation functions, which have the advantages of lower complexity and better robustness to noise compared with the leaky ReLU functions. The specific structure is shown in Fig. 4.

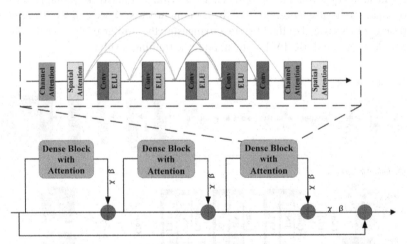

Fig. 4. RRADB structure

The SRGAN high-level architecture is retained in the generator model to increase training stability. The network structure is shown in Fig. 5. Residual and dense networks make up the advanced network structure. Entirety features are extracted using residual networks, and local features are extracted using dense convolutional layers. A residual dense network integrates the features of a network, combining global and local features to form continuous memory mechanisms. It is implemented by connecting the features extracted from the original residual dense network with the layers of the current residual dense network. In addition, the generator structure uses progressive upsampling, which is achieved by convolution at the subpixel level. Subpixel convolution has the advantage of having a large field of perception and providing a large amount of information and more accurate features.

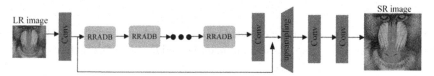

Fig. 5. The structure of the generator network

4.2 Discriminator Network Structure

We extend an added discriminator into a GAN discriminant network, which forms a dual discriminator to stabilize the training process. When training the discriminator, the parameters cannot be changed, and when the discriminator performance exceeds a certain level, the parameters of the discriminator must also be fixed. This alternating training leads to an unbalanced training process; for example, it is not possible to know when to switch from using the discriminator to the generator when training. To overcome the condition of imbalance, we introduce the idea of dual discriminators [24], replacing one discriminator with a dual discriminator, and alternating training with the generator. The dual discriminator is composed of two discriminators D_1 and D_2.

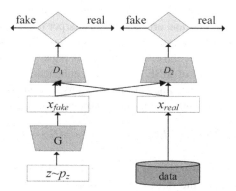

Fig. 6. Flow chart of dual discriminators

For training, data sample space x is first given, and if x is in the distribution of the actual data, then $D_1(x)$ scores high and $D_2(x)$ scores low, and vice versa. As with GAN, the overall learning ability of the network can be better improved by training D_1, D_2

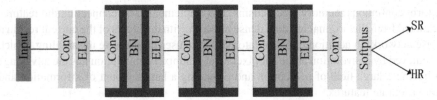

Fig. 7. Discriminator D_1 and D_2 network structure

and G alternately; consequently, generated images can be better visualized. The specific process is shown in Fig. 6, and the discriminator D_1 and D_2 network structures are shown in Fig. 7.

The loss function formula for the discriminator can be expressed in Eq. (4).

$$
\begin{aligned}
L_{D_1} &= \alpha \times E_{x \sim P_{data}}[\ln D_1(x)] + E_{z \sim P_z}\{-D_1[G(z)]\}, \\
L_{D_2} &= E_{x \sim P_{data}}[-D_2(x)] + \beta \times E_{z \sim P_z}\{\ln D_2[G(z)]\},
\end{aligned}
\tag{4}
$$

Where α and β are hyperparameters. $E_{x \sim P_{data}}$ describes the expected value for real data, $E_{z \sim P_z}$ describes the expected value in generated data, and P_{data} and P_z denote the data distribution of real and generated data, respectively.

4.3 Loss Function

Rather than the MSE loss, in this paper, we employ the more robust Charbonnier loss to evaluate the similarity of the obtained image with the actual image and use TV loss to smooth the training results. SRGAN uses MSE loss as content loss. MSE loss belongs to L2 loss, which can punish large losses, but it can do nothing about small losses and is not as effective as L1 loss. In fact, the L2 loss does not take into account the image quality, causing a lack of high-frequency details and resulting in overly smooth textures, which is unsatisfactory. In contrast, L1 loss requires more time to reconstruct, despite the ability to remove speckles and artifacts using sharper boundary reconstruction as a cost function. Charbonnier loss can better handle outliers. Therefore, Charbonnier loss is more supervised and more robust. Therefore, this paper uses Charbonnier loss as a content loss function instead of MSE loss, which enhances the supervisory ability of image reconstruction and the accuracy of image reconstruction. The PSNR values of the reconstructed results are also increased. The Charbonnier loss function L_C is defined as:

$$
L_C\left(I_{HR}, \hat{I}_{HR}\right) = \alpha E_{LR, HR \sim P_{data}(LR, HR)} \rho(I_{HR} - G_\theta(I_{LR})),
\tag{5}
$$

where α is a hyperparameter, $\rho(x) = \sqrt{x^2 + \varepsilon^2}$ is the Charlsonian penalty function, and ε is a constant. In this paper, the value $\varepsilon = 10^{-6}$ is used.

The content loss is recalculated using eigenvalues of the VGG-16 network before activation, as shown in Eq. (6).

$$
L_{VGG/i,j}^{SR} = \frac{1}{W_{i,j}H_{i,j}} \sum_{x=1}^{W_{i,j}} \sum_{y=1}^{H_{i,j}} \left(\phi_{i,j}\left(I^{HR}\right)_{x,y} - \phi_{i,j}\left(G_{\theta_G}\left(I^{LR}\right)\right)_{x,y}\right)^2,
\tag{6}
$$

where $\phi i, j$ denotes the map of features obtained from the j-th layer of convolution after the i-th layer of maximum pooling in the VGG network. Wi, j and Hi, j denote the size of the map of features. The adversarial loss used in this paper is the same as that used by SRGAN, as shown in Eq. (7).

$$L_{Gen}^{SR} = \sum_{n=1}^{N} - \log D_{\theta_D}\left(G_{\theta_G}\left(I^{LR}\right)\right), \tag{7}$$

where $D_{\theta_D}(G_{\theta_G}(I^{LR}))$ denotes the probability that reconstructed image $D_{\theta_D}(G_{\theta_G}(I^{LR}))$ is an HR image.

During image SR reconstruction, noise in the image may have a significant impact on the restoration results. At this time, we need to add some regular terms to keep the image smooth. The common regular terms include TV loss. To suppress the noise and smooth the image, this paper adds TV loss. As shown in Eq. (8).

$$L_{TV} = \delta \|\hat{y}\|_{TV}, \tag{8}$$

where δ is the hyperparameter.

In summary, the designed loss perception function in this paper is shown in Eq. (9).

$$L^{SR} = L_C + L_{VGG/i,j}^{SR} + L_{Gen}^{SR} + L_{TV}, \tag{9}$$

5 Experiment Results and Analysis

This section introduces the implementation details, evaluation metrics, experimental dataset, and results of the experiments. For a more comprehensive evaluation and analysis of the proposed method performance, in the case of a $4\times$ factor, in addition to the original SRGAN, other outstanding methods are compared in the experiments.

5.1 Implementation Details

The input image size is cropped to 88 × 88, the number of iteration rounds is 200, the training batch size is 64, the validation batch size is 1 and the Adam optimizer is used to optimize the network with a learning rate of 0.0002.

5.2 Datasets and Evaluation Metrics

We use DIV2K [25] as the training set. For a more accurate comparison with other SR methods, test datasets include three of the most commonly used test datasets: Set5 [26], Set14 [27] and BSD100 [28]. The LR images used in the experiments are generated by double triple interpolation by a factor of $4\times$.

The objective evaluation metrics are PSNR (in dB) and SSIM (in the range of 0 to 1), respectively. PSNR is a measure of image quality after denoising or noise addition, and SSIM is used to determine how similar two images are. Larger PSNR and SSIM values indicate better image quality and clarity; smaller PSNR and SSIM values indicate image blurriness and poor visual quality. The specific equations for PSNR and SSIM are shown in (10) and (11).

$$MSE = \frac{1}{H \times W} \sum_{i=1}^{H} \sum_{j=1}^{W} (X(i,j) - Y(i,j))^2,$$

$$PSNR = 10 \cdot \log_{10} \left(\frac{(2^n - 1)^2}{MSE} \right),$$

(10)

where MSE denotes the mean square error between the current image X and the original image Y, H is the image height, W is the image width, and n is a constant.

$$SSIM(X, Y) = \frac{(2u_X u_Y + C_1)(2\sigma_{XY} + C_2)}{(u_X^2 + u_Y^2 + C_1)(\sigma_X^2 + \sigma_Y^2 + C_2)},$$

(11)

Where μ_X and μ_Y are the average values of X and Y, σ_X and σ_Y are the standard deviations of X and Y, σ_{XY} denotes the covariance of X and Y, and C_1 and C_2 are constants.

5.3 Experimental Results and Analysis

SR reconstruction is performed on LR images of test datasets Set5, Set14, and BSD100 after the model training is complete. To test whether the proposed algorithm is effective and to compare subjective effects and objective evaluation values with classical methods such as bicubic, SRCNN [6], ESPCN [7], SRGAN [12] and ESRGAN [13], as well as References [14–16]. Among them, References [14–16] are all different improved methods based on SRGAN. For the fairness of the experiments, the training set used for the experiments of each SR reconstruction algorithm is the DIV2K datasets. Therefore, the test results of the experiment will be different from the data of the original paper. From a subjective perspective, three images of "baby", "butterfly" and "sculpture" are selected from the three datasets of Set5, Set14 and BSD100 to visualize some of the above methods. As shown in Figs. 8, 9 and 10. As seen from Figs. 8, 9 and 10, the visual effect of the reconstructed image by bicubic is very unclear and poorly reconstructed. Both the SRCNN reconstructed image and the ESPCN reconstructed image are clearer than the bicubic reconstructed image. Despite the improvement in their PSNR values, the visual effect is still blurry, and artifacts also appear. The overall effect of SRGAN and ESRGAN reconstruction is better than the overall effect of the three reconstructions, bicubic, SRCNN and ESPCN. However, the visual effect of SRGAN is still blurred, and the image details are fewer, which is not as good as ESRGAN. The visual effect of ESRGAN reconstruction is the best, but the PSNR values of "baby" and "sculpture" are the lowest, which also indicates that a low PSNR value cannot necessarily mean that the visual effect of the image is poor. Our method is similar to ESRGAN; the generated images closely resemble the original images. Comparing this method to others, its PSNR and SSIM are the highest; only the "butterfly" image value is smaller than that of ESRGAN. The loss function is replaced by Charbonnier loss instead of MSE. The GAN adversarial network is used to distinguish the natural HR image from the reconstructed

HR image, which will make the trained result visually closer to the natural HR image. This will reduce the PSNR because it is lower in the case of greater magnification and less smoothing. However, it is most realistic visually because too much smoothing will make the edges of objects inside the image look blurred. For this reason, the improved method will have a low PSNR value, but this is not a common phenomenon. In general, a low value does not necessarily mean a bad result. It is significantly better than other methods in terms of texture details of reconstruction, resulting in a better visual effect and richer texture details.

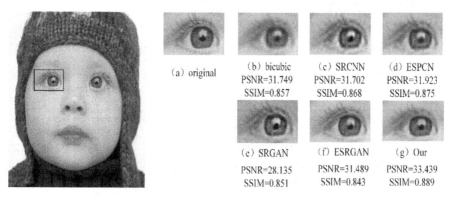

Fig. 8. Comparison map of the reconstruction results of the image "baby" from Set 5

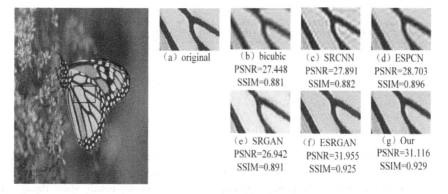

Fig. 9. Comparison map of the reconstruction results of the image "butterfly" from Set 14

The PSNR and SSIM values for different reconstruction methods are shown in Tables 1 and 2, respectively. From Tables 1 and 2 below, the PSNR and SSIM values for SRGAN are lowest, followed by the values of ESRGAN in the BSD100 datasets. Compared with SRGAN, PSNR has increased by an average of approximately 2.24 dB, and the SSIM value is improved by approximately 0.07 on average. Combining the data analysis in Table 1 and Table 2, compared to other algorithms, our method has the highest PSNR value and SSIM value.

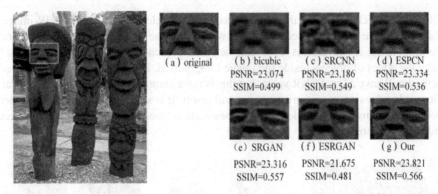

(a) original   (b) bicubic   (c) SRCNN   (d) ESPCN
PSNR=23.074   PSNR=23.186   PSNR=23.334
SSIM=0.499   SSIM=0.549   SSIM=0.536

(e) SRGAN   (f) ESRGAN   (g) Our
PSNR=23.316   PSNR=21.675   PSNR=23.821
SSIM=0.557   SSIM=0.481   SSIM=0.566

Fig. 10. Comparison map of the reconstruction results of the image "sculpture" from BSD100

Table 1. PSNR comparison was performed for distinct SR methods under distinct test sets

Datasets	Scale	Bicubic	SRCNN [6]	ESPCN [7]	SRGAN [12]	ESRGAN [13]	Reference [14]	Reference [15]	Reference [16]	Ours
Set5	×4	28.405	28.588	29.139	28.549	30.459	29.640	29.510	29.278	31.182
Set14	×4	25.634	26.009	26.337	25.688	26.285	26.730	26.443	25.791	27.811
BSD100	×4	25.952	26.077	26.268	25.193	25.322	25.240	25.884	25.486	27.155

Table 2. SSIM comparison was performed for distinct SR methods under distinct test sets

Datasets	Scale	Bicubic	SRCNN [6]	ESPCN [7]	SRGAN [12]	ESRGAN [13]	Reference [14]	Reference [15]	Reference [16]	Ours
Set5	×4	0.811	0.813	0.831	0.816	0.851	0.856	0.852	0.841	0.879
Set14	×4	0.711	0.735	0.739	0.703	0.714	0.735	0.738	0.766	0.776
BSD100	×4	0.667	0.699	0.697	0.655	0.651	0.663	0.701	0.687	0.723

5.4 Ablation Study

To make the proposed method more convincing, an ablation study can also be performed to demonstrate the performance of the RRDB module, dual discriminator module, CBAM module and Charbonnier module. The experimental results show the effect of these four different modules on the reconstruction effect. The number of iterations and the amplification factor are consistent with the experiments in this paper. Table 3 and Table 4 record the comparison results of PSNR values and SSIM values in ablation experiments, respectively. From Table 3 and Table 4, we can see that the reconstruction of the model with the addition of one of these four modules is better than that of SRGAN. The PSNR and SSIM values are also higher than those of the SRGAN. In contrast, using any one module is not as effective as using all four modules together.

As seen from Fig. 11 and Fig. 12, the PSNR and SSIM values are improved by adding different modules to the SRGAN, respectively, but the visual effect is not very good. It can be concluded that only one module has less effect on the original network, but

Table 3. Comparison of PSNR of different module experiments under test set values

method	Set5	Set14	BSD100
SRGAN (Baseline)	28.549	25.688	25.193
+CBAM	28.768	26.108	25.568
+Charbonnier	28.869	26.214	25.475
+RRDB	29.468	26.546	26.462
+Dual Discriminator	29.064	26.114	26.224
+CBAM + Charbonnier + RRDB + Dual Discriminator (Ours)	31.182	27.811	27.155

Table 4. Comparison of SSIM of different module experiments under test set values

method	Set5	Set14	BSD100
SRGAN (Baseline)	0.816	0.703	0.655
+CBAM	0.831	0.736	0.689
+Charbonnier	0.833	0.739	0.694
+RRDB	0.832	0.741	0.697
+Dual Discriminator	0.821	0.735	0.682
+CBAM + Charbonnier + RRDB + Dual Discriminator (Ours)	0.879	0.776	0.723

the simultaneous addition of these four different modules to the original network structure produces a very significant effect. Ablation experiments confirm that the proposed method is effective and feasible.

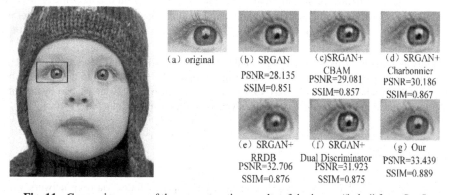

Fig. 11. Comparison map of the reconstruction results of the image "baby" from Set 5

The improvement method of this article first removes the BN layer in the generator part, and on the basis of removing the BN layer, replaces the original basic block in the

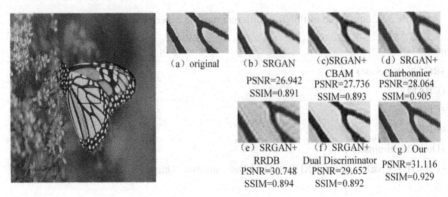

Fig. 12. Comparison map of the reconstruction results of the image "butterfly" from Set 14

generator with a residual dense block, which is the same as ESRGAN. However, this article also adds a CBAM attention mechanism to the residual dense block and replaces all the leaky ReLU activation functions with ELU activation functions. Experiments have shown that this improvement is useful for SRGAN, and the same reason can be shown that this improvement is also useful for ESRGAN. The rest will continue in subsequent experiments to draw better conclusions.

6   Conclusions

In this paper, an improved GAN image SR reconstruction method is proposed. In the generator part, the RRDB module is introduced without the BN layer. The attention module is fused with the RRDB module to form the RRADB module, which increases the high-frequency information of the images. Charbonnier loss is used instead of MSE loss to increase reconstruction accuracy. Finally, the training results are smoothed using TV loss. Experimental results indicate that the proposed method can achieve higher PSNR and SSIM values than other methods, and reconstconstructed images show better texture details and have improved visual effects.

Acknowledgments. This work was supported in part by the Basic Scientific Research Project of Liaoning Provincial Department of Education under Grant No. LJKQZ2021152; in part by the National Science Foundation of China (NSFC) under Grant No. 61602226; and in part by the PhD Startup Foundation of Liaoning Technical University of China under Grant No. 18-1021.

References

1. Xie, C., Zeng, W.L., Lu, X.B.: Fast single-image superresolution via deep network with component learning. IEEE Trans. Circuits Syst. Video Technol. **29**(12), 3473–3486 (2019)
2. Chen, G.Q., He, J., Luo, S.C.: Improved CycleGAN-based superresolution recovery algorithm for video surveillance faces. Comput. Appl. Res. **38**(10), 3172–3176 (2021)
3. Jiang W J, Luo X S, Dai Q X.: Research on superresolution reconstruction of remote sensing images based on adversarial network. Comput. Eng. Appl. **56**(21), 199–203 (2020)

4. Li, X.X., Xie, X., Li, B., et al.: Application of generative adversarial networks in medical image processing. Comput. Eng. Appl. **57**(18), 24–37 (2021)
5. Jia H D, Chen X A, Han Z, et al.: Nonconvex Nonlocal Tucker Decomposition for 3D Medical Image Super-Resolution. Front. Neuroinform. **16** (2022)
6. Dong, C., Loy, C.C., He, K.M., et al.: Image superresolution using deep convolutional networks. IEEE Trans. Pattern Anal. Mach. Intell. **38**(2), 295–307 (2016)
7. Shi, W.Z., Caballero, J., Huszár, F., et al.: Real- time single image and video superresolution using an efficient subpixel convolutional neural network. In: Proceedings of the IEEE Conference on Computer Vision and Pattern Recognition, Las Vegas, Washington, pp. 1874–1883 (2016)
8. Kim, J., Lee, J.K., Lee, K.M.: Accurate image superresolution using very deep convolutional networks. In: Proceedings of the IEEE Conference on Computer Vision and Pattern Recognition, pp. 1646–1654 (2016)
9. Lim, B., Son, S., Kim, H., et al.: Enhanced deep residual networks for single image superresolution. In: Proceedings of the IEEE Conference on Computer Vision and Pattern Recognition Workshops, pp. 136–144 (2017)
10. Lai, W.S., Huang, J.B., Ahuja, N., et al.: Deep Laplacian pyramid networks for fast and accurate superresolution. In: Proceedings of the IEEE Conference on Computer Vision and Pattern Recognition, pp. 624–632 (2017)
11. Ledig, C., Theis, L., Huszár, F., et al.: Photo-realistic single image superresolution using a generative adversarial network. In: Proceedings of the 2017 IEEE Conference on Computer Vision and Pattern Recognition, Honolulu, 21–26 July 2017, pp. 105–114. IEEE Computer Society, Washington (2017)
12. Wang, X., et al.: ESRGAN: enhanced super-resolution generative adversarial networks. In: Leal-Taixé, L., Roth, S. (eds.) ECCV 2018. LNCS, vol. 11133, pp. 63–79. Springer, Cham (2019). https://doi.org/10.1007/978-3-030-11021-5_5
13. Gao, H., Chen, Z., Huang, B., et al.: Image super-resolution based on conditional generative adversarial network. IET Image Proc. **14**(13), 3006–3013 (2020)
14. Zun, X.L., Zhong, H.J., Xing, L.R.: Multiscale generative adversarial networks for image superresolution algorithms. Sci. Technol. Eng. **20**(13), 5217–5223 (2020)
15. Peng, Y.F., Zhang, P.J., Gao, Y., Zi, L.L.: Single-image superresolution reconstruction of generative adversarial networks with fused attention. Adv. Lasers Optoelectron. **58**(20), 182–191 (2021)
16. Liu, G.Q., Liu, J.F., Zhu, D.H.: Image superresolution reconstruction algorithm based on generative adversarial network. Liq. Cryst. Display **36**(12), 1720–1727 (2021)
17. Goodfellow, I.J., Pouget-Abadie, J., Mirza, M., et al.: Generative adversarial nets. In: Proceedings of the Annual Conference on Neural Information Processing Systems, pp. 2672–2680. Curran Associates, Red Hook/Montreal (2014)
18. He, K., Zhang, X., Ren, S., et al.: Deep residual learning for image recognition. In: Proceedings of the IEEE Conference on Computer Vision and Pattern Recognition, pp. 770–778 (2016)
19. Radford, A., Metz, L., Chintala, S.: Unsupervised representation learning with deep convolutional generative adversarial networks. arXiv:1511.06434 (2015)
20. Arjovsky, M., Chintala, S., Bottou, L.: Wasserstein GAN. arXiv:1701.07875 (2017)
21. Isola, P., Zhu, J.Y., Zhou, T., et al.: Image-to-image translation with conditional adversarial networks. In: IEEE Conference on Computer Vision and Pattern Recognition, pp. 1125–1134 (2017)
22. Lim, B., Son, S., Kim, H., et al.: Enhanced deep residual networks for single image superresolution. In: 2017 IEEE Conference on Computer Vision and Pattern Recognition Workshops (CVPRW), Honolulu, HI,USA, pp. 1132–1140 (2017)

23. Woo, S., Park, J., Lee, J.Y., Kweon, I.S.: CBAM: convolutional block attention module. In: Ferrari, V., Hebert, M., Sminchisescu, C., Weiss, Y. (eds.) ECCV 2018. LNCS, vol. 11211, pp. 3–19. Springer, Cham (2018). https://doi.org/10.1007/978-3-030-01234-2_1

24. Nguyen, T.D., Le, T., Vu, H., et al.: Dual discriminator generative adversarial nets. In: Proceedings of the 27th International Conference on Neural Information Processing Systems, Long Beach, CA, USA, NIPS (2017)

25. He, L., Cheng, J.H., Jim, Z.Y., et al.: Multilevel perceptual residual convolutional networks for single image hypersegmentation reconstruction. Chin. J. Graph. Graph. 26(04), 776–786 (2021)

26. Lei, P.C., Liu, C., Tang, J.G., et al.: Hierarchical feature fusion attention network image superresolution reconstruction. Chin. J. Graph. 25(09), 1773–1786 (2020)

27. Jiang, Y.N., Li, J.H., Zhao, J.L.: A superresolution reconstruction algorithm for images based on generative adversarial networks. Comput. Eng. 47(03), 249–255 (2021)

28. Sun, C., Kou, K.H., Lu, J.W., et al.: Research on image superresolution methods based on wavelet deep networks. Appl. Res. Comput. 37(S1), 380–384 (2020)

Real-World Superresolution by Using Deep Degradation Learning

Rui Zhao[1(✉)], Junhong Chen[2], and Zhen Zhang[3]

[1] Institute of Special Environments Physical Sciences, Harbin Institute of Technology (Shenzhen), Shenzhen, China
zhaorui2020@hit.edu.cn
[2] Sun Yat-sen University, Guangzhou, China
caily3@mail2.sysu.edu.cn
[3] Shanghai Institute of Aerospace Electronic Technology, Shanghai, China

Abstract. Most current deep convolutional neural networks can achieve excellent results on a single image superresolution and are trained using corresponding high-resolution (HR) images and low-resolution (LR) images. Conversely, their superresolution performance in real-world superresolution tests is reduced because these methods create paired LR images by simply interpolating and downsampling HR images, which is very different from natural degradation. In this article, we design a new unsupervised framework conditioned by degradation representations of real-world hyperresolution problems. The approach presented in this paper consists of three stages: we first learn the implicit degradation representation from real-world LR images and then acquire LR images by shrinking the network, which will share similar degradation with real-world images. Finally, we make paired data of the generated real LR images and HR images for training the SR network. Our approach can obtain better results than the recent SR approach on the NTIRE2020 real-world SR challenge Track1 dataset.

Keywords: Super resolution · Contrastive learning · Image degradation

1 Introduction

Image superresolution is a problem that attempts to obtain a higher resolution image from a low-resolution quality image. In the last few years, DNN-based methods have achieved remarkable results of impressive visual quality, which mainly concentrate on building complicated network architectures to enhance various metrics in existing datasets. Most methods use simple interpolation operations to downsample HR images to construct paired training data. Despite the effectiveness and convenience of this operation, these methods have not considered the uncertainty of real-world degradation. Some approaches model the degradation by the ideal downsampling method:

$$Y_{lr} = (X_{hr} \otimes k) \downarrow_s + n \tag{1}$$

The Support Plan for Core Technology Research and Engineering Verification of Development and Reform Commission of Shenzhen Municipality (number 202100036).

where Y_{lr} and X_{hr} indicate the LR image and HR image, respectively $X_{hr} \otimes k$ means the HR image is blurred with the kernel k, $\downarrow_s$ means the image will be downsampled by scale factor s times, and the parameter n indicates the natural noise. According to this degradation model, we can easily obtain paired training data. Alternatively, it is also difficult to predict the real-world complicated blur kernel. Some recent studies have proposed GAN-based kernel estimation methods to generate realistic images. Based on paired data, the goal of the SR model is to minimize the average cost of images in the dataset:

$$\arg \min_{F} \frac{1}{N} \sum_{i=1}^{N} Loss_{sr}(F(y_i), x_i) \qquad (2)$$

where $F(.)$ refers to an SR model and $Loss_{sr}$ is a total loss function. If we valid the SISR model with the same downsampling dataset, the results are not unexpected. Once testing the model with the real-world image, the SR images are of poor quality. It is clear that the image downscaled by the bicubic operation does not have the same degradation as the real-world images. Therefore, creating an LR image similar to real-world degradation is a very useful problem that must be addressed. To this end, enlightened by the popularity of contrastive learning in computer vision, we promote a novel unsupervised-based superresolution approach that uses degradation representations to assist the downsampling network in generating realistic LR images and constructing paired input data for the superresolution network. Specifically, we assume that the images in the same dataset have approximate degradation. Consequently, the degradation of a real-world low resolution image will be approximate to other images in a dataset and far from high resolution images, as shown in Fig. 1.

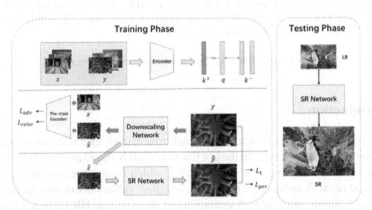

Fig. 1. The overview map of our proposed method in this paper. First, we train the degradation learning network, as shown by the yellow arrows. Given a trained encoder, the downscaling network is trained by employing adversarial and color losses, depicted by blue arrows. The SR network is optimized to obtain the high-resolution images in the third step, assisted by paired data $(\hat{x}, y)$ created in our downscaling network. (Color figure online)

Additionally, we develop an LR image generator to create realistic and paired LR images, which are the input of the superresolution network. By this means, our model is available for real-world degradation patterns instead of interpolations (e.g., nearest neighbor). To test the validity of our approach, we perform many experiments on the NTIRE2020 real-world SR challenge Track1 dataset. The results of experiments show that our method outperforms most of the current methods. Finally, we perform an ablation study to show the significance of the degradation learning module and downsampling network.

2 Related Work

2.1 Real-World Superresolution

In most previous superresolution studies, datasets with paired images are usually obtained by downsampling HR images with fixed operations. The SRCNN [1] was the first to apply a convolutional neural network to superresolution, and then various LR-to-HR reconstruction networks were developed to enhance SR performance. However, these models can only achieve good results on clean datasets because the model has not been trained with blurry or noisy image data. This is obviously different from real-world images, which often carry serious noise and blur. Cai et al. [2, 3] collected paired photos from the real world with a special camera directly. However, making such a dataset requires considerable manpower and material resources. To solve the problem of real-world superresolution, some research attempts to solve the SR problem without using paired training datasets. Lugmayr et al. [4] designed a downsample network to generate images with degradation and then used them to train an upsample network. Yuan et al. [5] developed a cycle framework to train degradation and superresolution networks concurrently. Since these aforementioned methods regard degradation as input, a growing number of studies have started relying on predicting degradation for real-world SR. Hence, incorrect degradation estimation can lead to poor SR performance with respect to fidelity. To address this issue, Gu et al. [6] optimized the estimated degradation by iteratively comparing the SR image with the ground truth.

2.2 Contrastive Learning

Currently, there are two unsupervised representation learning methods, generative learning and contrastive learning. Generative learning methods usually rely on autoencoding of images and conduct representation learning to minimize the similarity of the output images and the ground truth images in the pixel features. As a result, most of them require expensive calculation costs. Instead, contrastive learning aims to make the output representations closer to the positive images and farther away from those negative ones. Chen et al. [7] proposed a novel framework, named SimCLR, which extracts representations using a variety of data augmentations and contrastive learning. After that, He et al. [8] developed MoCo and MoCo v2, using a momentum encoder and a memory bank to maintain consistent representations. In this paper, images in datasets that seem to take the same degradation are considered positive counterparts, and contrastive learning will learn to draw content-invariance degradation features.

3 PurPosed Method

3.1 Overview of the Unsupervised Framework

Our real-world SR framework component is shown in Fig. 1, which contains three parts: a degradation learning network, an LR image generator and a reconstruction model. The degradation learning network aims to train an encoder that can extract degradation features from real-world images and assist generators in obtaining more realistic LR images. Motivated by kernelGAN [9], our LR image generator is a linear network without any nonlinear activation, which downscales images only by convolution and subsampling. The generated LR image will be extracted representation by the degradation encoder to enable the degradation feature in it to be as similar as the real world. Specifically, given real-world images x as LR and unpaired high-resolution images y as input, the first stage of our framework is to obtain the real-world degradation representation q, k+, k− from two groups of images by contrastive learning, as the yellow arrow pointed out in Fig. 1. Once the initial degradation learning is finished, we use a linear network to generate paired LR images $\hat{x}$ by downsampling y. Then, we encode $\hat{x}$ to ensure that it should have the same degradation as x. This process is marked by the blue arrow in Fig. 1. Our final goal is to use the generated samples that are used to train SR models. The constructed method tries to recover corresponding HR images $\hat{y}$ and make it approximate y, as shown by the green arrow in Fig. 1. In the testing phase, we only use the SR network to obtain the SR image and evaluate their quality by calculating the PSNR and SSIM. Different from previous work [10], our real-world degradation is learned from the LR and HR datasets and the LR image generated by our downscaling network instead of a fixed operator.

3.2 Degradation Model

The degradation model is the second part, which learns to obtain more realistic LR images by an unsupervised method. We extract the degradation representations from LR images using a contrastive learning framework. We assume that the degradation representation in real-world domain images is similar and is distinguished from high-resolution domain images. We randomly select patches from real-world images as the query patch, and the positive patches come from the same dataset. Other patches extracted from HR images are regarded as the negative patches. Then, all the query, positive and negative patches are encoded as degradation representations by a convolutional network model, which is shown in Fig. 2(a). SimCLR [11] pointed out that the representations need to go through a three-layer fully connected network to obtain q, k^+, and k^-. Contrastive learning aims to make q and k^+ more similar and keep q away from k^-. Following MoCo [8], we use InfoNCE loss to measure the similarity, which can be formulated as follows:

$$\text{Loss}_q = -\log \frac{e^{q \cdot k^+/\tau}}{\sum_{i=0}^{k} e^{q \cdot k_1^-/\tau}} \tag{3}$$

in which K is the number of negative samples, $\cdot$ is the dot product and τ is a hyper-parameter. Previous contrastive learning methods [6] mentioned that a large number of negative samples is essential for the model to obtain a good representation. A queue with negative samples is maintained for learning in the degradation model. During the training phase, we first randomly extract N patches from real-world datasets and divide two patches per group. Then, these N patches are encoded into q_i, k_i^+ by the degradation encoder. For negative samples, we also encode the HR images into k_i^- to update the queue, where i denotes the position of the i-th image. We define the total loss as follows:

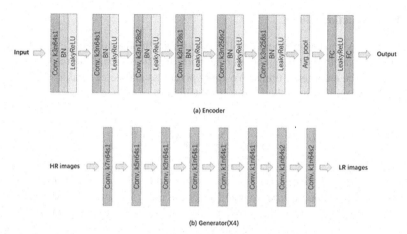

(a) Encoder

(b) Generator(X4)

Fig. 2. Architecture of Encoder and Generator Network. The 'k', 'n' and 's' in each layer indicate the kernel size, number of channels and stride size, respectively.

$$Loss_{deg} = \sum_{i=0}^{N/2} -log \frac{e^{q_t - k_i^+ / \tau}}{\sum_{j=0}^{k} e^{q_i - k_t^- / \tau}} \qquad (4)$$

where S is the size of the queue. Figure 2(a) shows the architecture of the degradation encoder. We adapt a similar encoder as the work of DASR [10]. Specifically, we assemble the convolutional layer with 3 filters, batch normalization (BN) [12] and LeakyReLU [13] layers with a negative slope of 0.1. Note that the average pooling output is embedding. The final multilayer perceptron consists of two fully connected networks and one LeakyReLU layer. The characters 'k', 'n' and 's' indicate the parameters of kernel size, numbers of different channels and the size of each stride. For instance, k7n128s2 means that there are 128 filters in the convolutional layer, the kernel size is 7, and the stride is 2. For the LR image generator, inspired by kernelGAN [9], we also design a linear model that does not contain any activation layer that is more in line with the degradation equation. This is consistent with the degradation model equation mentioned before, since downscaling by blur kernel is a linear operation applied to LR images. The generator architecture is shown in Fig. 2(b). There are eight convolutional layers with 64 channels each. The first three kernel sizes are 7, 5, and 3, and the rest are 1. The last two layers refer to the downscaling operator, whose scale factor is 4. The whole network can be regarded as a single convolutional layer with a 15 × 15 receptive field. The reason why we design a multilayer network instead of one convolutional layer is that gradient-based optimization is more efficient for deep linear networks than only one layer, as kernel-GAN [9] mentioned. We fine tune the pretrained encoder as a discriminator to ensure that the generated LR image can obtain the same degradation as the real-world image. The color loss and the adversarial loss aim to maintain the basic structure information of the original image.

3.3   Reconstruction Model

Based on SRGAN, we implement an SR model trained by paired data $\hat{x}, y$. The network adopts the architecture of the generator network in SRGAN, and the resolution of the SR image will be enlarged 4 times. We apply pixel loss and perceptual loss [14] during training. The pixel loss uses the L1 distance, which is calculated as:

$$Loss_1 = \frac{1}{S^2 WH} \sum_{i=0}^{sW} \sum_{j=0}^{sH} \left\| X_{i,j}^{hr} - F\left(Y_{i,j}^{lr}\right) \right\|$$
(5)

in which s is the scale factor, W, H is the width and height of the HR image and $I_{i,j}^{HR}$ describes the pixel value of the image. This is the most widely used loss function for image SR. However, there is the problem that solutions of the L1 regularized method often tend to lack high-frequency content, so we add a perceptual loss to obtain a sharper texture. The perceptual loss uses the inactive features of VGG-19 [15], which benefits the image vision quality:

$$Loss_{per} = \frac{1}{W_{i,j} H_{i,j}} \sum_{m=0}^{W_{i,j}} \sum_{n=0}^{H_{i,j}} \left(\phi_{i,j}\left(I^{HR}\right)_{m,n} - \phi_{i,j}\left(F\left(I^{LR}\right)\right)_{m,n} \right)^2$$
(6)

where $\phi_{i,j}$, $W_{i,j}$, and $H_{i,j}$ indicates the feature map created by the j-convolution before the i-maxpooling of the VGG-19 network. The total loss can be calculated as the weighted sum of these two different losses as follows:

$$Loss_{total} = \lambda_1 \cdot Loss_1 + \lambda_2 \cdot Loss_{per}$$
(7)

where λ_1 and λ_2 are set as 1 and 0.1.

4   Experiments

4.1   Training Data

We train the proposed model in the NTIRE-2020 dataset. There are 2650 degraded images that can be regarded as real-world images, 800 high-resolution images, and 100 paired validation images in Track 1. The 2650 degraded images are not paired with the 800 high-quality images and do not exist in the same image. We randomly select 800 images from the degraded image and 800 unpaired high-resolution images as training data during the degradation learning phase.

"0842" and "0896" from   (a) ground truth   (b) bicubic   (c) SRFBN   (d) ESRGAN   (e) EDSR   (f) ours
Track1 validation set

Fig. 3. Qualitative comparison with state-of-the-art blind methods on the NTIRE 2020 Real World SR challenge Track 1 validation set (SR scale ×4).

4.2 Training Details

As in Fig. 1, we split our training phase of the superresolution process into three parts. We first train the degradation learning network. During training, 64 h patches of size 192×192 and size 48×45 are images that are cropped from high-resolution images. Sixty-four LR patches of size 48×48 are randomly cropped from low-resolution images. In detail, we set τ and S in Eq. 4 to 0.06 and 8192. The model is optimized by the Adam optimizer, in which $\beta_1 = 0.5$ and $\beta_2 = 0.999$ without weight decay for 600 epochs. The learning rate is initialized as 1×10^{-3}. It will decrease 0.1 every 500 epochs. Given a trained encoder, we train the downscaling network with the same optimizer parameters for 100 epochs. Then, we train the reconstruction model and pretrained dowscaling network using generated paired LR-HR images for 300 epochs. The learning rates of the downscaling network and SR model are initialized to 1×10^{-4} and decreased by 0.5 every 100 epochs. The minibatch size of all networks is set to 32. Note that the patch size of the image and the network layers depend on the scale parameter, which is 4 in our experiments. We implement the proposed method on NVIDIA TITAN XP GPUs in the PyTorch platform, and it takes approximately two days and a half to train our model.

4.3 Training Details

We prove the effectiveness of our approach with current good methods in the same field: SRFBN [16], ESRGAN [17], EDSR [18], Impressionism [19] and DBPN [20]. Table 1 displays different parameters, such as the average PSNR, and SSIM values of the NTIRE2020 real-world SR Track1 validation set with different methods trained with clean LR images downscaled from HR images. Our methods outperform the previous methods. This shows that EDSR and SRFBN do not achieve good performance if the degradation is unknown in the training phase. The unsupervised ESRGAN enhances the noise and degradation, leading to poor quality of the SR image. Note that the Impressionism method makes more effort on so that the PSNR is lower than others. Our approach is much better than those other methods in both PSNR and SSIM, which may be because the model is trained on paired degradation image data. Several subjective results are illustrated in Fig. 3.

To validate the advantages of our model in solving real-world SR tasks, we apply the ESRGAN generator in our SR model, which is named RRDB.

Table 1. Quantitative results for the NTIRE 2020 real world SR challenge Track 1 validation dataset

Methods	PSNR	SSIM
Bicubic	25.48	0.680
EDSR	25.36	0.640
SRFBN	25.37	0.642
ESRGAN (Unsupervised)	19.04	0.242
Impressionism	24.82	0.662
DBPN	24.51	0.701
Ours	**25.50**	**0.738**

Table 2. Quantitative results for the NTIRE 2020 Real World SR challenge Track 1 validation dataset, comparing the ESRGAN (Supervised) and ours (RRDB)

Methods	ESRGAN (Supervised)	Ours (RRDB)
PSNR/SSIM	24.74/0.695	24.98/0.6873

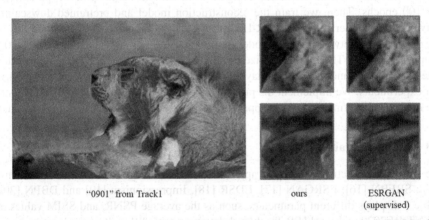

"0901" from Track1   ours   ESRGAN
                                                              (supervised)

Fig. 4. Qualitative comparison between ESRGAN (Supervised) and our method (RRDB) for Track 1 (SR scale ×4).

Table 2 shows a comparison with the supervised ESRGAN. The supervised ESRGAN is trained with real paired data provided by the NTIRE 2020 official baseline. We change the SR network into the RRDB but keep the same setting of the loss functions. The whole training process is also the same. As shown by the experimental results, our framework outperforms the supervised ESRGAN in PSNR and makes the SSIM value close to it. The quality results are illustrated in Fig. 4.

5   Conclusion

We propose a novel framework assisted by image degradation learning. In contrast to existing methods that downscale HR images to obtain LR images by a fixed operation, we acquire degradation representations of real-world LR images that assist the downsampling network in generating LR images with the consistent domain using contrastive learning. This assists us in obtaining more realistic and paired image data for the later reconstruction network. Experiments on NTIRE2020 datasets show the effectiveness of our approach.

References

1. Dong, C., Loy, C.C., He, K., Tang, X.: Image superresolution using deep convolutional networks. IEEE Trans. Pattern Anal. Mach. Intell. **38**(2), 295–307 (2015)
2. Cai, J., Zeng, H., Yong, H., Cao, Z., Zhang, L.: Toward real-world single image superresolution: a new benchmark and a new model. In: Proceedings of the IEEE/CVF International Conference on Computer Vision, pp. 3086–3095 (2019)
3. Xu, X., Ma, Y., Sun, W.: Toward real scene superresolution with raw images. In: Proceedings of the IEEE/CVF Conference on Computer Vision and Pattern Recognition, pp. 1723–1731 (2019)
4. Lugmayr, A., Danelljan, M., Timofte, R.: Unsupervised learning for real-world superresolution. In: 2019 IEEE/CVF International Conference on Computer Vision Workshop (ICCVW), pp. 3408–3416. IEEE (2019)
5. Yuan, Y., Liu, S., Zhang, J., Zhang, Y., Dong, C., Lin, L.: Unsupervised image superresolution using cycle-in-cycle generative adversarial networks. In: Proceedings of the IEEE Conference on Computer Vision and Pattern Recognition Workshops, pp. 701–710 (2018)
6. Gu, J., Lu, H., Zuo, W., Dong, C.: Blind superresolution with iterative kernel correction. In: Proceedings of the IEEE/CVF Conference on Computer Vision and Pattern Recognition, pp. 1604–1613 (2019)
7. Chen, X., Fan, H., Girshick, R., He, K.: Improved baselines with momentum contrastive learning. arXiv preprint arXiv:2003.04297 (2020)
8. He, K., Fan, H., Wu, Y., Xie, S., Girshick, R.: Momentum contrast for unsupervised visual representation learning. In: Proceedings of the IEEE/CVF Conference on Computer Vision and Pattern Recognition, pp. 9729–9738 (2020)
9. Bell-Kligler, S., Shocher, A., Irani, M.: Blind superresolution kernel estimation using an internal-GAN. arXiv preprint arXiv:1909.06581 (2019)
10. L. Wang, Y. Wang, X. Dong, Q. Xu, J. Yang, W. An, and Y. Guo, "Unsupervised degradation representation learning for blind superresolution. In: Proceedings of the IEEE/CVF Conference on Computer Vision and Pattern Recognition, pp. 10 581–10 590 (2021)
11. Chen, T., Kornblith, S., Norouzi, M., Hinton, G.: A simple framework for contrastive learning of visual representations. In: International Conference on Machine Learning. PMLR, pp. 1597–1607 (2020)
12. Ioffe, S., Szegedy, C.: Batch normalization: accelerating deep network training by reducing internal covariate shift. In: International Conference on Machine Learning. PMLR, pp. 448–456 (2015)
13. Maas, A.L., Hannun, A.Y., Ng, A.Y., et al.: Rectifier nonlinearities improve neural network acoustic models. In: Proceedings of the ICML, vol. 30, no. 1, p. 3. Citeseer (2013)

14. Johnson, J., Alahi, A., Fei-Fei, L.: Perceptual losses for real-time style transfer and super-resolution. In: Leibe, B., Matas, J., Sebe, N., Welling, M. (eds.) ECCV 2016. LNCS, vol. 9906, pp. 694–711. Springer, Cham (2016). https://doi.org/10.1007/978-3-319-46475-6_43

15. Simonyan, K., Zisserman, A.: Very deep convolutional networks for large-scale image recognition. arXiv preprint arXiv:1409.1556 (2014)

16. Li, Z., Yang, J., Liu, Z., Yang, X., Jeon, G., Wu, W.: Feedback network for image super-resolution. In: Proceedings of the IEEE/CVF Conference on Computer Vision and Pattern Recognition, pp. 3867–3876 (2019)

17. Wang, X., et al.: ESRGAN: enhanced super-resolution generative adversarial networks. In: Leal-Taixé, L., Roth, S. (eds.) ECCV 2018. LNCS, vol. 11133, pp. 63–79. Springer, Cham (2019). https://doi.org/10.1007/978-3-030-11021-5_5

18. Lim, B., Son, S., Kim, H., Nah, S., Lee, K.M.: Enhanced deep residual networks for single image superresolution. In: Proceedings of the IEEE Conference on Computer Vision and Pattern Recognition Workshops, pp. 136–144 (2017)

19. Ji, X., Cao, Y., Tai, Y., Wang, C., Li, J., Huang, F.: Real-world superresolution via kernel estimation and noise injection. In: Proceedings of the IEEE/CVF Conference on Computer Vision and Pattern Recognition Workshops, pp. 466–467 (2020)

20. Haris, M., Shakhnarovich, G., Ukita, N.: Deep back-projection networks for superresolution. In: Proceedings of the IEEE Conference on Computer Vision and Pattern Recognition, pp. 1664–1673 (2018)

Probability Loop Closure Detection with Fisher Kernel Framework for Visual SLAM

Ge Zhang$^{(\boxtimes)}$ (ID), Qian Zuo (ID), and Hao Dang (ID)

Henan University of Chinese Medicine, Zhengzhou 450000, Henan, China
`zhangge@hactcm.edu.cn`

Abstract. A typical approach to describe an image in loop closure detection for visual SLAM is to extract a set of local patch descriptors and encode them into a co-occurrence vector. The most common patch encoding strategy is known as bag-of-visual-words (BoVW) representation, which consists of clustering the local descriptors into visual vocabulary. The distinctiveness of images is difficult to represent since most of them contain similar texture information, which may lead to false positive results. In this paper, the vocabulary is used as a whole by adopting the Fisher kernel (FK) framework. The new representation describes the image as the gradient vector of the likelihood function. The efficiently computed vectors can be compressed with a minimal loss of accuracy using product quantization and perform well in the task of loop closure detection. The proposed method achieves a higher recall rate with 100% precision in loop closure detection compared with state-of-the-art methods, and the detection on bidirectional loops is also enhanced. vSLAM systems may perceive the environment more efficiently by constructing a globally consistent map with the proposed loop closure detection method, which is potentially valuable for applications such as autonomous driving.

Keywords: Fisher kernel · Loop closure detection · visual SLAM

1 Introduction

Simultaneous localization and mapping (SLAM) is regarded as the fundamental technology in autonomous robots and related sectors [37]. It is attracting increasing attention in industry and academic fields [26]. Visual SLAM (vSLAM), which uses a visual sensor as the only information source, has developed extensively during the past decades [10]. The loop closure detection (LCD) module recognizes the previously visited places to reduce the accumulated deviation during pose estimation, which is crucial in consistent map construction [48]. It is easy to realize with expensive odometry equipment but still remains an open problem with visual sensors only.

BoVW [28] is a representative approach to characterize texture features into visual words, which is utilized in detecting loop closures in vSLAM [13]. This

Y. Wang et al. (Eds.): ICPCSEE 2022, CCIS 1628, pp. 219–239, 2022.
https://doi.org/10.1007/978-981-19-5194-7_17

approach clusters the extracted feature descriptors into a set of visual words that compose the visual vocabulary. The clustering process can be realized in many aggregation approaches, such as k-means [46] and the information bottle-neck (IB) algorithm [16]. Image pairs are extracted for feature descriptors and matched into visual words to co-occurrence vectors, which is recognized as quantization in computer vision [42]. The vectors are evaluated to similarity scores, and the potential image pairs captured from a loop closure are found. The vocabulary construction provides meaningful summarization of all extracted features with identical length and achieves promising results in LCD [50].

Although accurate, the BoVW-based approach has its own limitations. Aggregation and quantization perform hard assignment while assigning points to clusters. Our previous work focuses on the aggregation technique and adopts the sequential information bottleneck (sIB) algorithm in vocabulary construction [16], but the operation still assumes spherical clustering with equal probability of a particular cluster. A large number of feature descriptors are represented by several cluster centers in aggregation. As a result, the accuracy is low when the training image set and testing set share very few similar texture features. Moreover, it is usually the case that hundreds of words appear 0 times before encountering a matched word. Consequently, the co-occurrence vectors are usually very sparse, and the computational cost in matching is relatively high [32]. To reduce the computational cost of extracting and matching features, the use of local binary descriptors gradually emerges in embedded platform development [44]. However, it is still incapable of action when the newly emerged features are not included in the training image set.

Our goal is to recognize the previously visited places during the motion period. Similar to content-based image retrieval technology, supervised and unsupervised deep neural networks yield substantial gain in various benchmarks, especially place recognition tasks [32]. Specifically, convolutional neural networks (CNNs) are widely adopted in vSLAM for sequence-based matching procedures to handle viewpoint and condition variance [3]. Although increased model size and computational cost tend to improve the quality in most tasks, they suffer from serious low parameter count problems in LCD. Our previous work [23] also adopts CNN to remove the semantic inconsistency in matching. However, the time complexity is too high to adopt in vSLAM systems with real-time requirements. Another trend combines local and global image features through histogram and keypoint matching [20] with classical computer vision technology. They have achieved better and faster performance in scenarios with illumination and texture variations, but there is no proper solution to detection on bidirectional loops [1], which is frequently encountered in vSLAM systems.

To cope with the issues that exist in the BoVW-based method and deep learning approach, we propose a novel LCD method via the Fisher kernel (FK) framework for the patch encoding strategy. The FK framework-based approach represents the probability of a feature descriptor that belongs to a particular "visual word" in vocabulary. It is possible that the system may describe features that are not contained in training images. The idea is to characterize the descriptors from the training image set with a gradient vector derived from a Gaussian mixture model (GMM) [41]. First, the feature descriptors extracted from the training image set are clustered with the k-means algorithm to generate the GMM, and the parameter k in the GMM is also defined in this step. A "probability visual vocabulary" contains expected information of the first order, and variance information of the second order is generated by our method. Second, an image that is potentially being encoded is extracted for descriptors and converted to probability density function (pdf) representation [27]. The Fisher vector (FV) of this image is computed with the descriptor set and GMM prior parameters. Histogram generation requires less computational cost since FV is a denser matrix than the co-occurrence vector of BoVW-based approaches. Third, the current image and loop candidate in the loop detection task are converted to a high-dimensional histogram vector by mapping to the FV set generated in the previous steps for similarity comparison. The potential loop closures are detected after comparing the similarities and returned to the mapping module after a verification stage. Experiments demonstrate that our method detects loops more accurately with less time consumption, which facilitates vSLAM systems to reduce the accumulated error in environment perception and correct estimated poses in case of needs.

The contributions of this study can be summarized as follows:

- We introduce the FK framework into vocabulary construction in LCD to focus on the distinctive features of training images. The pdf representation of visual features contributes to better recognition of similar images, which facilitates loop closure recognition.
- More accurate characterization of images with expected and variance information during image representation with FV creates a denser matrix during the matching process for reduced time consumption in LCD.
- The LCD method constructed with FV provides the possibility of detecting bidirectional loops, which enables the autonomous driving system to recognize previously visited places in opposite directions and extends the application field of the vSLAM system.

2   Related Works

Typical appearance-based BoVW LCD methods use Euclidean distance or other metrics to measure the similarity between two co-occurrence vectors, which are generated by measuring the similarity between the extracted features and visual vocabulary. This technology has been successfully applied in practical vSLAM systems [29,30] by taking the binary robust features of the BRIEF descriptor [11] and the efficient computation feature of FAST feature detector [38]. However, the generated feature descriptor is not able to characterize the spatial features of images. As a result, DBoW2 and DBoW3 [33] adopt the robust ORB descriptor [39] in feature extraction. This approach is highly efficient and is both scale and rotation invariant in the implementation of the vSLAM system due to the delicate designs of the ORB descriptor. However, the hard assignment between feature points and clusters during clustering makes new newly emerged features uncharacterizable. Soft assignment [18,35] or sparse coding [8,45] as well as the use of deep learning technologies [22,23,50] are proposed to cope with robust and efficiency issues. However, they either provide unacceptable performance in the detection of bidirectional loops or are time-consuming.

There are two different directions in addressing the problems of high time complexity and low accuracy. To reduce the time complexity, binary local descriptors are implemented to represent the image features. They are much faster to extract and can also be matched faster by using Hamming distance. The related literature [5,49] reports that computation of ORB is an order of magnitude faster than the computation of the SURF descriptor and two orders faster than the SIFT descriptor. However, although binary local descriptors are compact, the mapping process is difficult to scale up to a large vocabulary since the association between images and descriptors is usually very large [43]. Fisher vectors (FV) and vectors of locally aggregated descriptors (VLAD) are typical ways to improve aggregation methods [25]. They are able to analyze the local descriptors, such as SIFT and SURF, contained in an image. The vectors treat the local descriptors as global ones by creating statistical summaries. They have the qualities of effectiveness power from local descriptors and flexibility of global descriptor. The characterization process is a soft transform rather than hard assignment, and the frequency counts are transferred to probabilities. It is possible to characterize the features that are not contained in the training set [12], which makes it possible to detect loop closures from different orientations. In addition to improvements in feature extraction and quantization, dimensionality reduction techniques, such as principal component analysis (PCA) [21], have been proposed to reduce the storage requirement. Similar technologies, such as product quantization compression [19,24] and binary codes [36], also contribute to storage complexity reduction. The similarity matching scales well in very large databases in the tasks of image retrieval and classification, but there is no implementation in LCD application. Concerning the effectiveness and accuracy, we implement it in the task of LCD.

3   Methodology

The traditional BoVW-based approach clusters the feature descriptor with the k-means algorithm, and the feature descriptor is represented by the nearest cluster center. This approach only focuses on the quantity of key words, which is a 0-order statistic. This representation loses considerable useful information, and the computational cost is relatively high due to sparse matched features. The FV representation also includes 1st-order expectation and 2nd-order covariance data, and the vectors are almost dense, resulting in a more efficient matching operation in LCD. The computational cost is also reduced significantly since the creation of a histogram that measures the similarity of two images requires less computation than typical BoVW-based approaches. The construction of the FV and similarity measurement, as well as the LCD, will be discussed in the following sections.

3.1 Fisher Vector Generation

The generation of FV is explained in a threefold manner in this section. First, we describe the underlying principle of the FK framework. Then, we introduce FK to image representation for LCD. Finally, the approximation on the pdf is discussed to calculate the gradient of the parameters, which are used to characterize images.

The FV is essentially a gradient vector of the likelihood function that characterizes an image. Every pdf of the feature descriptor is a weighted representation by k GMM clusters in the FK framework, which means that the image is indirectly represented by k GMM clusters. The partial derivative of GMM represents the distribution of Gaussian models in this image, and the image can be converted to vector representation, which is similar to the co-occurrence vector in the BoVW-based approach. The physical significance of the gradient is the parameter tuning process in data fitting. The vectors that represent the most feature descriptors are found in this process. The likelihood function is related to statistical model parameters, which is a probability distribution.

Sample Preprocessing. An image in the training set needs to be preprocessed for the preparation of likelihood function creation. Suppose an image X is preprocessed and extracted for T feature descriptors; it can be formally represented as:

$$X = \{x_t, t = 1, 2..., T\}. \tag{1}$$

Since a single feature descriptor x_t is independently and identically distributed (i.i.d.) [2] in this image, the parameter θ of the known descriptor set X can be estimated as:

$$p(X; \theta) = \prod_{t=1}^{T} p(x_t; \theta). \tag{2}$$

The multiplication of multiple decimal numbers may cause floating-point underflow, and we can convert the multiplication to addition on Eq. (2) by performing a logarithmic operation on the two sides:

$$L(X; \theta) = ln[p(X; \theta)] = \sum_{t=1}^{T} ln[p(x_t; \theta)], \tag{3}$$

where θ is the feature parameter of descriptor x_t. Now, the concretization calculus of Eq. (3) can be performed, but there is no concrete representation on $p(X; \theta)$. Since the feature descriptors are i.i.d. in this image, the pdf of this image is calculated as:

$$L(X; \theta) = \sum_{t=1}^{T} ln[p(x_t; \theta)]. \tag{4}$$

GMM is chosen for clustering as the functional model, which is required to perform approximation calculation to this Eq. (4). The model makes use of the

GMM distribution to attempt every possible approximation of the pdf, which is used to model the distribution of all extracted features from an image. The FV can encode the gradients of the likelihood function with respect to the GMM parameters.

Gaussian Mixture Model Clustering. The GMM parameters denote the first-order moments of the features, which are recognized as expected information. The FV can denote the second-order moments of the features, which is variance information. The number of models is limited to a certain range, and every Gaussian model is called a "component". The pdf of GMM is calculated by linearly adding these components:

$$p(x_t; \theta) = \sum_{k=1}^{K} w_k N_k(x_t; \theta), \tag{5}$$

where

$$N_k(x_t; \theta) = \frac{e^{-\frac{1}{2}(x_t - u_k)^T \sum^{-1}(x_t - u_k)}}{(2\pi)^{\frac{D}{2}} (|\sum|)^{\frac{1}{2}}}. \tag{6}$$

Note that w_k denotes the weighting vector, u_k denotes the mean vector and $\sum$ denotes the covariance deviation vector. Since we have $\sum_{k=1}^{K} w_k = 1$, the equations above can be summarized as:

$$L(X; \theta) = \sum_{t=1}^{T} ln[p(x_t; \theta)] = \sum_{t=1}^{T} ln[\sum_{k=1}^{K} w_k \frac{e^{-\frac{1}{2}(x_t - u_k)^T \sum^{-1}(x_t - u_k)}}{(2\pi)^{\frac{D}{2}} (|\sum|)^{\frac{1}{2}}}], \tag{7}$$

where D denotes the dimensionality of the feature vector and $|\sum|$ means apply the determinant operator to summation.

EM Iteration. Since there is an addition operation inside the logarithm, we cannot find the maximum value directly. Therefore, we perform the EM iteration discussed in [47]. The approximation is carried out as follows:

E-step. Estimate every probability value that a particular data point is generated by a certain component, which is called the "Occupancy Probability". According to the Bayes formula, the probability of each data point x_t being generated by component k is given as:

$$\gamma(t, k) = \frac{w_k N_k(x_t; u_k, \sum_k)}{\sum_{j=1}^{K} w_j N_j(x_t; u_j, \sum_j)}. \tag{8}$$

Because u_k and $\sum_k$ are unknown in this calculation, we can estimate them by supposing they equal to the values from the previous iteration or the default value in the first iteration.

M-step. Estimate the parameter of every component. Suppose the $\gamma(t, k)$ in the previous step is the correct one that denotes the probability of x_k being generated by component k, and the value γ is also regarded as the contribution brought by the component in the generation of these data. Because each component is a standard Gaussian distribution, we can calculate the parameter θ according to maximum likelihood easily by finding the partial derivative of parameters $\theta = \{w_k, u_k, \sum_k\}$ in Eq. (7).

Pdf Approximation. According to the discussion above, we can approximate the probability density function in finite GMM. For example, the $p(x_t, \theta)$ in Eq. (3) can be calculated as:

$$p(x_t; \theta) = \sum_{k=1}^{K} w_k N_k(x_t; \theta), \tag{9}$$

and

$$N_k(x_t; \theta) = \frac{e^{-\frac{1}{2}(x_t - u_k)^T \sum^{-1}(x_t - u_k)}}{(2\pi)^{\frac{D}{2}}(|\sum|)^{\frac{1}{2}}}, \text{where} \sum_{k=1}^{K} = 1. \tag{10}$$

Therefore, the embodiment expression of Eq. (3) can be found by substituting Eq. (9) and Eq. (10) as:

$$L(X; \theta) = ln[p(X; \theta)] = \sum_{t=1}^{T} ln[\sum_{k=1}^{K} w_k N_k(x_t; \theta)]. \tag{11}$$

The estimation of parameter θ can be achieved by separately estimating the weighting vector (w_k), mean vector (u_k) and covariance deviation vector $(\sum_k)$. The solution to the likelihood function is achieved by finding the solution to $\{w_k, u_k, \sum_k\}$. This is solved by performing partial derivative on θ in GMM as:

$$\frac{\partial L(X; \theta)}{\partial \theta} = \frac{\partial L(X; \theta)}{\partial w_k}, \frac{\partial L(X; \theta)}{\partial u_k}, \frac{\partial L(X; \theta)}{\partial \sum_k}. \tag{12}$$

The optimized parameters are calculated with the method discussed in [40]:

$$\frac{\partial L(X; \theta)}{\partial w_k} = \sum_{t=1}^{T} [\frac{\gamma_t(k)}{w_k} - \frac{\gamma_t(1)}{w_1}], \tag{13}$$

and

$$\frac{\partial L(X; \theta)}{\partial u_k} = \sum_{t=1}^{T} \gamma_t(k)[\frac{x_t - u_k}{\sum_k}], \tag{14}$$

and

$$\frac{\partial L(X; \theta)}{\partial \sum_k} = \sum_{t=1}^{T} \gamma_t(k)[\frac{x_t - u_k}{\sum_k^{\frac{3}{2}}} - \frac{1}{\sqrt{\sum}}] \tag{15}$$

where $\gamma_t(k) = \frac{w_k N_k(X; \theta)}{\sum_{k=1}^{K} w_k N_k(X; \theta)}$ and $k \geq 2$.

3.2 Probability Visual Vocabulary

The FV of an image is the collection of gradient vectors of the likelihood function. We propose applying FK to the visual vocabulary, where the visual words are represented by means of GMM. In practice, we extract ORB descriptors from images according to the characteristics of the training set and the requirement of application. To simplify the sampling operation, we resize all the images in the training set to the same resolution. The computation of FV is denoted by Algorithm 1.

Algorithm 1. Fisher vector generation process

Require: Local image descriptors $X = \{x_t \in \mathbb{R}^D, t = 1, ..., T\}$, Gaussian mixture model parameters $\lambda = \{w_k, u_k, \sum_k, k = 1, ..., K\}$
Ensure: normalized FV representation $g_\theta^X \in \mathbb{R}^{K(2D+1)}$
1: **function** COMPUTE STATISTICS
2:      **for** $k = 1 : K$ **do**
3:          according to the method described in [18], initialize accumulators as
4:          $S_k^0 \leftarrow 0,\ S_k^1 \leftarrow 0,\ S_k^2 \leftarrow 0$
5:          where the superscripts 0, 1 and 2 denote the order of zero, first and second statistics, respectively
6:      **end for**
7:      **for** $t = 1, ..., T$ **do**
8:          Compute $\gamma_t(k)$ using $\gamma_t(k) = \frac{w_k N_k(X;\theta)}{\sum_{k=1}^K w_k N_k(X;\theta)}$
9:          **for** $k = 1 : K$ **do**
10:              $S_k^0 \leftarrow S_k^0 + \gamma_t(k)$
11:              $S_k^1 \leftarrow S_k^1 + \gamma_t(k)x_t$
12:              $S_k^2 \leftarrow S_k^2 + \gamma_t(k)x_t^2$
13:          **end for**
14:      **end for**
15: **end function**
16: **function** COMPUTE FV SIGNATURE
17:      **for** $k = 1 : K$ **do**
18:          $g_{w_k}^X = (S_k^0 - Tw_k)/\sqrt{w_k}$
19:          $g_{u_k}^X = (S_k^1 - u_k S_k^0)/(\sqrt{w_k}\sum_k)$
20:          $g_{\sum_k}^X = (S_k^2 - 2u_k S_k^1 + (u_k^2 - \sum_k^2)S_k^0)/(\sqrt{2w_k}\sum_k^2)$
21:      **end for**
22:      $g_\theta^X = (g_{w_1}^X, \cdots, g_{w_K}^X, g_{u_1}^{X'}, \cdots, g_{u_K}^{X'}, g_{\sum_1}^{X'}, \cdots, g_{\sum_k}^{X'})$
23: **end function**
24: **function** APPLY ℓ_2−NORMALIZATION
25:      $g_\theta^X = g_\theta^X / \sqrt{g_\theta^{X'} g_\theta^X}$
26: **end function**
27: **return** g_θ^X

For every single Gaussian in the probability visual vocabulary, w_k denotes the relative frequency of the word k, u_k denotes the mean of the word and $\sum_k$

denotes the variation around the mean. The local feature descriptors set X can be represented by the vector:

$$g_\theta^X = \frac{1}{T}\nabla_\theta ln u_\theta(X), \tag{16}$$

where θ denotes both the set of parameters of u and the estimate of parameters for simplicity. The notation ∇ denotes the gradient computation on the logarithm of the likelihood function. Following the parameter selection in [34], we choose u_θ as the GMM:

$$u_\theta(x) = \sum_{i=1}^{K} w_i u_i(x), \tag{17}$$

where we recognize the model u_θ as a probabilistic visual vocabulary that is trained offline with the training image set through EM iteration. Upon feature extraction, a soft assignment is performed on all the clusters, and the sum of the weights from this assignment is utilized to represent the images. We relate the traditional BoVW representation to Fisher kernels since our method also considers the 0-th order statistics, which is word counting in BoVW representation. The Fisher kernel also considers first-order and second-order statistics, which are denoted by Eqs. (14) and (15), respectively. For a typical vocabulary of size N, the BoVW representation generates an N-dimensional histogram, while the full gradient representation generates a vector of $(2 \times D + 1) \times N - 1$ dimensions. The dimensionality of our method is approximately 100 times higher than that of BoVW. In the concept of the GMM model, BoVW representation only considers the quantity of a single Gaussian model, which can be understood as zero order information. FV representation involves expected information of the first order and covariance information of the second order, which facilitates the algorithm acquiring more precise loop closures.

After acquiring the vectors of the current image and loop candidate, the similarity between them can be assessed by calculating the similarity between two vectors. Suppose there are the current image I_c from the camera input and the loop candidate I_f from the potentially revisited place to be compared. They are converted to FV representation as g_c and g_f, respectively. The similarity between two images can be evaluated by calculating the L1-score between them as:

$$s(g_c, g_f) = 1 - \frac{1}{2}\left|\frac{g_c}{|g_c|} - \frac{g_f}{|g_f|}\right|, \tag{18}$$

where $|\cdot|$ stands for the L1-norm, and the calculation will result in a positive correlation between score and similarity. The image pair with a higher value of s is potentially a true loop closure. The image pair with a certain level of similarity can be passed to the verification stage. It involves three separate stages and will be discussed in the following sections.

3.3 Loop Closure Detection

Originating from classic computer vision applications, such as image classification and retrieval, detection on loop closures is achieved through comparison between the FV of the current image and several other FVs of loop candidates. However, it is not sufficient in LCD since the input frames are consecutive in the vSLAM application. The pair of current images and loop candidates with the highest similarity may come from adjacent frames but not the revisited place. The threshold of the similarity score is crucial in the design of the LCD method, and we define a similarity metric in the first part as follows. A temporal consistency check is implemented in our actual implementation to avoid the adjacent frames competing with each other. A geometric verification is also defined to eliminate the erroneous detected loop closures as many as possible.

Similarity Score Normalization. Depending on the probability visual vocabulary and the image frames, the similarity score may vary from 0 to greater than 10. The similarity scores of the two images may lie in completely different ranges with different probability visual vocabularies. We adopt a normalization method to convert the score between 0 and 1 when comparing the two vectors. We perform a normalization operation to obtain the possible highest score in a particular sequence:

$$\eta(\boldsymbol{g}_c, \boldsymbol{g}_{f_j}) = \frac{s(\boldsymbol{g}_c, \boldsymbol{g}_{f_j})}{s(\boldsymbol{g}_c, \boldsymbol{g}_{f-\Delta\tau})}, \tag{19}$$

where $s(\boldsymbol{g}_c, \boldsymbol{g}_{f_j})$ denotes the scores from the potential matches, $s(\boldsymbol{g}_c, \boldsymbol{g}_{f-\Delta\tau})$ is the expected score of the previous image pair and $\boldsymbol{g}_{f-\Delta\tau}$ is the computed FV from the previous time interval. According to the reduction above, the highest similarity may come from the match of I_f and its adjacent frame $I_{f-\Delta\tau}$ from the previous time interval $\Delta\tau$. As a result, the scores are converted to a float number between 0 and 1 by calculating η. Extremely low $s(\boldsymbol{g}_c, \boldsymbol{g}_{f-\Delta\tau})$ on the denominator will cause a very high value of η. These image pairs are rejected, and only matches that pass the threshold can move to the next stage.

Temporal Consistency. The similarity between adjacent images can be higher than those from revisited places. A true loop may only occur in the time interval of 10 s in a typical vSLAM application. In a typical video sequence, the capture frequency 30 Hz in the NTSC standard, which means 60 pairwise consistent frames are captured in 2 s. We ignore the adjacent frames 5 s before and after the current image in the temporal consistency check to avoid images that are close in time from competing with each other. The temporal value can also be changed according to the characteristics of the scene and the sensitivity requirement. A temporal threshold α is defined in this stage to filter the image pairs with similarity scores lower than a certain level. Suppose current image I_c and loop candidate $I_{f\prime}$ are represented by matched vectors $<\boldsymbol{g}_c, \boldsymbol{g}_{f\prime}>$. The similarity score is normalized as $\eta(\boldsymbol{g}_c, \boldsymbol{g}_{f\prime})$. The program checks if $\eta(\boldsymbol{g}_c, \boldsymbol{g}_{f\prime}) \geqslant \alpha$ and forward the passed image pairs to the geometrical verification stage. In the probability LCD, we set $\alpha = 0.43$ in the evaluation since this threshold may accept most of the potential real loop closures.

Geometrical Verification. Due to the unordered nature of FV representation of images, two completely different image frames can be converted to the same FV and recognized as a loop. A geometrical verification procedure is introduced to eliminate falsely detected loops by comparing the geometrical structure of the feature descriptor from two images. Specifically, an epipolar constraint is defined on the geometrical structure to determine the unchanged location of the feature point.

Although the FV representation with probabilistic visual vocabulary contains more information than BoVW representation, the 1st-order and 2nd-order data have no spatial information available. Consequently, two images from completely different places may have very high similarity scores. We perform geometrical verification between them to evaluate the actual coordinates of the features.

Suppose several loop candidates pass the temporal consistency check and are accepted as potential loop closures. We calculate the fundamental matrix supported by 12 correspondences between the current image and loop candidate with the RANSAC algorithm [31]. The probability visual vocabulary is used to approximate the nearest feature neighbors to check whether the matched features come from the same part of the image frames. For more details of geometrical verification, please refer to [13]. The LCD process is complete after this validation, and the whole process is summarized as in Algorithm 2.

Algorithm 2. Loop closure detection with geometrical verification

---

Require: consecutive frames $\boldsymbol{F}$
Ensure: loop closure LC, potential loop closure LC^*
1: initialize $LC \leftarrow \{\}$, $LC^* \leftarrow \{\}$
2: **if** there is a new frame in $\boldsymbol{F}$ **then**
3:     select the latest frame as the current image I_c
4:     $\boldsymbol{g}_c \leftarrow$ build_Fisher_vector (I_c)
5:     **for** every single previous frame I_f in $\boldsymbol{F}$ **do**
6:         $\boldsymbol{g}_l \leftarrow$ build_Fisher_vector (I_f)
7:         $s(\boldsymbol{g}_c, \boldsymbol{g}_f) \leftarrow$ similarity_measurement $(\boldsymbol{g}_c, \boldsymbol{g}_f)$
8:         **if** $s(\boldsymbol{g}_c, \boldsymbol{g}_f) \geqslant \alpha$ **then**
9:             $LC^* \leftarrow \{< I_c, I_f >\} \cup LC^*$
10:        **end if**
11:        **for** each element LC_i in LC^* **do**
12:            **if** geometrical_verification (LC_i) **then**
13:                $LC \leftarrow LC_i \cup LC$
14:            **end if**
15:        **end for**
16:    **end for**
17: **end if**
18: **return** LC

---

4 Results and Discussion

In this section, we carry out experiments to validate the improvement of our method with FVs in terms of LCD accuracy by means of the contrast relationship between precision and recall against the state-of-the-art LCD methods. Specifically, we compare our method with DBoW3 [33], which is employed in various feature-based vSLAM systems. The competitors also include PREVIeW [4] and iBoW-LCD [15], which represent the latest advancement of the appearance-based LCD method. As a continual tracing and research on LCD, our previous work "SLCD" [23] and other deep learning-based approaches [32] are also involved in some tests. An ablation study is also carried out to analyze the impact of the "LCD" module on the vSLAM system on the pose estimation accuracy.

First, we compare the precision-recall performance of our method with four competitors on the KITTI [17] and Bovisa [7] datasets, which are widely used to simulate the vSLAM application in horizontal movement. Second, we evaluate the LCD accuracy in the EuRoC MAV [9] dataset, which is captured with horizontal and vertical movements. Third, we study the detection of bidirectional loops by five methods on the Malaga dataset [6], which contains several revisits of the same places with different orientations. It is extremely important in autonomous driving applications since vehicles usually travel in an identical way in different directions. For a more detailed explanation of bidirectional loops, please refer to [16]. Finally, we study the effect of our method in actual pose estimation with a real-time vSLAM system. The LCD module in ORB-SLAM2 is substituted with our method to evaluate the improvement compared with DBoW3. The evaluation can be recognized as an ablation study on the LCD module in the vSLAM system.

4.1 Dataset and Preprocessing

The KITTI dataset contains visual data captured during road movement, which is ideal for the evaluation of autonomous driving applications. The Bovisa dataset has similar characteristics but more sophisticated textures and illumination conditions. Selected sequences from the two datasets are utilized in the evaluation of five methods in 2D motion of vSLAM application. The EuRoC dataset contains grayscale visual measurement sequences collected in a machine hall building, which is equipped on a drone and contains movement in any direction. It is ideal for evaluating the method in a 3D environment. The Malaga dataset contains six sequences captured in different places and lengths, and we utilize the sequences with bidirectional loops for evaluation. The attributes of the four datasets are summarized in Table 1.

The four datasets provide trajectory ground truth from laser radar, GPS, and inertial measurement unit (IMU) data. Therefore, the benchmark of loop detection can be calculated from the trajectory ground truth with the method in [14]. The data collected by physical hardware contain precise location and pose information, which can be converted for the benchmark of evaluating the performance of LCD. The calculated data also play an important part in the

Table 1. Characteristics of experimental datasets

Dataset	Dynamic	Overlap	Images	Image size
KITTI00	Low	Medium	4541	1241×736
KITTI02	Low	Small	4661	1241×376
KITTI05	Slight	Tiny	2761	1241×376
KITTI06	None	Tiny	1101	1241×376
KITTI07	Medium	Medium	1101	1241×376
KITTI09	Low	Small	1591	1241×376
Bovisa04	Low	Medium	34173	640×480
Bovisa06	Medium	Small	32240	640×480
Bovisa07	Low	Tiny	26825	640×480
Machine Hall 01	Low	Medium	2700	752×480
Machine Hall 03	Low	Medium	2700	752×480
Machine Hall 05	Medium	Large	2273	752×480
Malaga Parking 6L	Low	Large	6948	1024×768

calculation of precision and recall indexes, which is crucial in evaluation and will be discussed in the next section.

4.2   Evaluation Metrics

A typical LCD approach is evaluated in terms of precision and recall rate, which are two inversely proportional values. To calculate the two indexes, we collect the number of correct matches as true positives (TP), the incorrect matches as false positives (FP), and matches that the system erroneously discards as false negatives (FN). Precision and recall can be calculated as:

$$Precision = \frac{TP}{TP + FP}, Recall = \frac{TP}{TP + FN}. \tag{20}$$

The two indexes have a negative correlation, and we need to improve the recall rate while holding precision as high as possible. Because the vSLAM system requires zero FP result in most cases, we only focus on the maximum recall rate with 100% precision. The correctly discarded matches in the system are called true negative (TN) results, which are not involved in the calculation. The evaluation is also performed in terms of precision-recall curves, which implies the trend of precision with recall rate increase. We plot the two values in a coordinate system with precision as the vertical coordinate and recall as the horizontal coordinate. The dots in the system are then smoothly connected to form a precision-recall curve. The PR curves are frequently used to reproduce the tendency of results in LCD.

4.3 2D Motion

We conduct a set of experiments on the KITTI and Bovisa datasets to evaluate the performance with horizontal movement. The two datasets are frequently used to simulate the 2D motion estimation of vSLAM systems, which is recognized as the autonomous driving scenario simulation. LCD is crucial in reducing the accumulated error during mapping. The results from the six algorithms are shown in Table 2; literature [32] does not make the source code available for their DNN model, and we put their published results in comparison directly. The other dashed lines indicate that the relevant method does not achieve 100% precision detection in the test.

Table 2. Recall rate at 100% precision of the five algorithms for the KITTI and Bovisa datasets. Entries in bold indicate the best performing result for each dataset.

Dataset	Algorithms					
	DNN in [32]	SCLD [23]	DBoW3 [33]	PREVIeW [4]	iBoW-LCD [15]	Our method
KITTI00	95.99	94.32	**99.63**	93.14	86.75	91.6
KITTI02	90.44	88.74	83.33	–	71.52	**91**
KITTI05	87.06	**92.55**	87.32	88.27	87.55	84.37
KITTI06	–	87.79	88.58	–	74.69	**90.13**
KITTI07	–	85.87	82.12	–	83.67	**86.45**
KITTI09	–	90.34	–	–	–	**96.68**
Bovisa04	–	30.66	22.2	30.59	**31.8**	25.7
Bovisa06	–	27.68	14.72	**32.41**	30.72	19.85
Bovisa07	–	40.59	37.55	39.75	41.66	**42.38**

The data suggest that our method consistently performs better in most of the sequences of 2D motion, while the competitors have advantages under certain extreme conditions. In the test on the KITTI dataset, our method yields 100% precision in all 6 sequences and maintains the highest recall rate in KITTI 02/06/07/09. All six algorithms perform well in KITTI00, which is captured in low dynamic motion and contains the most loop closures. DBoW3 has a slightly higher recall rate at 100% precision in KITTI00, while the other five algorithms also gain a relatively high recall rate at 100% precision. There is only one place where loop closure occurs in KITTI05, and many border trees with mixed texture details emerge. The repeated texture feature is very difficult for appearance-based methods, and SLCD achieves the highest recall rate in this sequence. Our method performs the best in the remaining four sequences, and the results are extremely attractive in KITTI07/09. The possible reason is that FV contains a higher level of features and can characterize the distinctive textures, which are crucial in identifying the revisited places.

In the test on the Bovisa dataset, the performances of the six algorithms decrease simultaneously due to larger visual displacement and higher dynamic level. iBoW-LCD utilizes a dynamic island mechanism to detect loops and achieves a recall rate of 31.8% at 100% precision in Bovisa04. PREVIeW detects nearly 1/3 of the loops in Bovisa06, which has isolate loop closures over a very

Table 3. Loop closure counts of TP, FP and FN results from five algorithms in the MH01, MH03, and MH05 sequences in the EuRoC dataset. Entries in bold indicate the best performing result for each sequence

Sequence	Length/Duration	Avg.vel./angular vel	Algorithm	TP	FP	FN
MH01	80.6 m/182 s	0.44 m/s 0.22 rad/s	SLCD	72	95	55
			DBoW3	64	97	43
			PREVIeW	89	108	18
			iBoW-LCD	**93**	111	47
			our method	91	105	36
MH03	130.98 m/132 s	0.99 m/s 0.29 rad/s	SCLD	85	116	44
			DBoW3	62	137	67
			PREVIeW	77	120	52
			iBoW-LCD	87	104	58
			our method	**90**	132	66
MH05	97.6 m/111 s	0.88 m/s 0.21 rad/s	SCLD	57	176	58
			DBoW3	65	132	73
			PREVIeW	48	196	75
			iBoW-LCD	71	106	65
			our method	**76**	115	48

long moving distance. Our method performs the best in Bovisa07 due to the advantage in extracting distinguishable features with FV characterization. This sequence contains the most sophisticated texture features, and our method eliminates the disturbance from irrelevant interference.

4.4   3D Motion

We conduct a set of experiments on the EuRoC MAV dataset to evaluate the performance in horizontal and vertical movements. This dataset contains 11 sequences captured by an MAV flying around a machine hall and two office rooms, which can be recognized as a 3D motion scenario. The sequences are categorized into easy, medium and difficult groups by investigating the speed, illumination and scene texture complexity variations during captioning. We choose three sequences from machine halls to simulate the detection performance in 3D motion, which is usually seen in applications with flying objects, e.g., drones.

Since the method proposed in [32] does not support detection in a 3D motion scenario, the corresponding result is ignored in this test. Limited by the available data, we put the TP and FP results out of the total detection for the five algorithms directly in Table 3. The left three columns denote the name of sequences and their attributes, e.g., the duration of capture and average velocity of movement. The latter three columns denote the quantity of TP, FP and FN loop closures for each algorithm. The TN index can be extremely large in long video sequences and is ignored since it does not contribute to the final judgment.

The results suggest that our method not only detects more TP results with increased difficulty but also generates fewer FP loops during LCD. The video frames in the MH01 sequence are captured in a bright scene with good texture, and the five algorithms perform roughly the same. iBoW-LCD has slightly more TP loops, and our method has the fewest FN results. The MH03 sequence is captured with higher speed, and there are fewer repeated frames from one place. iBoW-LCD and PREVIeW detect fewer loops since there is little chance to group "image islands" in this sequence. Our result decreases accordingly, but we still maintain 90 TP loops in this scene. SLCD generates the fewest FN loops in this video sequence. MH05 has an even darker illumination condition and identical fast motion. The five algorithms detect even fewer loops and generate more FP matches. Our method still finds 76 TP results according to the benchmark with only 115 FP results. The other four competitors have lower recall rates with lower precision, and our method maintains a high recall rate with better image characterization by expect and covariance information. We believe that the algorithm with the deep learning model is overfitting in this sequence and has relatively poor performance.

4.5 Bidirectional Loops

We collect precision-recall data to study the detection on bidirectional loops of the four algorithms. The publicly available source code of PREVIeW only generates the number of detections and is not suitable to draw the curve. The detection on bidirectional loops helps the vSLAM system in many ways, such as relocalization, mapping accuracy improvement and more consistent registration results. The Malaga dataset contains six sequences with detailed calibration information, and we choose parking-6L and campus-2L as the test scenarios for LCD. The two datasets are captured in a parking lot and campus, which have lengths of 1222 m and 2150 m, respectively. The vehicle moves in a round way where the bidirectional loops account for nearly 1/3 of the total loop closures. Refer to Fig. 1 for the PR curves of the parking-6L and campus-2L dataset.

The curves suggest that the precision maintains a longer time of 100% with the recall rate increase of our method. DBoW3 uses a traditional method of feature extraction and clustering, which creates many redundant visual words, and the LCD is greatly affected. iBow-LCD has a similar mechanism, and neither of them can achieve 100% precision with 100% recall. It is obvious that the bidirectional loops are not detected by the two methods. The deep learning approach SLCD can detect some of the loops in opposite positions due to the inconsistency removal features. Our method based on FV is able to maintain 100% precision even with a recall rate higher than 0.5, which means that more than 1/3 of the bidirectional loops are being detected by us. Actually, our method can achieve a recall rate of 0.51 in parking-6L and 0.53 in campus-2L. The corresponding recall rate does not exceed 0.4 without detection on bidirectional loops and is often recognized as unsuitable in autonomous driving applications.

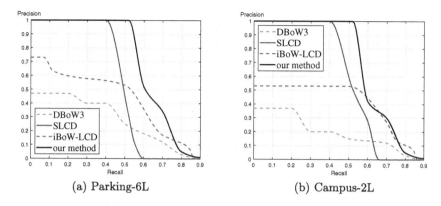

Fig. 1. Precision-recall curves of DBoW3 and proposed method for the Malaga dataset. (Color figure online)

4.6 Ablation Study

As a module of the vSLAM system, LCD serves as the optimizer of pose estimation and 3D registration. The final goal is to reduce the accumulated error during mapping and construct consistent results. We conduct a set of experiments to evaluate the performance of our method in a real vSLAM system. Specifically, we choose ORB-SLAM as the baseline and substitute the LCD module for our method. Since ORB-SLAM adopts DBoW3 as the LCD module, it is recognized as the ablation study between DBoW3 and our method on the accuracy of estimated trajectories. Refer to Fig. 2 for the comparison of the two methods in selected sequences from the KITTI and EuRoC datasets.

Blue lines denote the trajectories generated by the original ORB-SLAM system, green lines denote the improved ORB-SLAM with our method, and black lines denote the ground truth. The plotted trajectories suggest that the vSLAM system is able to acquire globally consistent poses with our method. In the comparison of trajectories from KITTI02, the ORB-SLAM system encounters several drifts during mapping, and our method gives timely relocalization to correct the drift. As a result, we can see that the green line better matches the ground truth. The blue and green lines match the ground truth in KITTI05 since there are fewer loops to be detected. Our method does not have an obvious advantage in this sequence. ORB-SLAM with its original LCD module experiences significant drift in KITTI06. The blue line scratches out of the road at the top roundabout. In contrast, the green trajectory matches the ground truth and forms a nearly perfect round way. The reason is that our method can detect bidirectional loops and correct the drift in the downbound trip. We also see the same phenomenon in three sequences from the EuRoC dataset. It is obvious that LCD is important in environment perception in the vSLAM system, and constructing the vocabulary with FV contributes to reducing the accumulated errors.

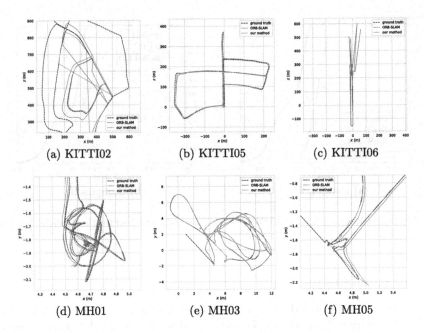

(a) KITTI02   (b) KITTI05   (c) KITTI06

(d) MH01   (e) MH03   (f) MH05

Fig. 2. Comparison of the estimated trajectories of ORB-SLAM with its original LCD module and ours; some of the trajectories are magnified. (Color figure online)

5   Conclusions

In this paper, we propose the FV representation as an alternative to BoVW approaches, which are commonly adopted in the LCD for vSLAM systems. In the FV framework, a probabilistic visual vocabulary is learned offline from a training image set, and feature descriptors from images are computed for their deviations from the vocabulary. The representation is recognized as the gradient vector with respect to the parameters of the vocabulary, which we choose as the Gaussian mixture with diagonal covariance. Our LCD method constructed on FV offers a more complete representation of the images since it encodes both the probabilistic count of occurrences and higher order data related to its visual word distribution in the vocabulary. The more compact and efficient representation facilitates more robust detection of loop closures. The consistent improvement in the recall rate with 100% precision and detection on bidirectional loops is able to improve the accuracy of semantic environment perception in vSLAM, which provides potential map construction approaches for autonomous driving. We will investigate low time complexity approaches of environment perception for vSLAM by leveraging different feature extraction algorithms and sensors in the future.

References

1. Arroyo, R., Alcantarilla, P.F., Bergasa, L.M., Yebes, J.J., Gamez, S.: Bidirectional loop closure detection on panoramas for visual navigation. In: Proceedings of IEEE Intelligent Vehicles Symposium (IV), pp. 1378–1383 (2014)
2. Arshad, S., Kim, G.W.: Role of deep learning in loop closure detection for visual and Lidar SLAM: a survey. Sensors **21**(4), 1243 (2021)
3. Bai, D., Wang, C., Bo, Z., Xiaodong, Y.I., Yang, X.: CNN feature boosted SeqS-LAM for real-time loop closure detection. Chin. J. Electron. **27**(3), 488–499 (2018)
4. Bampis, L., Amanatiadis, A., Gasteratos, A.: Fast loop-closure detection using visual-word-vectors from image sequences. Int. J. Robot. Res. **37**(1), 62–82 (2018)
5. Bay, H., Tuytelaars, T., Van Gool, L.: SURF: speeded up robust features. In: Leonardis, A., Bischof, H., Pinz, A. (eds.) ECCV 2006. LNCS, vol. 3951, pp. 404–417. Springer, Heidelberg (2006). https://doi.org/10.1007/11744023_32
6. Blanco, J.L., Moreno, F.A., González, J.: A collection of outdoor robotic datasets with centimeter-accuracy ground truth. Auton. Robot. **27**(4), 327–351 (2009)
7. Bonarini, A., Burgard, W., Fontana, G., Matteucci, M., Sorrenti, D.G., Tardos, J.D.: RAWSEEDS: robotics advancement through web-publishing of sensorial and elaborated extensive data sets. In: Proceedings of International Conference on Intelligent Robots and Systems Workshop on Benchmarks in Robotics Research (ICIRS) (2009)
8. Boureau, Y.L., Bach, F., Lecun, Y., Ponce, J.: Learning mid-level features for recognition. In: Computer Vision & Pattern Recognition, pp. 2559–2566 (2010)
9. Burri, M., et al.: The EuRoC micro aerial vehicle datasets. Int. J. Robot. Res. **35**(10), 1157–1163 (2016)
10. Cadena, C., et al.: Past, present, and future of simultaneous localization and mapping: toward the robust-perception age. IEEE Trans. Rob. **32**(6), 1309–1332 (2016)
11. Calonder, M., Lepetit, V., Strecha, C., Fua, P.: BRIEF: binary robust independent elementary features. In: Daniilidis, K., Maragos, P., Paragios, N. (eds.) ECCV 2010. LNCS, vol. 6314, pp. 778–792. Springer, Heidelberg (2010). https://doi.org/10.1007/978-3-642-15561-1_56
12. Dong, H., Yang, L., Wang, X.: Robust semi-supervised support vector machines with Laplace kernel-induced correntropy loss functions. Appl. Intell. **51**(21), 1–15 (2021)
13. Galvez-López, D., Tardos, J.D.: Bags of binary words for fast place recognition in image sequences. IEEE Trans. Rob. **28**(5), 1188–1197 (2012)
14. Gao, X., Zhang, T.: Unsupervised learning to detect loops using deep neural networks for visual SLAM system. Auton. Robot. **41**(1), 1–18 (2017)
15. Garcia-Fidalgo, E., Ortiz, A.: iBoW-LCD: an appearance-based loop closure detection approach using incremental bags of binary words. IEEE Robot. Autom. Lett. **99**, 3051–3057 (2018)
16. Ge, Z., Xiaoqiang, Y., Yangdong, Y.: Loop closure detection via maximization of mutual information. IEEE Access **7**, 124217–124232 (2019)
17. Geiger, A., Lenz, P., Stiller, C., Urtasun, R.: Vision meets robotics: the KITTI dataset. Int. J. Robot. Res. **32**(11), 1231–1237 (2013)
18. Van Gemert, J.C., Veenman, C.J., Smeulders, A.W., Geusebroek, J.M.: Visual word ambiguity. IEEE Trans. Pattern Anal. Mach. Intell. **32**(7), 1271–1283 (2010)
19. Gray, R.M., Neuhoff, D.L.: Quantization. IEEE Trans. Inf. Theory **44**(6), 2325–2383 (1998)

20. Guclu, O., Can, A.B.: Fast and effective loop closure detection to improve SLAM performance. J. Intell. Rob. Syst. **93**(3), 495–517 (2019)
21. Gupta, A., Barbu, A.: Parameterized principal component analysis. Pattern Recogn. **78**, 215–227 (2018)
22. Han, D., Li, Y., Song, T., Liu, Z.: Multi-objective optimization of loop closure detection parameters for indoor 2D simultaneous localization and mapping. Sensors **20**(7), 1906 (2020)
23. Haosheng, C., Ge, Z., Yangdong, Y.: Semantic loop closure detection with instance-level inconsistency removal in dynamic industrial scenes. IEEE Trans. Ind. Inf. **17**(3), 2030–2040 (2021)
24. Hervé, J., Matthijs, D., Cordelia, S.: Product quantization for nearest neighbor search. IEEE Trans. Pattern Anal. Mach. Intell. **33**(1), 117–128 (2010)
25. Jégou, H., Douze, M., Schmid, C., Perez, P.: Aggregating local descriptors into a compact image representation. In: Proceedings of the IEEE Computer Society Conference on Computer Vision and Pattern Recognition, pp. 3304–3311, July 2010
26. Klein, G., Murray, D.: Parallel tracking and mapping for small AR workspaces. In: IEEE International Symposium on Mixed and Augmented Reality (ISMAR), pp. 1–10 (2008)
27. Ksibi, S., Mejdoub, M., Amar, C.B.: Deep salient-Gaussian Fisher vector encoding of the spatio-temporal trajectory structures for person re-identification. Multimed. Tools Appl. **78**(2), 1583–1611 (2018). https://doi.org/10.1007/s11042-018-6200-5
28. Kwon, H., Yousef, K.M.A., Kak, A.C.: Building 3D visual maps of interior space with a new hierarchical sensor fusion architecture. Robot. Auton. Syst. **61**(8), 749–767 (2013)
29. Labbe, M., Michaud, F.: Appearance-based loop closure detection for online large-scale and long-term operation. IEEE Trans. Rob. **29**(3), 734–745 (2013)
30. Latif, Y., Cadena, C., Neira, J.: Robust loop closing over time for pose graph SLAM. Int. J. Robot. Res. **32**(14), 1611–1626 (2013)
31. Marr, D., Hildreth, E.: Theory of edge detection. In: Proceedings of the Royal Society of London, vol. 207, pp. 187–217 (1980)
32. Memon, A.R., Wang, H., Hussain, A.: Loop closure detection using supervised and unsupervised deep neural networks for monocular SLAM systems. Robot. Auton. Syst. **126**, 103470 (2020)
33. Mur-Artal, R., Tardós, J.D.: Fast relocalisation and loop closing in keyframe-based slam. In: Proceedings of IEEE International Conference on Robotics and Automation (ICRA), pp. 846–853 (2014)
34. Perronnin, F., Dance, C.R.: Fisher kernels on visual vocabularies for image categorization. In: Proceedings of the IEEE Computer Society Conference on Computer Vision and Pattern Recognition (CVPR), pp. 1–8 (2008)
35. Perronnin, F., Dance, C., Csurka, G., Bressan, M.: Adapted vocabularies for generic visual categorization. In: Leonardis, A., Bischof, H., Pinz, A. (eds.) ECCV 2006. LNCS, vol. 3954, pp. 464–475. Springer, Heidelberg (2006). https://doi.org/10.1007/11744085_36
36. Perronnin, F., Yan, L., Sánchez, J., Poirier, H.: Large-scale image retrieval with compressed fisher vectors. In: Proceedings of the IEEE Computer Society Conference on Computer Vision and Pattern Recognition, pp. 3384–3391 (2010)
37. Pire, T., Fischer, T., Civera, J., Cristóforis, P.D., Berlles, J.J.: Stereo parallel tracking and mapping for robot localization. In: Proceedings of International Conference on Intelligent Robots and Systems (ICIRS) (2015)

38. Rosten, E., Drummond, T.: Machine learning for high-speed corner detection. In: Leonardis, A., Bischof, H., Pinz, A. (eds.) ECCV 2006. LNCS, vol. 3951, pp. 430–443. Springer, Heidelberg (2006). https://doi.org/10.1007/11744023_34
39. Rublee, E., Rabaud, V., Konolige, K., Bradski, G.: ORB: an efficient alternative to sift or SURF. In: International Conference on Computer Vision, pp. 2564–2571 (2012)
40. Safarinejadian, B., Mozaffari, M.: A distributed averaging-based evidential expectation-maximization algorithm for density estimation in unreliable sensor networks. Measurement **165**, 108–162 (2020)
41. Sánchez, J., Perronnin, F., Mensink, T., Verbeek, J.: Image classification with the fisher vector: theory and practice. Int. J. Comput. Vis. **105**(3), 222–245 (2013)
42. Sünderhauf, N., et al.: Place recognition with convnet landmarks: viewpoint-robust, condition-robust, training-free. In: Robotics: Science and Systems, p. 296 (2015)
43. Tonellotto, N., Gotta, A., Nardini, F.M., Gadler, D., Silvestri, F.: Neural network quantization in federated learning at the edge. Inf. Sci. **575**(4), 417–436 (2021)
44. Uchida, Y., Sakazawa, S.: Image retrieval with fisher vectors of binary features. In: Proceedings of IAPR Asian Conference on Pattern Recognition (ACPR), pp. 1–11 (2017)
45. Wang, J., Yang, J., Kai, Y., Lv, F., Huang, T.S., Gong, Y.: Locality-constrained linear coding for image classification. In: Computer Vision & Pattern Recognition, pp. 3360–3367 (2010)
46. Yan, X., Ye, Y., Qiu, X., Yu, H.: Synergetic information bottleneck for joint multi-view and ensemble clustering. Inf. Fusion **56**, 15–27 (2020)
47. Yang, Y., Mémin, E.: Estimation of physical parameters under location uncertainty using an ensemble2-expectation-maximization algorithm. Q. J. R. Meteorol. Soc. **145**(719), 418–433 (2019)
48. Younes, G., Asmar, D., Shammas, E., Zelek, J.: Keyframe-based monocular SLAM: design, survey, and future directions. Robot. Auton. Syst. **98**, 67–88 (2017)
49. Zhou, W., Zhang, L., Gao, S., Lou, X.: Gradient-based feature extraction from raw Bayer pattern images. IEEE Trans. Image Process. **30**, 5122–5137 (2021)
50. Zhu, Z., Xu, X., Liu, X., Jiang, Y.: LFM: a lightweight LCD algorithm based on feature matching between similar key frames. Sensors **21**(13), 4499 (2021)

A Complex Background Image Registration Method Based on the Optical Flow Field Algorithm

Zhentao Liu, Lei Xu$^{(\boxtimes)}$, and Shiyao Jiang

Peng Cheng Laboratory, Shenzhen, China
{liuzht,xul06,jiangshy}@pcl.ac.cn

Abstract. An effective nonrigid image registration method is developed based on the optical flow field (OFF) framework for the complex registration of structure images. In our method, a new force is modeled and integrated into the original optical flow equation to jointly drive the motion direction of pixels. At any point in the offset field, in addition to the force generated by the OFF model derived from local gradient information to drive the pixels in the floating image to infiltrate into the reference pixel set, a new "guiding force" derived from the global grayscale overall trend in a given neighborhood system helps the pixels to more properly spread into the corresponding reference pixel set, particularly when the gradient field of the reference image is unstable. In the experiment, a data set containing several images with complex structures was employed to validate the performance of our registration model. The test results show that our method can quickly and efficiently register complex images and is robust to noise in images.

Keywords: Target tracking · Image registration · Complex background · High noise

1 Introduction

With the development of computer science and artificial intelligence, intelligent processing of images and videos is applied in an increasing number of fields [1]. Due to the wide application of target positioning and tracking technology, the requirements of computers for image processing in different fields also increase [2]. Image registration technology can obtain the change information of the target's position space and shape at any time point during different feature change phases between the image sequences [3]. In the field of target tracking, image registration can be used to reduce the errors in the relative positions of different frames caused by the shaking of the shooting equipment and to obtain the deformation information of the target between different frames [4]. By utilizing different registration strategies, image registration technology is widely applied

This work is supported in part by the National Key Research and Development Program of China under Grant no. 2020YFB1806403.

in numerous target tracking application fields [5]. At present, image registration technology has a relatively complete system and framework and has become increasingly mature. It is mainly divided into rigid-based methods and nonrigid-based methods [6–9]. Rigid-based registration methods utilize rotation and translation-based DOF transformation to map the corresponding location of the reference image to the corresponding location of the moving image. In some fields, there are changes in the target position and its own structure of images with complex structures, which leads to the limitation of rigid image registration in processing images with complex structures. Nonrigid registration can extract different features in the image and reflect the spatial and deformation information of the image at the same time [10], which is a research hot spot in the field of object tracking.

Existing nonrigid-based image registration approaches can be split into two categories: feature-based categories and pixel-based categories. Under the framework of the two categories above, registration methods for different kinds of images have been studied [11–15]. However, the methods under the two above frameworks still have difficulties when dealing with image sequences with complex intensity changes in the background. Feature-based algorithms are effective for handling images with noise by matching the morphology features and local texture features in registered images with the same features in reference images. However, potential problems of feature extraction probably occur during the deformable image matching stage due to the various target attribute changes caused by the stress of image structure deformation. In addition, additional time is required to complete the feature extraction step. Grayscale-based algorithms can directly use pixel information, but grayscale-based approaches, e.g., the sum of squared intensity differences (SSD) or the sum of absolute differences (SAD), are susceptible to pixels with drastic gray value changes, such as noise.

The optical flow field (OFF) is one of the most commonly used frameworks for nonrigid image registration due to its linear computational complexity and accessibility [16]. Many image registration methods based on optical flow fields have been studied [17–19]. In Ref. [17], an 'active force' along with an adaptive force strength adjustment was integrated into the original optical flow equation to accelerate the algorithm during the iterative process. In Ref. [18], the optical flow field algorithm can be generalized to make image-driven locally adaptive regularization that maintains the linear complexity and usability of the original approach simultaneously. Ref. [19] proposed an optical flow field-based nonrigid image registration method for handling medical images. By combining the image orientation field with the traditional optical flow field model to estimate the distortion, the direction information can be better used, and the deformation model can be well simplified. However, these methods are suitable for processing images with homogeneous intensity distributions. For images with complicated structures, the registration may obtain unexpected results since the unstable grayscale changes. In this paper, an image registration method integrating multiple driving forces based on the OFF framework is studied for processing images with complex structures. To address the low signal-to-noise ratio of poor-quality images, the proposed method constructs a new driving force and integrates it into the original OFF equation. The deformation field at any point in the deformation field will generate a guiding force originating from the grayscale variation trend in the local Moore neighborhood system. Based on the guiding force, the proposed method can drive pixels in moving images to infiltrate into the relevant reference images more appropriately, especially when the gradient information of images with complex background changes is unstable. In our experiments, a series of complex images are applied to evaluate the efficiency of the proposed method. Experimental results indicate that the proposed method is effective, robust, real-time and automatic for registering images with complex background changes.

2 The Proposed Method

The traditional optical flow field-based algorithm uses the local gradient information of the pixels in the static image to drive the pixels in the moving image to diffuse into the corresponding static image. In the original optical flow algorithm, the optical flow equation is applied to model the optical flow force. For a given point p of the image, the estimated displacement required to match the corresponding points can be expressed as:

$$|v_c| = \frac{[M(p) - S(p)]\nabla S(p)}{|\nabla S(p)|^2 + [M(p) - S(p)]^2} \tag{1}$$

where $S(p)$ is the intensity value of p in static image S and $M(p)$ is the intensity value of p in image M. $v_c = (v_i, v_j)$ and i, j are the corresponding pixel coordinates of a given point in the static image and M moving image, respectively. In Eq. (1), $\nabla S(p)$ is seen as an "internal force" acquired from the local gradient information in the static image, reflecting the relationship between adjacent pixels in a static image. $M(p) - S(p)$ is seen as an external force used to reflect the interact and influence difference between static and dynamic images. As shown in Fig. 1, in the original optical flow algorithm, the displacement of the pixels will be pulled according to the direction of $\nabla S(p)$ forced by the optical flow field when $S(p) < M(p)$ or driven consistent with the direction of $-\nabla S(p)$ when $S(p) > M(p)$.

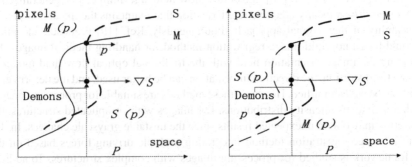

Fig. 1. Working principle of the optical flow algorithm

Equation (1) calculates and estimates the optical flow field force iteratively until the optical flow field converges to a stable state. However, since the original optical flow field force driving the deformation is only taken from the gradient of each pixel in the static image space, the local gradient information in the static image may change when the change trend of the grayscale is seriously unstable due to noise or complex background changes, which may cause errors when processing images with complex background changes based on the original optical flow field algorithm. The situation of the optical flow field when dealing with complex images is shown in Fig. 2.

In Fig. 2(a), the marked pixel $M(p)$ of M can obtain a good position matching result with the relative point in effect space S. The driving force of the original optical field is imposed on point p of the moving image in Fig. 2(b). From Fig. 2(b), we can see that due to unstable gradient variation, point p in effect space M is pushed to an unexpected position. In addition, since the displacement of moving pixels is computed according to the local image information only, a smoothing operation is required to suppress noise and maintain the geometric continuity of the deformed image.

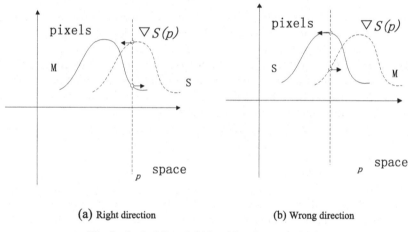

(a) Right direction   (b) Wrong direction

Fig. 2. Optical flow field in addressing complex images

In the original OFF-based algorithm, in each iteration, the optical flow equation needs to be accompanied by a Gaussian filter with a variance σ^2 for regularizing the deformation field. However, when σ^2 is too small, the efficiency of suppressing noise may be low. If σ^2 is too large, the preservation of image detail may not be effective. Previous studies modeled the optical flow field forces by incorporating different types of additional forces into the optical flow equations. However, these studies still have problems in handling image scenes with low signal-to-noise ratios. When gradient operations are performed on local information, the registration process can also be inefficient, i.e., unstable in both static and moving images. To help the moving space to infiltrate into the corresponding static space more efficiently, an additional guiding force is modeled and combined with the original optical flow field equation. The re-estimated optical flow field force is defined as:

$$|v_c'| = \frac{\left[M(p) - S(p)\right] \cdot \nabla S'(p)}{|\nabla S'(p)|^2 + \left[M(p) - S(p)\right]^2} \tag{2}$$

$$\nabla S'(p) = \nabla S^{(t)}(p) + \theta^{(t-1)}(p) \tag{3}$$

$$\theta^{(t)} = \text{mean}\left[\sum_{I=h}^{h+r} S^{(t)}(p_I) - \sum_{I=h-r}^{h} S^{(t)}(p_I)\right] \tag{4}$$

where h is the gray level of point p in position (i, j). $h \pm r$ is the gray level of $(i, j \pm r)$ or $(i \pm r, j)$, where r is the radius and t is the number of iterations. $\theta(p)$ is a guiding force that considers the overall trend of the gray value variation of pixels in a specified effect scope. In Eq. (2), the term $M(p) - S(p)$ is the extrinsic force for controlling the offset of $M(p)$, and $\nabla S'(p)$ is the new internal force obtained by integrating the original internal force $S(p)$ and the guiding force $\theta(p)$.

3   Evaluation Functions

To validate the efficiency of the registration performance, three similarity measurement functions, the sum of squared difference (SSD) and correlation coefficient (CC), are used. The squared difference and normalized mutual information (MI_N) are utilized to statistically evaluate the registration efficiency. SSD is one of the most general similarity measures computed by subtracting pixel intensities in the squared neighborhood between static and warped images. Let S be the pixel set of the static image and D be the pixel set of the corresponding deformed image. The index SSD is defined as:

$$S_{SD} = \frac{\sqrt{\sum_{i=1}^{m}\sum_{j=1}^{n}\left[S(i,j) - D(i,j)\right]^2}}{m \times n} \tag{5}$$

When the ratio of SSD is low, the similar difference registered between deformed images and static images is small. The correlation coefficient is utilized to count the linear correlation of the matching degree between images S and D, representing the extension of the overlapping area between two corresponding images. If two images are more similar, the value of the correlation coefficient is closer to 1, meaning that the registration result is better. CC is defined as:

$$C_C = \frac{\sum_{S(i,j)}\sum_{D(i,j)}\left[S(i,j) - \overline{S}(i,j)\right]\left[D(i,j) - \overline{D}(i,j)\right]}{\sqrt{\sum_{S(i,j)}\left[S(i,j) - \overline{S}(i,j)\right]^2} \cdot \sqrt{\sum_{D(i,j)}\left[D(i,j) - \overline{D}(i,j)\right]^2}} \tag{6}$$

where $\overline{S}(i,j)$ represents the mathematical expectation of $S(i,j)$ and $\overline{D}(i,j)$ represents the mathematical expectation of $D(i,j)$. Normalized mutual information can qualitatively measure how well one image matches another by registering the joint histogram of the grayscale distribution between two images. Normalized mutual information is defined as:

$$M_I = H(S) + H(D) - H(S, D) = \sum_{s,d} p(s, d) \log \frac{p(s, d)}{p(s) \cdot p(d)} \tag{7}$$

where $p(s)$ is the entropy of image S, $p(d)$ is the entropy of image D, and $p(s, d)$ is the joint entropy of image S and image D. In this paper, the mutual information MI_N is normalized to facilitate the data statistics.

4 Experimental Results

To evaluate the effect of the proposed optical flow field diffusion model, the proposed method is first applied to a group of aerial images with complex structures captured from an aerial video with 10 fps. The registration performance of our method is compared with the registration performance of the original optical flow field registration method and the active force-based optical flow field registration (AFO) method [17]. The experimental results are shown in Fig. 3.

Figure 3(a) is the original image captured from the aerial video at the inception, and the relative reference image captured from the same video after 0.5 s is shown in Fig. 3(b). The registration result is shown in Fig. 3(c) by using the original optical flow field model. It can be seen from Fig. 3(c) shows that many mosaic-like speckle areas are generated in the result, and the registration result appears to be not smooth. This is because the estimated optical flow force of the original optical flow field model originated from the local gradient information of individual pixels. Due to noise in natural scene images with complex background changes, the moving direction of moving objects is confused because the gradient change trend on both sides of the noise point is the same. In this situation, the moving objects may be driven in opposite directions toward the corresponding static objects, leading to trivial regions or mosaics being wrongly generated.

The registration result generated by using the AFO method is shown in Fig. 3(d). The mosaic-like speckle areas are greatly reduced. The registration result generated by using the integrated driving force of the optical flow field is shown in Fig. 3(e). Compared with Fig. 3(d), the registration performance of the AFO method appears smoother than that of the proposed method. However, some regions in the registration result of the AFO method do not match well with the reference image, while the details of the original image and the geometric continuity of the deformed image are well reserved by the proposed method. This is because by applying the guiding force, the unsteady force at speckle noise points can be corrected according to the global grayscale trend during the iterative process. Therefore, the proposed method is insensitive to noise. Without utilizing any image smoothing operation, the image details can be reserved well. In addition, since the optical flow field force is integrated by the local image information and the global intensity trend information at any point in the deformation field, the geometric continuity of the original image can also be reserved.

To statistically verify the registration performance of different methods, for each instance, we calculate the SSD between image S and image M: S_{SDsm} and the SSD between image S and image D: S_{SDdm} generated by using different registration approaches. A normalized threshold is used as the similarity criterion, i.e., the similarity between S_{SDsm} and S_{SDdm} is calculated as $P_{\text{SD}} = S_{\text{SDsd}}/S_{\text{SDsm}}$. P_{SD} closer to 0 indicates that the difference between the registered image and the static image is smaller than that of the difference between the moving image and the static image, implying that the registration is effective.

(a) Moving image   (b) Reference image

(c) Result of the original OFF method   (d) Result of the AFO method

(e) Results of the proposed method

Fig. 3. The registration results of different methods

For the images in the database, the mean P_{SD} values of the original optical flow method and the proposed method are 1.02 and 0.57, respectively. This indicates that the result obtained by the original OFF registration method has a large deviation compared with the moving image. Compared with the traditional OFF registration method, the proposed method has a lower P_{SD} value, indicating that the deformation result of the proposed method is better. To evaluate the overall performance of different methods, the registration model based on the original OFF method and our method are applied to the data set with the same test samples. The statistical results with three validating functions are shown in Table 1. Table 1 shows that the SSD value of our method is lower, and the values of CC and MI_N are simultaneously higher than those of the original OFF method. This demonstrates that the registration result of our method is more accurate.

Table 1. Table captions should be placed above the tables.

Methods	SSD	CC	MI_N
Original OFF	0.87	0.74	0.68
AFO	0.56	0.82	0.75
The proposed method	0.45	0.92	0.82

The registration process of the original OFF method and the proposed method are executed on a computer with a CPU of 4 cores and a memory of 8G RAM. The computational time for performing nonrigid image deformation depends on the image size and the similarity degree between the moving image and the corresponding static image. Compared to the OFF method, the proposed method takes less time to acquire the deformation because the proposed method does not use any smoothing operation to suppress noise during each iteration.

5 Conclusion

To improve the efficiency of deformable image registration with complicated structures and signal-to-noise ratios, an effective deformable image registration technique for images is proposed. To address low signal-to-noise ratios, the proposed method adds a new guiding force to the original optical flow field equation to drive pixels in moving images to diffuse into corresponding reference image objects more effectively. Compared with the original OFF registration methods, the proposed method can deform the images more effectively by introducing the new modeled force.

References

1. Zhang, Y., Liu, Y., Cheng, H., Li, Z., Liu, C.: Fully multi-target segmentation for breast ultrasound image based on fully convolutional network. Med. Biol. Eng. Comput. **58**(9), 2049–2061 (2020). https://doi.org/10.1007/s11517-020-02200-1
2. Liu, Y., Chen, Y., Han, B., Zhang, Y.T., Zhang, X.T., Su, Y.X.: Fully automatic Breast ultrasound image segmentation based on fuzzy cellular automata framework. Biomed. Signal Process. Control **40**, 433–442 (2018)

3. Zitova, B., Flusser, J.: Image registration methods: a survey. Image Vis. Comput. **21**(11), 977–1000 (2003)
4. Kashyap, M., Bhattacharya, M.: Medical image registration using extremal region based interest points. In: IEEE International Conference on Computer Graphics. IEEE (2016)
5. Alam, F., Rahman, S.U.: Intrinsic registration techniques for medical images: a state-of-the-art review. JPMI: J. Postgrad. Med. Inst. **30**(2), 119–132 (2016)
6. Jianchao, Y.: Image registration based on both feature and intensity matching. In: IEEE International Conference on Acoustics (2001)
7. Fookes, C.B., Bennamoun, M.: Rigid and nonrigid image registration and its association with mutual information: a review. Image Registration (2002)
8. Alam, F., Rahman, S.U., Hassan, M., et al.: An investigation toward issues and challenges in medical image registration. J. Postgrad. Med. Inst. **31**(3), 202–214 (2017)
9. Wu, Y., Ma, W., Gong, M., et al.: A novel point-matching algorithm based on fast sample consensus for image registration. IEEE Geosci. Remote Sens. Lett. **12**(1), 43–47 (2017)
10. Schnabel, J.A., Tanner, C., Castellano-Smith, A.D., et al.: Validation of nonrigid image registration using finite-element methods: application to breast MRI images. In: IEEE Trans. Med. Imaging **22**(5), 238–247 (2003)
11. Chen, H.M., Varshney, P.K.: Mutual information-based CT-MR brain image registration using generalized partial volume joint histogram estimation. IEEE Trans. Med. Imaging **22**(9), 1111–1119 (2003)
12. Hill, D.L., Batchelor, P.G., Holden, M., et al.: Medical image registration. Phys. Med. Biol. **46**, R1–R45 (2001)
13. Chen, C.J., Chang, R.F., Moon, W.K., Chen, D.R., et al.: 2-D ultrasound strain images for breast cancer diagnosis using nonrigid subregion registration. Ultrasound Med. Biol. **32**(6), 837–846 (2006)
14. Pluim, J.P.W., Maintz, J.B.A., Viergever, M.A.: Mutual-information-based registration of medical images: a survey. IEEE Trans. Med. Imaging **22**(8), 986–1004 (2003)
15. Yang, D.S., Li, H., Low, D.A., et al.: A fast inverse consistent deformable image registration method based on symmetric optical flow computation. Phys. Med. Biol. **53**, 6143–6165 (2008)
16. Liu, Y., Cheng, H.D., Huang, J.H., et al.: An effective nonrigid registration approach for ultrasound image based on Demons algorithm. J. Digit. Imaging **26**(3), 521–529 (2013)
17. Wang, H., Dong, L., Daniel, D.J., et al.: Validation of an accelerated 'Demons' algorithm for deformable image registration in radiation therapy. Phys. Med. Biol. **50**, 2887–2905 (2005)
18. Cahill, N.D., Noble, J.A., Hawkes, D.J.: A Demons algorithm for image registration with locally adaptive regularization. In: Yang, G.Z., Hawkes, D., Rueckert, D., Noble, A., Taylor, C. (eds.) MICCAI 2009. LNCS, vol. 5761, pp. 574–581. Springer, Heidelberg (2009). https://doi.org/10.1007/978-3-642-04268-3_71
19. Lan, S., Guo, Z., You, J.: Nonrigid medical image registration using image field in Demons algorithm. Pattern Recognit. Lett. **125**, 98–104 (2019)

Collaborative Learning Method for Natural Image Captioning

Rongzhao Wang and Libo Liu[✉]

School of Information Engineering, Ningxia University, Yinchuan, China
liulib@nxu.edu.cn

Abstract. We propose a collaborative learning method to solve the natural image captioning problem. Numerous existing methods use pretrained image classification CNNs to obtain feature representations for image caption generation, which ignores the gap in image feature representations between different computer vision tasks. To address this problem, our method aims to utilize the similarity between image caption and pix-to-pix inverting tasks to ease the feature representation gap. Specifically, our framework consists of two modules: 1) The pix2pix module (P2PM), which has a share learning feature extractor to extract feature representations and a U-net architecture to encode the image to latent code and then decodes them to the original image. 2) The natural language generation module (NLGM) generates descriptions from feature representations extracted by P2PM. Consequently, the feature representations and generated image captions are improved during the collaborative learning process. The experimental results on the MSCOCO 2017 dataset prove the effectiveness of our approach compared to other comparison methods.

Keywords: Image captioning · Pix2pix inverting · Collaborative learning

1 Introduction

The goal of image captioning is to generate a meaningful sentence from a given image. In recent years, increasing attention has been given to vision-language tasks, which involve two domains: computer vision [10] and natural language processing [6]. Image captioning is one of the vision-language tasks, as it needs to obtain the global and local entities in the scene and their attributes and spatial and logical relationships to generate natural language descriptions, which makes it a very challenging task. Recently, image captioning also has some applications, such as aiding visually impaired people in perceiving information from the scene and helping people manage massive video and image information produced by the internet. Furthermore, image captioning is also crucial for artificial intelligence to interact with surroundings.

Numerous image caption frameworks have been proposed and can be classified into template-based [9,15,16], retrieval-based and neural-based [26,30]. With the rapid

This work was supported by grant of no.61862050 from the National Nature Science Foundation of China and no.2020AAC03031 from Natural Science Foundation of Ningxia, China.

ⓒ The Author(s), under exclusive license to Springer Nature Singapore Pte Ltd. 2022
Y. Wang et al. (Eds.): ICPCSEE 2022, CCIS 1628, pp. 249–261, 2022.
https://doi.org/10.1007/978-981-19-5194-7_19

growth of computer vision and natural language processing, an increasing number of neural-based methods have been proposed and have achieved remarkable performance. These frameworks treat image captioning as a translation task from vision to text rather than from one language to another. Inspired by machine translation frameworks, pretrained classification CNNs are used for extracting a tensor as a feature representation from an image, and language models such as LSTM [12] and GRU [7] are used for decoding the vector and generating meaningful sentences.

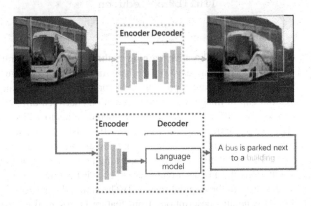

Fig. 1. Illustration of the commonalities between pix2pix inverting and encoder-decoder-based image caption methods. These two methods both need to extract feature representations from a given image and then upsample them to the corresponding image or decode them to a meaningful language description, which includes both object and spatial information.

Although these neural-based "encoder-decoder" frameworks [26,29,30,33] use pretrained image classification CNNs, for example, GoogLeNet [24], VGG [23] and Dense-Net [13] as their encoder, and have achieved impressive performance, there is a gap in representations between vision and language tasks: the features obtained from CNNs are not suitable for image captioning task for some reason. For example, the classification model pays more attention to differences in the features extracted from different images because they need to distinguish them, but the common features between different images are also important to the image caption task (different objects may share the same actions, attributes and relationships). Furthermore, the spatial information that is crucial for caption generation may be lost during the feature extraction process. As a result, the feature representation gap between the image caption task and classification task should be taken into account.

We suppose that a method that can extract feature representations better for the image caption task may bring further performance improvement. Therefore, we construct a framework using a collaborative learning method to ease the feature representation gap between the image caption and classification tasks. Our proposed model combines a pix2pix module and a natural language generation module. The feature extracted by the pix2pix module is used for generating both natural language descriptions and regenerating the original image. Our main idea is that the image caption task and pix2pix inversion task both aim to decode the features extracted from the original image (as shown in Fig. 1). Meanwhile, the language description and image generated from these two tasks have common semantics and spatial information. To this end, the quality of image captioning could be improved during the collaborative learning process.

Specifically, the main works of this paper are summarized below:

1. We propose an image captioning framework that uses a collaborative learning method to achieve feature representations with rich semantics and information. Compared with other conventional encoder-decoder-based frameworks, we take into account the feature representation gap between the classification task and image caption task so that the image captioning can be improved during the collaborative learning process.
2. We utilize a pix2pix module that has a U-Net [21] architecture in our method. We modified it to extract feature representations more suitable for the image caption task and our collaborative learning framework. The experimental results qualitatively and quantitatively demonstrate the efficiency of our method.

2 Related Work

Image captioning always treats themselves as a translation task and uses a language model to generate natural language descriptions [20, 26, 30]. Different from machine translation, image captioning requires an extra feature extraction module to obtain a feature vector from a given image. The feature vector should preserve sufficient information that makes the model generate meaningful sentences. However, having the ability to gain an effective feature vector and translate it to natural language description is difficult. Numerous existing encoder-decoder methods only use a pretrained classification model to extract image features, which is insufficient. Recently, some approaches, e.g., Bottom-Up, VLP, Oscar and PTBD [2, 17, 28, 34], have used detectors to preextract region features of salient objects, even bounding boxes and object tags, which further improves the quality of the generated captions. However, these approaches are difficult to train end-to-end due to their preextraction process, and more labeled information is needed. In contrast, our method extracts image features during the training step and only uses the image data and their corresponding descriptions as input.

Collaborative learning approaches construct share modules to achieve representations that are more informative and have been extensively employed in deep learning areas, such as image classification [4, 31, 35], unsupervised domain adaptation [32] and keyframe selection [22]. These methods excavate the same concept from different environments by training an unsupervised model, but they only learn semantic features. Guan et al. [11] proposed a collaborative learning framework for application in the stylegan embedding task, which requires pixel-level inversion. However, as far as we know, a collaborative learning method based on image captioning and pix2pix inverting tasks has not been used for tackling the image captioning task. Instead, our framework first uses a collaborative learning method that introduces a pix2pix inverting module to improve image captioning. Therefore, our method can boost the extracted feature representations during the training stage and can be trained more efficiently.

3 Method

The aims of our approach are to obtain more informative feature representations and generate descriptions with richer semantics from an image. To this end, we construct a collaborative learning framework, which is detailed in Fig. 2. This framework consists of two parts: a pix2pix inverting module and a natural language generation module. Given a real image, the feature extraction part of the pix2pix inverting module extracts the feature vector, which is used for regenerating the original image and decoding to the language description.

In this section, we introduce the pix2pix inverting module first (Sect. 3.1) and then describe the natural language generation module (Sect. 3.2). Finally, we illustrate our collaborative learning framework and detail its final objective function (Sect. 3.3).

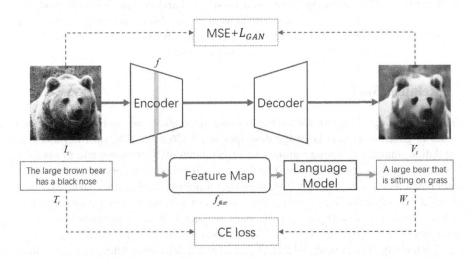

Fig. 2. The main structure of our collaborative learning framework for image captioning. The encoder obtains the latent code and feature map from the given image first, which are sent to the remaining parts of U-Net and a language model to regenerate the original image and language description, respectively. The generated image and language description supervise the optimization of the encoder and the language model using MSE, adversarial and cross-entropy loss. The blue arrow indicates the pipeline of P2PM, and the orange arrow indicates the pipeline of NLGM.

3.1 P2PM: Pix2Pix Inverting Module

The P2PM in our framework has a U-net architecture similar to that of the pix2pix framework in [14] but has some differences: 1) We use a pretrained image classification CNN as the feature extraction part of the P2PM, which is shown in the orange dashed box of Fig. 3. We utilize lower-level feature representations from classification CNNs such as the work in [30] to learn a correspondence between the features and their region positions. 2) We only preserve part of the skip connections because the first goal of our P2PM is to invert the given image to the original image, rather than to another style image or image segmentation, and to make the feature extraction process preserve more informative features. Overmuch skip connections will make our decoder obtain excessive semantic and spatial information from the mirror downsampling layers, which will reduce the feature extraction ability of the encoder. Therefore, through preliminary experiments we preserve three skip connections in our U-net architecture.

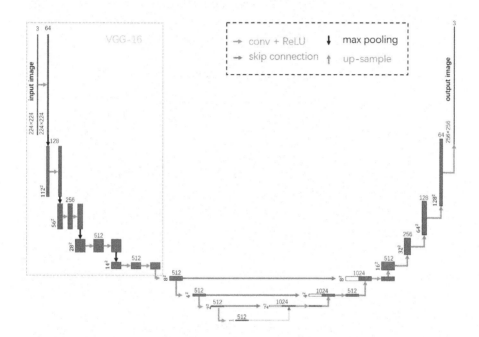

Fig. 3. Illustration of the details of P2PM, which includes a feature extractor, an encoder and a decoder. Each box represents a multichannel feature map. White boxes represent feature maps copied from the mirror downsampling layers. We replace the first five layers of U-Net with pretrained VGG-16 as a feature extractor (conv_1 to conv_5 in the orange dashed box) and only build three skip connections for our P2PM.

As shown in Fig. 3, our P2PM is a U-net architecture. To implement U-net in our collaborative learning framework, we utilize the tensors from conv5_3 in the VGG-16 model to better obtain representations with more semantic and spatial information. Given an image I, the feature f_{feat} is defined as:

$$f_{feat} = E_{feat}(I; \theta), \tag{1}$$

where θ are the parameters of E_{feat}(conv_1 to conv_5 of VGG-16). After the feature map extraction process, f_{feat} is sent to U-net to downsample to the latent code and upsample to the original image. Specifically,

$$l = U_E(f_{feat}),$$
$$I_g = U_D(l), \tag{2}$$

where U_E is the encoder of U-net used to downsample the feature map. U_D is the decoder of U-net to upsample the latent code. l denotes the latent code, and I_g denotes the regenerated image.

The feature map will also be sent to the NLGM to generate a language description, which will be introduced in Sect. 3.2. However, different from other pix2pix frameworks, our aim is to regenerate the original image, so we only preserve three skip connections in U-Net to avoid the problem introduced in Sect. 3.1 paragraph 1.

3.2 NLGM: Natural Language Generation Module

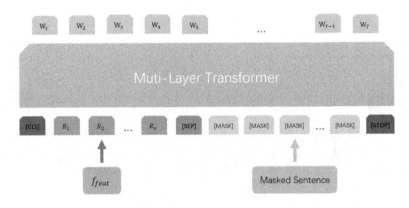

Fig. 4. Natural language generation module

We will illustrate the details of our natural language generation module below. To model the image caption generation, we build a transformer that uses pretrained variants: Unified VLP [34] as the initialization of our language model. The feature map f_{feat} that is extracted from VGG-16 is denoted as different image regions:

$$R = [R_1, ..., R_N] \in \mathbb{R}^{e \times N}, \tag{3}$$

where R_i denotes the region feature,

$$G = [P_1, ..., P_N] \in \mathbb{R}^{c \times N}, \tag{4}$$

where P_i denotes the position information. e denotes the embedding size, and $c = 5$ are five values that consist of the coordinates of each area (we treat the image as 98 different parts normalized between 0 and 1) and the ratio of the region area to the whole image. We set these values to the same between 0 and 1 because these regions have the same area. The descriptions are denoted as one-hot vectors S, which will be embedded with size s : $y_t \in \mathbb{R}(t \in 1, 2, ..., T)$, where T is the number of words. All the settings are the same as VLP except for the region object labels because we do not use an object detector to preextract the object tags. We replace it with a vector C. Masked Token Loss (MTL) is used for VLP pretraining,

$$\mathcal{L}_{MTL} = -\mathbb{E}_{(v,h)\sim\mathcal{D}} \log p(h_i \mid h_{\backslash i}, v), \tag{5}$$

which is similar to BERT. h is the discrete token sequence. h_i is the masked token with 15% probability. $h_{\backslash i}$ is the nonmasked token. At each iteration, h_i is predicted based on $h_{\backslash i}$ and image features v by minimizing the negative log-likelihood. We utilize the seq2seq objective for the training stage of our natural language generation module, which can be seen in Fig. 4.

3.3 Collaborative Learning Framework

From Fig. 2, we can see that given an image I, the encoder extracts its feature map and latent code. The latent code is upsampled to image I itself, and the feature map is translated to a natural language description. We use cross entropy loss in the natural language generation part. Specifically, the loss function is detailed below:

$$\mathcal{L}_{CE} = -\sum_{t=0}^{L-1} \log p_t(T_t), \tag{6}$$

The regenerated image from the U-net decoder supervises the encoder network and natural language generation module by using adversarial loss:

$$\mathcal{L}_{cGAN}(G, D) = \mathbb{E}_{x,y}[\log(D(x, y))] \\ + \mathbb{E}_{x,z}[\log(1 - D(x, G(x, z)))], \tag{7}$$

where G aims to minimize this objective, and D tries to maximize it:

$$G^* = arg \min_G \max_D \mathcal{L}_{cGAN}(G, D), \tag{8}$$

and MSE loss:

$$\mathcal{L}_{MSE} = \|I_i - V_i\|_2^2 \tag{9}$$

where I_i and V_i are the given image and reconstructed image, respectively. In summary, the total loss of the collaborative learning framework is:

$$\mathcal{L} = \mathcal{L}_{CE} + \beta_1 G^* + \beta_2 \mathcal{L}_{MSE} \tag{10}$$

where β_1 and β_2 are hyperparameters to control the contribution of cross entropy, adversarial loss and MSE loss.

4 Experiments

In this section, we set up a number of experiments to demonstrate the efficiency of our proposed method. Microsoft COCO caption evaluation tools are used for calculating the BLEU [19], METEOR [3], ROUGE [18], CIDEr [25] and SPICE [1] metrics. We will first summarize the dataset, metrics and implementation details in the experiments. Then, we compare and analyze the results of our image caption model.

4.1   Experimental Setup

Dataset. We set up our experiments on the MSCOCO 2017 dataset [5], which has 118,287 and 5000 images for training and validation, respectively. Each image is paired with 5 human annotated captions.

Evaluation Metrics. In addition, referring to the evaluation metrics of existing image caption methods, we choose many evaluation metrics to measure the effectiveness of our model, *e.g.*, BLEU, METEOR, ROUGE, CIDEr and SPICE. Bilingual evaluation understudy (BLEU) is a commonly used metric for the evaluation of machine translation that is based on n-gram precision. METEOR is based on the harmonic mean of unigram precision and recall with which the recall is weighted higher than precision. Different from the BLEU metric, the METEOR seeks correlation at the corpus level. ROUGE is a recall-oriented metric based on n-grams, word sequences, and word pairs. It mainly measures the recall of the generated description relative to the ground truth description. CIDEr is designed for evaluating image descriptions using human consensus, and it can more objectively evaluate the similarity between the predicted description and ground truth description. SPICE is a metric for image captioning tasks to solve the insufficiency of n-gram-based evaluation metrics.

Implementation Details. We utilize the pretrained VGG-16 to extract the feature map. Each feature map is represented as an $14 \times 14 \times 512$ dimensional tensor. Our language model is initialized with Unified VLP [34], which uses BERT-base [8] as the transformer backbone ($H = 768$), where H is the hidden size of the pretrained BERT-base model. The feature map f_{feat} is sent to the language model as 98 image regions to generate a natural language description. We transform the 196×512 feature map to the 98×1024 tensor as 98 region features because 98 is near 100, which is the number of inputs in the baseline image caption module we used in this paper. Meanwhile, f_{feat} is also sent to U-Net with only three skip connections and regenerates the original image with a size of 256×256.

4.2   Main Results

In this section, we present both qualitative and quantitative experiments to analyze our method with other comparison methods to verify the effectiveness of our collaborative learning framework. Our method is compared with a classic encoder-decoder image caption method [30] and some newly proposed methods [2,17,27,33,34,36]. All these methods use the same dataset split and are optimized based on cross-entropy loss.

The evaluation metrics of our framework and other comparison methods are shown in Table 1. Our framework outperforms other methods in most of the evaluation metrics on the MSCOCO dataset. Specifically, Bleu-4: 37.0, METEOR: 31.3, ROUGE_L: 58.2, CIDEr: 125.6. The results show the effectiveness of our collaborative learning method for natural image captioning.

We also use an ablation study to prove the efficiency of different losses in Table 2. We compare $\mathcal{L}_{CE}$, $\mathcal{L}_{CE} + G^*$ and $\mathcal{L}_{CE} + G^* + \mathcal{L}_{MSE}$. When we use $\mathcal{L}_{CE} + G^*$

Table 1. Experimental results on the MSCOCO2017 dataset.

Method	Bleu-4	METEOR	ROUGE_L	CIDEr	SPICE
Hard Attention [30]	24.8	23.8	52.1	94.2	18.6
CL-HRA [33]	35.8	27.1	–	109.8	–
Up-Down [2]	36.2	27.0	56.4	113.5	20.3
UP-Down+SG+RWS+WR [27]	36.6	28.0	56.9	116.9	21.3
XE Pre-Train+POS-SCAN [36]	36.5	27.9	–	114.9	20.8
Unified VLP [34]	36.5	28.4	–	116.9	21.2
Oscar [17]	36.5	30.3	–	125.3	**23.1**
Ours	**37.0**	**31.3**	**58.2**	**125.6**	23.0

$+ \mathcal{L}_{MSE}$, our method achieves the best performance. We suppose that the $\mathcal{L}_{CE}$ loss and the G^* loss make the encoder extract feature representations with rich semantic and spatial information. In addition, they boost the image caption by contrasting the pixelwise visual similarity between the given image and reconstructed image.

Table 2. Ablation experiments on the MSCOCO2017 dataset.

Different loss	Bleu-1	Bleu-2	Bleu-3	Bleu-4	METEOR	ROUGE_L	CIDEr	SPICE
$\mathcal{L}_{CE}$	74.8	59.0	45.8	35.5	29.8	57.6	120.5	22.6
$\mathcal{L}_{CE} + G^*$	75.7	59.9	46.3	35.8	30.3	57.8	120.8	22.7
$\mathcal{L}_{CE} + G^* + \mathcal{L}_{MSE}$	**77.3**	**62.0**	**48.2**	**37.0**	**31.3**	**58.2**	**125.6**	**23.0**

Moreover, we test the effect of different β_1 and β_2 in the final objective function. We set β_1 and β_2 to different values to balance the contribution of different loss functions in the experiments because during the training stage, the values of adversarial loss and MSE loss are in different ranges compared to cross entropy loss. Table 3 shows the impact of different β_1 and β_2.

Table 3. The effects of different values of β_1 and β_2.

β_1, β_2	Bleu-1	Bleu-2	Bleu-3	Bleu-4	METEOR	ROUGE_L	CIDEr	SPICE
$\beta_1 = 0.01, \beta_2 = 0.001$	74.5	58.0	44.3	35.0	28.8	57.1	119.4	22.5
$\beta_1 = 0.05, \beta_2 = 0.005$	75.5	60.1	45.7	35.4	30.1	57.5	122.9	22.8
$\beta_1 = 0.1, \beta_2 = 0.01$	**77.3**	**62.0**	**48.2**	**37.0**	**31.3**	**58.2**	**125.6**	**23.0**

Figure 5 lists some captions from some kinds of image caption methods and the corresponding given annotations. We find that for some simple background images, such as the second and forth images, our method and the comparison methods can all generate relevant and accurate descriptions. However, for some complex background images, such as the first and third images, our method can generate more accurate and informative image captions and always concentrates on the salient contents in the scene. Specifically, compared to the caption "a group of people standing next to each other" and "a man sitting next to a cow holding an umbrella" generated by Hard Attention and Oscar, "a young boy holding an umbrella next to a herd of cows" generated by our framework is more specific in some details and concentrates on the salient objects.

Images	Generated Captions	Ground Truth
	**Hard Attention:** a group of people standing next to each other. **Oscar:** a man standing next to a cow holding an umbrella. **Ours:** a young boy holding an umbrella next to a herd of cows.	① A child holding a flowered umbrella and petting a yak. ② A young man holding an umbrella next to a herd of cattle. ③ a young boy barefoot holding an umbrella touching the horn of a cow. ④ A young boy with an umbrella who is touching the horn of a cow. ⑤ A boy holding an umbrella while standing next to livestock.
	**Hard Attention:** a man riding skis down a snow covered slope. **Oscar:** a man riding skis down a snow covered slope. **Ours:** a man riding skis down a snow covered slope.	① A person skiing in an open area of snow. ② An older man is skiing down a snowy mountain. ③ An older man downhill skiing down a slope. ④ A man skiing down a snowy hill alone. ⑤ A man wearing ski equipment on a snowy slope.
	**Hard Attention:** a man sitting on a skate-board on a side walk. **Oscar:** a white airplane sitting on top of an airport tarmac. **Ours:** a large airplane parked on the runway at an airport.	① An airplane sits on the tarmac of an airport, with a disconnected boarding gate. ② Plane boarding passengers while at a fancy airport. ③ A plane sitting on a runway getting ready to be emptied. ④ This is a airplane on the runway of the airport. ⑤ A plan parked on the cement near a terminal.
	**Hard Attention:** a small airplane flying through the air. **Oscar :** a red with white plane flying in the sky. **Ours:** a red and white plane flying in the sky.	① A small plane flying through a cloudy blue sky. ② Red and white small plane on partially cloudy day. ③ A red and white biplane in a blue, cloudy sky. ④ A plane that is flying in the air. ⑤ A small lightweight airplane flying through the sky.

Fig. 5. Caption generation results on the MSCOCO dataset.

5   Conclusion

We constructed a collaborative learning framework for image captioning in this paper. Unlike the works before, our method aims to utilize the similarity between image caption and pix-to-pix inverting tasks to ease the feature representation gap. Specifically, the framework in this paper consists of two modules: a pix-to-pix module (P2PM) and a natural language generation module (NLGM). The first module encodes the image into feature representations and then decodes them into the original image. The second module generates descriptions from feature representations extracted by P2PM. Consequently, the feature representations are improved during the collaborative learning process. The experimental results prove the effectiveness of our collaborative learning framework over other comparison methods.

Our motivation is that the image caption task and pix2pix inverting task both aim to decode the representations obtained from the given image. Therefore, the natural language descriptions and images generated from these two tasks have common semantics and spatial information. As a result, the quality of image captioning could be improved during the collaborative learning process. However, our method is mainly optimized based on cross-entropy loss, so we only compared our method with methods optimized based on cross-entropy loss. Therefore, the self-critical sequence training method [20] should be considered in future works.

References

1. Anderson, P., Fernando, B., Johnson, M., Gould, S.: SPICE: semantic propositional image caption evaluation. In: Leibe, B., Matas, J., Sebe, N., Welling, M. (eds.) ECCV 2016. LNCS, vol. 9909, pp. 382–398. Springer, Cham (2016). https://doi.org/10.1007/978-3-319-46454-1_24

2. Anderson, P., et al.: Bottom-up and top-down attention for image captioning and visual question answering. In: Proceedings of the IEEE Conference on Computer Vision and Pattern Recognition (CVPR), June 2018

3. Banerjee, S., Lavie, A.: METEOR: an automatic metric for MT evaluation with improved correlation with human judgments. In: Proceedings of the ACL Workshop on Intrinsic and Extrinsic Evaluation Measures for Machine Translation and/or Summarization, pp. 65–72 (2005)

4. Chen, T., Kornblith, S., Norouzi, M., Hinton, G.: A simple framework for contrastive learning of visual representations. In: III, H.D., Singh, A. (eds.) Proceedings of the 37th International Conference on Machine Learning. Proceedings of Machine Learning Research, vol. 119, pp. 1597–1607. PMLR, 13–18 July 2020. https://proceedings.mlr.press/v119/chen20j.html

5. Chen, X., et al.: Microsoft COCO captions: data collection and evaluation server. arXiv preprint arXiv:1504.00325 (2015)

6. Chowdhury, G.G.: Natural language processing. Ann. Rev. Inf. Sci. Technol. **37**(1), 51–89 (2003)

7. Chung, J., Gulcehre, C., Cho, K., Bengio, Y.: Empirical evaluation of gated recurrent neural networks on sequence modeling. In: NIPS 2014 Workshop on Deep Learning, December 2014 (2014)

8. Devlin, J., Chang, M.W., Lee, K., Toutanova, K.: BERT: pre-training of deep bidirectional transformers for language understanding. arXiv preprint arXiv:1810.04805 (2018)

9. Farhadi, A., et al.: Every picture tells a story: generating sentences from images. In: Daniilidis, K., Maragos, P., Paragios, N. (eds.) ECCV 2010. LNCS, vol. 6314, pp. 15–29. Springer, Heidelberg (2010). https://doi.org/10.1007/978-3-642-15561-1_2

10. Forsyth, D.A., Ponce, J.: Computer Vision: A Modern Approach. Pearson, London (2012)

11. Guan, S., Tai, Y., Ni, B., Zhu, F., Huang, F., Yang, X.: Collaborative learning for faster stylegan embedding. arXiv preprint arXiv:2007.01758 (2020)

12. Hochreiter, S., Schmidhuber, J.: Long short-term memory. Neural Comput. **9**(8), 1735–1780 (1997)

13. Huang, G., Liu, Z., Van Der Maaten, L., Weinberger, K.Q.: Densely connected convolutional networks. In: Proceedings of the IEEE Conference on Computer Vision and Pattern Recognition, pp. 4700–4708 (2017)

14. Isola, P., Zhu, J.Y., Zhou, T., Efros, A.A.: Image-to-image translation with conditional adversarial networks. In: Proceedings of the IEEE Conference on Computer Vision and Pattern Recognition (CVPR), July 2017
15. Kulkarni, G., et al.: Baby talk: understanding and generating simple image descriptions. In: CVPR 2011, pp. 1601–1608 (2011). https://doi.org/10.1109/CVPR.2011.5995466
16. Li, S., Kulkarni, G., Berg, T., Berg, A., Choi, Y.: Composing simple image descriptions using web-scale n-grams. In: Proceedings of the Fifteenth Conference on Computational Natural Language Learning, pp. 220–228 (2011)
17. Li, X., et al.: OSCAR: object-semantics aligned pre-training for vision-language tasks. In: Vedaldi, A., Bischof, H., Brox, T., Frahm, J.-M. (eds.) ECCV 2020. LNCS, vol. 12375, pp. 121–137. Springer, Cham (2020). https://doi.org/10.1007/978-3-030-58577-8_8
18. Lin, C.Y.: ROUGE: a package for automatic evaluation of summaries. In: Text Summarization Branches Out, pp. 74–81. Association for Computational Linguistics, Barcelona, July 2004. https://www.aclweb.org/anthology/W04-1013
19. Papineni, K., Roukos, S., Ward, T., Zhu, W.J.: BLEU: a method for automatic evaluation of machine translation. In: Proceedings of the 40th Annual Meeting of the Association for Computational Linguistics, pp. 311–318 (2002)
20. Rennie, S.J., Marcheret, E., Mroueh, Y., Ross, J., Goel, V.: Self-critical sequence training for image captioning. In: Proceedings of the IEEE Conference on Computer Vision and Pattern Recognition (CVPR), July 2017
21. Ronneberger, O., Fischer, P., Brox, T.: U-net: convolutional networks for biomedical image segmentation. In: Navab, N., Hornegger, J., Wells, W.M., Frangi, A.F. (eds.) MICCAI 2015. LNCS, vol. 9351, pp. 234–241. Springer, Cham (2015). https://doi.org/10.1007/978-3-319-24574-4_28
22. Sheng, L., Xu, D., Ouyang, W., Wang, X.: Unsupervised collaborative learning of keyframe detection and visual odometry towards monocular deep SLAM. In: Proceedings of the IEEE/CVF International Conference on Computer Vision (ICCV), October 2019
23. Simonyan, K., Zisserman, A.: Very deep convolutional networks for large-scale image recognition. arXiv preprint arXiv:1409.1556 (2014)
24. Szegedy, C., et al.: Going deeper with convolutions. In: Proceedings of the IEEE Conference on Computer Vision and Pattern Recognition, pp. 1–9 (2015)
25. Vedantam, R., Lawrence Zitnick, C., Parikh, D.: CIDEr: consensus-based image description evaluation. In: Proceedings of the IEEE Conference on Computer Vision and Pattern Recognition, pp. 4566–4575 (2015)
26. Vinyals, O., Toshev, A., Bengio, S., Erhan, D.: Show and tell: a neural image caption generator. In: Proceedings of the IEEE Conference on Computer Vision and Pattern Recognition, pp. 3156–3164 (2015)
27. Wang, L., Bai, Z., Zhang, Y., Lu, H.: Show, recall, and tell: image captioning with recall mechanism. In: Proceedings of the AAAI Conference on Artificial Intelligence, vol. 34, no. 07, pp. 12176–12183, April 2020. https://doi.org/10.1609/aaai.v34i07.6898, https://ojs.aaai.org/index.php/AAAI/article/view/6898
28. Wang, R., Liu, L.: Paint to better describe: learning image caption by using text-to-image synthesis. In: 2021 IEEE International Conference on Dependable, Autonomic and Secure Computing, International Conference on Pervasive Intelligence and Computing, International Conference on Cloud and Big Data Computing, International Conference on Cyber Science and Technology Congress (DASC/PiCom/CBDCom/CyberSciTech), pp. 958–964. IEEE (2021)
29. Xia, P., He, J., Yin, J.: Boosting image caption generation with feature fusion module. Multimed. Tools Appl. **79**(33), 24225–24239 (2020)
30. Xu, K., et al.: Show, attend and tell: neural image caption generation with visual attention. In: International Conference on Machine Learning, pp. 2048–2057 (2015)

31. Zhang, W., Tople, S., Ohrimenko, O.: Dataset-level attribute leakage in collaborative learning. arXiv preprint arXiv:2006.07267 (2020)
32. Zhang, W., Ouyang, W., Li, W., Xu, D.: Collaborative and adversarial network for unsupervised domain adaptation. In: Proceedings of the IEEE Conference on Computer Vision and Pattern Recognition (CVPR), June 2018
33. Zhang, W., Tang, S., Su, J., Xiao, J., Zhuang, Y.: Tell and guess: cooperative learning for natural image caption generation with hierarchical refined attention. Multimed. Tools Appl. **80**(11), 16267–16282 (2020). https://doi.org/10.1007/s11042-020-08832-7
34. Zhou, L., Palangi, H., Zhang, L., Hu, H., Corso, J.J., Gao, J.: Unified vision-language pretraining for image captioning and vqa. In: AAAI, pp. 13041–13049 (2020)
35. Zhou, Y., et al.: Collaborative learning of semi-supervised segmentation and classification for medical images. In: Proceedings of the IEEE/CVF Conference on Computer Vision and Pattern Recognition (CVPR), June 2019
36. Zhou, Y., Wang, M., Liu, D., Hu, Z., Zhang, H.: More grounded image captioning by distilling image-text matching model. In: Proceedings of the IEEE/CVF Conference on Computer Vision and Pattern Recognition (CVPR), June 2020

Visual Analysis of the National Characteristics of the COVID-19 Vaccine Based on Knowledge Graph

Yong Feng[1], Ning Zhang[1], Hongyan Xu[1(✉)], Rongbing Wang[1], and Yonggang Zhang[2]

[1] College of Information, Liaoning University, Shenyang 110036, China
xuhongyan@lnu.edu.cn
[2] Key Laboratory of Symbolic Computation and Knowledge Engineering of Ministry of Education, Jilin University, Changchun 130012, China

Abstract. The aim is to construct a country-dimension knowledge graph of COVID-19 vaccines from the information of COVID-19 vaccines and to analyze the leading countries of vaccine R&D by combining the advantages of easy operation and intuitive feeling of knowledge graph visualization, to provide a reference for Chinese vaccine R&D departments and international cooperation. In this paper, through data collection, based on entity extraction and relationship construction, a knowledge graph of country dimensions was established by specifying the central vaccine R&D countries and vaccine distribution, and multidimensional microdata such as word frequency and betweenness centrality were combined to analyze the national characteristics of the COVID-19 vaccine. The analysis of the knowledge graph of the country dimension of the COVID-19 vaccine shows that countries with robust technology and economies, such as the US and China, choose to develop vaccine distribution independently, countries with advanced economies, such as Saudi Arabia, decide to purchase vaccine distribution, and less developed countries, such as South Africa and Latin America, need international aid for vaccines or purchase low-cost vaccines. This paper constructs the correlation between nodes and nodes of the COVID-19 vaccine with the help of a knowledge graph, systematically and comprehensively reveals the research mainstay and distribution model of the COVID-19 vaccine from the national level, and provides rationalized suggestions for international cooperation in vaccine R&D in China.

1 Introduction

The outbreak of the COVID-19 virus reached a cumulative total of 200 million infections and more than 5 million deaths in 2020 [1], with some recovered patients experiencing sequelae [2], including pulmonary fibrosis, altered/loss of sense of smell or taste,

This work is partly supported by the Social science planning foundation of Liaoning province of China (Grant No. L21BGL026), the Key Laboratory of Symbolic Computation and Knowledge Engineering of Ministry of Education, Jilin University (Grant No. 93K172018K01) and the General project of scientific research funds of Liaoning Provincial Department of education (Grant No. LJKZ0085).

and posttraumatic stress disorder. The academic research on COVID-19 is in a rapid progress [3]. Countries worldwide have created protective barriers against the spread of the virus by encouraging or forcing residents to wear masks, entering the country for quarantine, and getting vaccinated. With the increase in vaccination rates for the COVID-19 vaccine, the incidence of individual infections, severe illnesses, and deaths has gradually decreased. This has played an essential role in the prevention and control of outbreaks worldwide. However, there are currently several problems: on the one hand, viruses are mutating faster than the development of COVID-19 vaccines, making vaccine development more difficult. On the other hand, vaccine resources are not properly allocated.

The knowledge graph is simple to operate and graphically intuitive. An increasing number of scholars have applied it for analysis in various fields, such as the evolution of internet opinion features and its visualization analysis based on the topic graph [4]. The visualization of basketball news is combined with knowledge graphs [5]. A visual analysis of Chinese corporate disclosure research based on knowledge graph [6], but visualization analysis of COVID-19 vaccine features based on knowledge graphs is not yet standard. Using intuitive visualization of knowledge graphs, we can analyze the significant R&D countries and vaccine distribution methods, promote the cooperation of vaccine R&D departments in different countries, reduce the difficulty of vaccine R&D and promote the rational distribution of vaccine resources.

This paper collects data through the Web of Science platform, extracts entities from the data to construct schema models, uses entity and relationship models to build a knowledge graph, and visualizes and analyses the characteristics of COVID-19 vaccines through knowledge graphs. From the country-based knowledge graph analysis, the central R&D countries are clarified, and the R&D country relationships are sorted out. China should strengthen cooperation in vaccine R&D with countries such as the UK, US, and Germany, learn from the experience of mRNA vaccines, and choose to aid vaccines in less developed regions to promote a rational distribution of vaccine quantities.

2   Related Studies

The most common visualization technologies include HTML + CSS [7], SVG [8], and WebGL [9], of which HTML + CSS is convenient and does not rely on third-party code but cannot visualize data and is relatively cumbersome to draw pictures. SVG makes up for the ability of HTML + CSS to draw graphics and can visualize data but requires too many SVG elements for complex graphics and consumes performance. WebGL is powerful but difficult to use and makes full use of GPU parallel computing power. COVID-19 vaccine characterization requires visualization techniques that are intuitive and simple to use, so the commonly used visualization techniques do not meet the needs of this study.

In 2012 Google introduced the concept of Google Knowledge Graph [10], which is essentially a semantic network that stores real-world entities and entities and relationships between entities and attributes in the form of structured triples <entity, relationship, entity> and <entity, attribute, attribute value>, and connects different kinds of information together through a network of relationships to achieve a structured semantic description of the real objective world [11]. Graphs have been widely used in many areas, such as semantic search, intelligent question and answer, assisted language understanding and assisted extensive data analysis [12].

With the maturity of knowledge graph technology, the application of knowledge graphs in the medical field has been extensively developed. Hou M [13] conducted a comprehensive analysis of the medical knowledge graph architecture, construction technology, and application status, which supported the construction of medical knowledge graph-related technology and implementation processes. Hot topics in the medical field use knowledge graphs to provide decision support [14, 15] for disease diagnosis, and the current research hotspot in the medical area is the COVID-19 virus. Several domestic institutions have released several COVID-19 virus knowledge graphs, and Openkg has shared several COVID-19-related knowledge graphs [16–18] covering diagnosis and treatment, clinical and material aspects, but little has been seen in the COVID-19 vaccine.

In summary, given the number of countries developing COVID-19 vaccines and the complex relationships, the knowledge graph is good at sorting out complex node relationships, and it is also applicable to the analysis of the COVID-19 vaccine. The evolution of COVID-19 vaccine features is reflected in the logical relationships between events, and by analyzing the logical connections between nodes represented by countries, patterns of COVID-19 vaccine features can be discovered. Therefore, this paper uses the knowledge graph as a tool to visually analyze the characteristics of the COVID-19 vaccine by constructing a knowledge graph of the COVID-19 vaccine countries, which can strongly support the deep reading of the COVID-19 vaccine news.

3   Construction of the COVID-19 Vaccine Knowledge Graph

In the era of epidemic, finding appropriate paths and methods for storytelling of COVID-19 vaccine data has important theoretical and practical significance for sorting out the impact of the COVID-19 vaccine and giving full play to the value of epidemic data assets [19]. Knowledge graph gives full play to the role of information technology, such as big data, to provide information and intelligence support for emergency management of major public health emergencies and modernization of social risk management [20].

The knowledge graph stores the entities and relationships in the COVID-19 vaccine data, forming a semantic relationship network with entities as nodes and relationships as edges and discovering the characteristics of the COVID-19 vaccine through analysis of the relationship network. This paper proposes a model for constructing the knowledge graph of COVID-19 vaccines, as shown in Fig. 1. The data of papers in the database are crawled by web crawlers, and the crawled content includes fields such as research countries, researchers, and research results. The entity extraction technique is used to obtain relevant entities from the collected field information, and the relationships between entities are constructed through the association between data. The knowledge graph of COVID-19 vaccines is constructed based on the extracted entities and the constructed relationships. The country-based knowledge graph of the COVID-19 vaccine is constructed by specifying the main force of COVID-19 vaccine research and the distribution method. The following section focuses on the critical aspects of the knowledge graph construction model: data acquisition, entity extraction to construct a relationship model, and knowledge graph establishment.

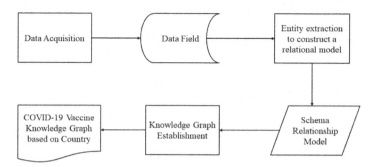

Fig. 1. Construction model of the COVID-19 vaccine knowledge graph

3.1 Data Acquisition

In this study, the Web of Science platform was selected as the source database, covering various database literature such as SCI, SSCI, AHCI, and CPCI-S. This study determined the SCI literature source as the data source, which is more representative. In this paper, we searched for the "COVID-19 vaccine/2019-nCOV vaccine" to ensure that we collected as much data as possible on the COVID-19 vaccine. The time interval selected was [2019, 2021], and a total of 13417 COVID-19 vaccine data were collected.

This study is based on a deduplication program written in Python and the ETL tool to d-duplicate the COVID-19 vaccine data, and the deduplication process is shown in Fig. 2. The COVID-19 vaccine data source is loaded into the ETL, the data source is previewed, the Python script component is connected to the data source component, and the deduplication is completed, resulting in 7827 COVID-19 vaccine data remaining.

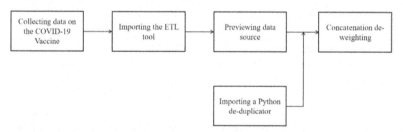

Fig. 2. Date deduplication process

3.2 Entity Extraction to Construct a Relational Model

In this study, structured data extraction of the collected COVID-19 vaccine data will be carried out by a Python program to obtain the corresponding fields as entity information. The schema model of COVID-19 vaccine knowledge and the representation scheme of the learning will be developed from the entities such as country, institution, researcher, the results, keywords and the relationships and attributes between the entities, and the detailed COVID-19 vaccine entities and the relationships between them are shown in Fig. 3.

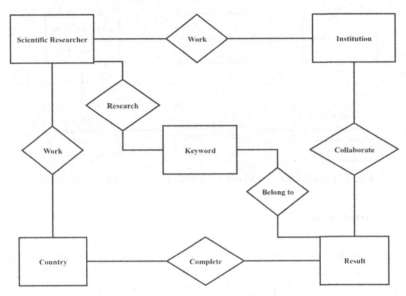

Fig. 3. Entity and relational data model of the COVID-19 vaccine knowledge graph

3.3 Knowledge Graph Establishment

The construction of the knowledge graph of the COVID-19 vaccine involves knowledge modeling, data acquisition, knowledge fusion, and knowledge inference [21]. Based on the characteristics of COVID-19 vaccine data, the knowledge graph is constructed top-down by developing the COVID-19 vaccine data ontology and Schema model, and its related technology and implementation process are shown in Fig. 4.

(1) COVID-19 vaccine knowledge schema model. The modeling process analyzes various scientific research entities in the data of COVID-19 vaccine abstract entity types, such as the country of research and development of the COVID-19 vaccine, institution (the unit or institution to which the results belong), researchers, the results, keywords (knowledge points) and the relationship between each entity type, and constructs the schema model of the COVID-19 vaccine knowledge graph, which conceptually describes the relationship between concepts and concepts and attributes related to COVID-19 vaccine representation.

(2) Heterogeneous data cleaning. Data from multisource heterogeneous systems were abstracted for initial data cleaning to form a normalized data format. The data in this study were mainly derived from cumulative vaccine papers in the COVID-19 field. For the cleaned and processed heterogeneous multisource data, a graph database was selected as the storage solution for the platform.

(3) Knowledge graph construction, knowledge mining, and inference. By organizing the acquired data in the form of a network, a globally unified knowledge identification system was formed through techniques such as fusion, disambiguation, and alignment of scientific research entities; a combination of manual and machine

learning techniques was used to fuse and align different national COVID-19 vaccine entities; a graph database was used to complete the storage and management of COVID-19 vaccine information; symbolic representation or vector representation of COVID-19 vaccine information was used; and computational inference and replenishment of potential relationships between COVID-19 vaccine research entities was used.

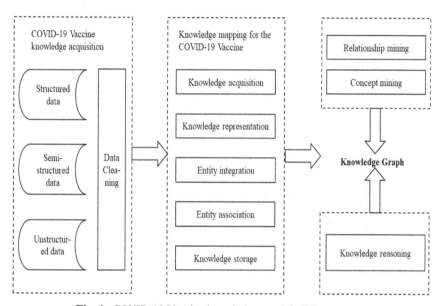

Fig. 4. COVID-19 Vaccine knowledge graph building process

4 Visual Analysis of the National Characteristics of the COVID-19 Vaccine

The COVID-19 vaccine knowledge graph is formed by connecting relationships between country entities, where larger nodes represent more occurrences of keywords and darker purple outlines of nodes the greater the medium. The country-based knowledge graph of COVID-19 vaccines makes it possible to visualize the significant countries developing COVID-19 vaccines. The results are shown in Fig. 5.

The importance of institutional nodes is judged by frequency and betweenness centrality. Table 1 shows that the top ten countries for COVID-19 vaccine R&D are the USA, PEOPLES R CHINA, ENGLAND, INDIA, ITALY, GERMANY, CANADA, AUSTRALIA, SAUDI ARABIA, and FRANCE.

Betweenness centrality is an index used to measure centrality, which refers to the ratio of the shortest path passing through a point and connecting the two points to the total number of shortest path lines between two points in the network [22]. The greater

Fig. 5. COVID-19 vaccine knowledge graph based on country

Table 1. Top ten nodes in frequency ranking

Serial number	Nodes	Frequency	Year
1	USA	2578	2020
2	PEOPLES R CHINA	806	2020
3	ENGLAND	757	2019
4	INDIA	643	2020
5	ITALY	541	2020
6	GERMANY	426	2020
7	CANADA	351	2020
8	AUSTRALIA	308	2019
9	SAUDI ARABIA	285	2020
10	FRANCE	256	2020

the betweenness of a node, the stronger and more critical [23] the node is, and the more pronounced the purple outline is in the diagram. The betweenness centrality of a node is calculated by Eq. (1).

$$C_B(v) = \sum_{s \neq v \neq t \in V} \frac{\sigma_{st}(v)}{\sigma_{st}} \tag{1}$$

where $C_B(v)$ denotes the betweenness centrality of node v, and $\sigma_{st}(v)$ indicates the number of shortest paths from node s to node t that pass through node v, and σ_{st} denotes the shortest path from node s to node t. The top 10 nodes in terms of mediating ability were obtained by Eq. (1), and the results are shown in Table 2.

Table 2 shows that the top ten countries with the most cooperation in the COVID-19 vaccine are BURKINA FASO, MOZAMBIQUE, GEORGIA, SLOVENIA, LITHUANIA, U ARAB EMIRATES, MALI, PARAGUAY, BAHRAIN, and GABON.

Table 2. Top ten nodes of betweenness centrality ranking

Serial number	Nodes	Betweenness centrality	Year
1	BURKINA FASO	0.89	2020
2	MOZAMBIQUE	0.87	2020
3	GEORGIA	0.86	2021
4	SLOVENIA	0.71	2020
5	LITHUANIA	0.43	2020
6	U ARAB EMIRATES	0.41	2020
7	MALI	0.34	2021
8	PARAGUAY	0.25	2020
9	BAHRAIN	0.22	2020
10	GABON	0.22	2020

There are currently 292 COVID-19 vaccines under development globally, of which 184 are in the preclinical phase, and 108 have entered clinical trials. The competition and cooperation between countries in the vaccine development process is not encouraging (see Fig. 5), and the major countries in vaccine development are clearly the USA, China, and the UK.

According to the frequency of words (see Table 1), among the significant R&D countries, except for China, India, and Saudi Arabia, all the other seven countries are developed countries with muscular economic strength and research capabilities. In the medical direction, the United States has strong backing for its medical system, with multiple vaccines advancing at the same time, including the famous Moderna vaccine and Pfizer-BioNTech COVID-19 vaccine, while the oral COVID-19 particular drug will soon be available; the UK has successfully developed the AstraZeneca vaccine, with muscular economic strength and scientific technology to achieve universal free medical care, and uses the strategy of mixing different varieties of COVID-19 vaccines to achieve better vaccination effects; Germany relies on its great Germany relies on its abundant medical resources, and eight vaccine projects are being promoted together, three of which are favored by the capital market, and vaccine R&D is developing in an echelon. Italy, Canada, Australia, and France have built up their national immunization defenses by developing vaccines in-house and purchasing vaccines with substantial financial resources. China and India stand out from the rest of the developing world, relying on their economic and technological development to complete their vaccine development. Among them, China has developed multiple vaccines at the same time, and vaccines such as the Sinopharm COVID-19 vaccine and CoronaVac vaccine have helped China form a robust immune barrier and successfully break the vaccine blockade in Europe and the US, selling vaccines to many countries around the world. Wealthy Saudi Arabia has chosen to collaborate with other countries to conduct clinical trials of vaccines while purchasing vaccines from other countries.

The vast majority of neighboring countries cooperate in the fight against the COVID-19 epidemic but are also influenced by geopolitics such as religion. As seen, by the centrality of betweenness (see Table 2), four underdeveloped African countries, Burkina Faso, Mozambique, Mali and Gabon, chose to collaborate with other African countries, receiving vaccine donations from different countries and purchasing vaccines for vaccination as their economic capacity allowed. Georgia in West Asia and Lithuania in Eastern Europe collaborate with West Asia and Eastern Europe to buy vaccines for immunization. Slovenia in Central Europe also cooperates with Eastern Europe to purchase the COVID-19 vaccine. Paraguay and others in Latin America cooperate in the fight against the epidemic. The U Arab Emirates cooperates with wealthy Islamic countries, such as Bahrain, to purchase the COVID-19 vaccine for vaccination.

In summary, influenced by economic and technological factors, economically and technologically powerful countries such as the United States, Germany, and China develop their vaccines; wealthy countries such as Saudi Arabia and the U Arab Emirates conduct clinical trials and purchase vaccines by working with technologically powerful countries; underdeveloped countries located in Africa and Latin America have the most difficulty in obtaining vaccines and require significant assistance from other countries. According to statistics, wealthy countries, which account for 14% of the world's population, buy 53% of the world's COVID-19 vaccine, the African continent has received less than 2% of the global market and only 0.9% of the population in low-income countries has received at least one dose of COVID-19 vaccine [24].

Compared with the Datavrap technique for the visualization of COVID-19 vaccine country characteristics, the country characteristics of the datavrap are dynamic 3D visualization with richer data, but knowledge graph visualization is more reasonable for the representation of complex relationships between countries. Plotly library for visualization of COVID-19 vaccines does better for vaccination data, but for analysis of country relationships, it is not excellent. In analyzing country characteristics, using knowledge graph techniques is the best compared to other techniques.

5   Conclusions and Recommendations

In this paper, by constructing a knowledge graph of the new vaccine countries and visualizing the characteristics of the COVID-19 countries with the knowledge graph and data, the following conclusions were drawn: countries with muscular technology and economy, such as the United States, Germany, and China, chose to develop their vaccines; countries with prosperous economies, such as Saudi Arabia, chose to purchase vaccines and cooperate with developed countries with technology; and underdeveloped countries, such as Latin America and Africa, waited for international assistance for vaccines and purchase vaccines that are within their economic range. In light of the findings of this paper, the following recommendations are made China aids and sells vaccines to underdeveloped countries such as Latin America and Africa when there are sufficient vaccines to promote equitable distribution of vaccine quantities and, depending on the epidemic, conducts clinical studies there and obtains a large amount of scientific data. There is a gradual increase in the trend towards globalization of scientific research collaboration on COVID-19 vaccine [25]. In this paper, there are still some limitations in the research process, as the specific emotional characteristics of the COVID-19 vaccine, such as the attitude of the public toward vaccination, have not been analyzed, and the analysis of straightforward emotional attitudes is also a vital part of the analysis of the characteristics of the COVID-19 vaccine. The combination of a knowledge graph and specific sentiment analysis of the COVID-19 vaccine will be further investigated in future work.

References

1. WHO Coronavirus (COVID-19) Dashboard. https://covid19.who.int/. Accessed 15 Sept 2021
2. Logue, J.K., Franko, N.M., Mcculloch, D.J., et al.: Sequelae in adults at 6 months after COVID-19 infection. JAMA Netw. Open **4**(2), e210830 (2021)
3. Xie, J., Li, W., Shen, H., Cheng, Y.: A study of credibility discriminators in the context of information explosion and information uncertainty - the COVID-19 outbreak as an example. J. China Soc. Sci. Tech. Inf. **40**(7), 714–724 (2021)
4. Chen, J., Xia, L., Liu, X.: The evolution of online public opinion features based on topic mapping and visualization analysis. Intell. Sci. **5**, 75–84 (2021)
5. Ji, N., Gao, Y., Zhao, Y., Yu, D., Chu, S.: Visualization of basketball news combined with the knowledge graph. J. Comput.-Aided Des. Graph. **6**, 838–846 (2021)
6. Tian, H., Jiang, C.: A visual analysis of Chinese corporate disclosure research based on knowledge graph. Inf. Sci. **38**(12), 98–104 (2020)
7. Xia, C., Zhang, Y.: New opportunities for mobile reading services in libraries: HTML5 and CSS3. Mod. Libr. Inf. Technol. (5), 17–25 (2012)
8. Li, Q., Xie, Z., Zuo, X., Wang, C.: Spatial information description and visual representation based on SVG. J. Surv. Mapp. **34**(1), 59–63 (2005)
9. Wang, X., Wei, S.: Research and application of 3D WebGIS platform with WebGL technology. Remote Sens. Inf. **34**(3), 134–138 (2019)
10. Amit, S.: Introducing the knowledge graph. Official Blog of Google, America (2012)
11. Amit, S.: Introducing the knowledge graph: things, not strings. https://www.blog.google/products/search/introducing-knowledge-graph-things-not/. Accessed 10 June 2020
12. Chinese Computer Society.CCF 2017-2018 China Computer Science and Technology Development Report. Machinery Industry Press, Beijing (2018)
13. Hou, M., Wei, R., Lu, L., et al.: A review of knowledge graph research and its application in the medical field. Comput. Res. Dev. **55**(12), 2585–2599 (2018)
14. Rotmensch, M., Halpern, Y., Climat, A., et al.: Learning a health knowledge graph from electronic medical records. Sci. Rep. **7**, 5994 (2017)

15. Bakal, G., Talari, P., Kakani, E.V., Kavuluru, R.: Exploiting semantic patterns over biomedical knowledge graphs for predicting treatment and causative relations. J. Biomed. Inform. **82**, 189–199 (2018)
16. Jiang, B., Yu, X., Li, K., et al.: Interactive visual analysis of the COVID-19 epidemic situation using geographic knowledge mapping. J. Wuhan Univ. **45**(6), 836–845 (2020)
17. Chen, X., Liu, J., Xu, L., et al.: Knowledge mapping of COVID-19 case activities in Zhengzhou City. J. Wuhan Univ. (Inf. Sci. Ed.) **45**(6), 816–825 (2020)
18. Xiang, J., Hu, H., Liu, Y., et al.: Construction of the COVID-19 material knowledge map. J. Wuhan Univ. (Sci. Ed.) **66**(5), 409–417 (2020)
19. Hou, X., Ying, J., Song, S.: Structural elements and processes of storytelling of epidemic data from an activity theory perspective. https://kns.cnki.net/kcms/detail/11.1762.G3.20220329.1644.004.html/. Accesses 29 Mar 2022
20. Wang, F., An, L., Huang, R., Wang, X., et al.: Scientific response and reflection in public health emergencies: graphic experts on the new crown epidemic. Libr. Intell. Knowl. (2), 4–14 (2020)
21. Wang, W., Li, N., Zheng, X.: Research on the key technology of research knowledge graph service platform for tobacco field. https://kns.cnki.net/kcms/detail/11.2985.TS.20210527.1101.002.html/. Accessed 27 May 2021
22. Freeman, L.C.: Set of measures of centrality based on betweenness. Sociometry **40**(1), 35–41 (1977)
23. Wang, W., et al.:Social Network Analysis: Methods and Practice. Machinery Industry Press, Beijing (2013)
24. Boao Forum for Asia releases Global COVID-19 vaccine Adoption Landscape Report. People's Daily (6) (2021)
25. Wang, F., Dong, J.: An interpretation of the research model of international cooperation on COVID-19 prevention and control from the perspective of network structure analysis. Intell. Explor. (3), 56–63 (2022)

Speech Recognition for Parkinson's Disease Based on Improved Genetic Algorithm and Data Enhancement Technology

Jing Qin[1], Tong Liu[1], Zumin Wang[1,2,3(✉)], Qijie Zou[1], Liming Chen[2],
and Chang Hong[3]

[1] Dalian University, Dalian 116000, China
{qinjing,zumingwang}@dlu.edu.cn
[2] Ulster University, Belfast, NIC 100166, UK
[3] Third People's Hospital of Dalian, Dalian 116000, China

Abstract. Parkinson's disease is one of the most destructive diseases to the nervous system. Speech disorder is one of the typical symptoms of Parkinson's disease. Approximately 90% of Parkin-son's patients develop some degree of speech disorder, which affects speech function faster than any other subsystem of the body. Screening Parkinson's disease by sound is a very effective method that has attracted a growing number of researchers over the past decade. Patients with Parkinson's disease could be identified by recording the sound signal of the pronunciation of words, extracting appropriate features and identifying the disturbance in their voices. This paper proposes an improved genetic algorithm combined with a data enhancement method for Parkinson's speech signal recognition. Specifically, the methods first extract representative speech signal features through the L1 regularization SVM and then enhance the representative feature data by the SMOTE algorithm. Following this, both original and enhanced features are used to train an SVM classifier for speech signal recognition. An improved genetic algorithm was applied to find the optimal parameters of the SVM. The effectiveness of our proposed model is demonstrated by using Parkinson's disease audio data set from the UCI machine learning library, and compared with the most advanced methods, our proposed method has the best performance.

Keywords: Parkinson's disease speech signal detection · Support vector machine · SMOTE algorithm · Genetic algorithm

1 Introduction

Parkinson's disease (PD) is a neurodegenerative disease related to the human nervous system. The loss of brain cells affects the body's ability to move [1–3]. PD is currently regarded as the second largest disease affecting humans, and the proportion of people with PD is increasing rapidly with the growing aging population worldwide, especially in Asian countries. The origin of the disorder begins with the decline of neurogenetic dopamine chemicals [4], and the main lesions are in the substantia nigra and striatum. As

274   J. Qin et al.

such, the information generated by the brain is unable to be sent to specific parts of the body, leading to various physiological symptoms in the patient. A previous investigation found that vocalization disorder is one of the main symptoms manifested 5 years before the diagnosis of PD. Sixty to 90% of PD patients suffer from speech disorders, such as slurred speech, slow speech [5] and other symptoms. Identifying dysphonic indicators plays an important role in early diagnosis to support a high-quality life for as long as possible. The later the disease is found, the more it will disturb the patient's normal life. In the later stage, it depends on the help of others to live a normal life [7].

Recently, scholars have used machine learning and deep learning methods to detect Parkinson's disease through speech signals. Different researchers have adopted different techniques, such as artificial neural networks, SVM, Dirichlet process mixture, random forest, fuzzy k-nearest neighbor (FKNN), and different mixture models, to improve classification accuracy.

In the field of using traditional machine learning methods as classifiers, in recent years, relevant researchers have considered the following experimental methods. For example, Vladimir Despotovic et al. [9] selected the best 5 features and applied the Gaussian process to obtain the best classification accuracy rate of 96.92%. Rohit Lamba et al. [10] recommended machine learning-related algorithms combined with the SMOTE algorithm for classification, and the final accuracy rate was approximately 95.58%. Aich, S, Younga, K et al. [18] proposed a principal component analysis (PCA) algorithm based on the original feature set and other nonlinear classifiers. The study gave the best random forest accuracy performance. At 96.87%, the author proposes that the dimensionality reduction of features can improve the classification effect of PD. Sakar et al. [11] proposed a decision support system for Parkinson's disease detection. A large number of speech signal processing methods are used to extract useful clinical information for PD patient classification and input it into a machine learning classifier. The adjustable Q factor wavelet transform (TQWT) was first used to extract features from speech signals. Feature selection is completed by the minimum redundancy-maximum correlation (mRMR) method, and classification is completed by six different classifiers. The best accuracy of 86% is achieved through the combination of the mRMR and SVM-RBF algorithms. Bibhuprasad Sahu et al. [7] proposed the chaos-mapped bat algorithm to optimize SVM parameters and obtained 98.24% accuracy through 10-fold cross-validation. Through the above literature, we found that relevant researchers did not consider the impact of redundant features on data classification, or the effect of feature selection was not good. Selecting reliable features that have a large amount of identification information remains one of the major challenges in PD speech signal recognition.

In the field of neural networks as classifiers, Sukhpal Kaur et al. [8] proposed using grid search to improve the performance of a specific framework, and a neural network model was developed and optimized to detect the early onset of Parkinson's disease, thereby setting and adjusting multiple hyperparameters to evaluate deep learning models. Grid search optimization includes three main stages, namely, the optimization of neural network model topology, hyperparameters and its performance. An accuracy of 89.23% was obtained. Olusola O. Abayomi-Alli, Robertas Damaševičius et al. [6] proposed an interpolation method to generate new data, expand the original data set, put all the data into a two-way LSTM for classification, and obtained an accuracy of 97.1%. Through the above literature, we found that relevant researchers did not consider the distribution of the categories of the original data set.

Through the analysis of the abovementioned literature, in view of the shortcomings of the models and methods proposed in the past, this article considers the following improvements:

- Due to the difficulty in obtaining medical data and the small amount of data, the use of deep learning methods makes it difficult to train the model. The classification effect of traditional machine learning algorithms is not good. Therefore, we propose using a data set expansion algorithm to address the problem of insufficient data set samples.
- Aiming at the problem that the genetic algorithm easily falls into a local optimal solution, we propose a better adaptive genetic algorithm that combines a new mutation point selection strategy. Improve the performance of genetic algorithms and apply them to the parameter optimization process of machine learning classifiers.

2 The Proposed Methods

This paper proposes a set of classification models. First, an effective feature selection algorithm is used to capture the most important feature from a among many feature sets. Next, effective methods are used to expand insufficient data sets. Finally, the traditional genetic algorithm is improved and used to find the most effective parameters of the classifier. The experimental results show that the above process improves the final classification accuracy of the classifier. The model process is shown in Fig. 1.

2.1 Method

Improved Adaptive Genetic Algorithm (Improved-GA)
For the classification steps in Fig. 1, we consider using the improved genetic algorithm to optimize the classifier. The following is a brief description of the genetic algorithm.

For the traditional genetic algorithm, the values of Pc (the crossover probability) and Pm (the mutation probability) are two important control parameters to change the optimization performance of the genetic algorithm. The value of Pc defines the degree of population diversity. The greater Pc is, the greater the diversity of the population, but the greater the probability that individuals with good genes will be filtered out. The value of Pm is the key factor in jumping out of the local optimal solution and finding the overall optimal solution. If the values of Pc and Pm are too small, it is not easy to produce new individuals, leading to stagnant evolution. Therefore, an excellent genetic algorithm mainly needs to improve the following two capabilities:

- The ability to converge to the best effect after finding the range that contains the optimal solution.
- Exploring new regions of the solution space to find the global optimal ability.

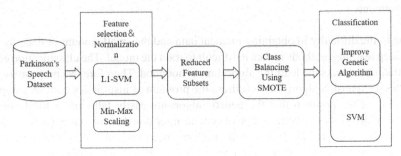

Fig. 1. The proposed method for hybrid Parkinson's speech signal recognition

In response to the above situation, Srinivas et al. [12] proposed an adaptive genetic algorithm. The main purpose of the adaptive genetic algorithm is to achieve the trade-off between search and randomness in different ways and automatically adjust the values of Pc and Pm according to the current fitness. The idea is to increase Pc and Pm when the population fitness is relatively concentrated; when the population fitness is relatively scattered, both Pc and Pm are reduced so that they conform to dynamic changes. The author proposes the following formula to automatically regulate Pc and Pm during operation.

$$Pc = \begin{cases} k_1(f_{max} - f')/(f_{max} - f_{ave}) & f' \geq f_{ave} \\ k_2 & f' < f_{ave} \end{cases} \tag{1}$$

$$Pm = \begin{cases} k_3(f_{max} - f)/(f_{max} - f_{ave}) & f \geq f_{ave} \\ k_4 & f < f_{ave} \end{cases} \tag{2}$$

f_{max} is the maximum fitness of the population; f_{avg} is the average fitness of the population; f' is the fitness of two crossed individuals with greater fitness; f is the fitness of the mutated individual; $0 < k_1, k_2, k_3, k_4 \leq 1$ is a random constant [12]. Researchers with different algorithms have put forward different ideas for improvement. Some researchers based on this, in order to speed up the optimization efficiency, In the selection operator, the introduced selection operator is combined with the optimal preservation strategy. To keep the population number constant during the genetic operation, the strategy of preserving parents is proposed. Based on previous research by relevant personnel, I associate the odd-even cross-point mutation based on the traditional odd-even crossover operator [15]. Through the odd-even point mutation, the difference between the new sample and the original sample is increased. The characteristics of the mutation probability are automatically adjusted in the iterative process, and the two are coordinated with each other. The optimal solution of the function to be sought is found in this process of declining and increasing. At the same time, it can be seen from the formula of the traditional adaptive genetic algorithm that although the poor individuals can be improved quickly, the high-quality genes carried in the poor individuals will also be lost, destroying the diversity of the population. Therefore, we consider a new formula that can nonlinearly and adaptively adjust the operation process of genetic evolution and the values of crossover probability Pc and mutation probability Pm according to the degree of concentration and dispersion of population fitness in evolution. The formula is as follows:

$$Pc = \begin{cases} k_1 \dfrac{\arctan\left(\frac{f_{ave}}{f_{max}}\right)}{\pi/2} & \arctan\left(\frac{f_{ave}}{f_{max}}\right) < \pi/8 \\[3mm] k_1\left(1 - \dfrac{\arctan\left(\frac{f_{ave}}{f_{max}}\right)}{\pi/2}\right) & \arctan\left(\frac{f_{ave}}{f_{max}}\right) \geq \pi/8 \end{cases} \tag{3}$$

$$Pm = \begin{cases} k_2\left(1 - \dfrac{\arctan\left(\frac{f_{ave}}{f_{max}}\right)}{\pi/2}\right) & \arctan\left(\frac{f_{ave}}{f_{max}}\right) < \pi/8 \\[3mm] k_2 \dfrac{\arctan\left(\frac{f_{ave}}{f_{max}}\right)}{\pi/2} & \arctan\left(\frac{f_{ave}}{f_{max}}\right) \geq \pi/8 \end{cases} \tag{4}$$

The meaning of the parameters is the same as above. For specific problems, we can also calculate the information entropy of the new population step by step when the initial random population of the algorithm is formed. The specific process is as follows:

- Three chromosomes are randomly generated as a population, and the information entropy function is called to calculate the population information entropy.
- Determine whether the population size meets the requirements. If it meets the requirements, go to step 4; otherwise, generate a chromosome, add it to the population, and calculate the information entropy again.
- If the information entropy of the new population increases compared to the previous information entropy, continue to step 2; if the information entropy decreases, delete this new chromosome from the population, and then continue to step 2.
- The cycle stops, and the initial population is obtained. Through this process, the initial population is more discrete in the sample space distribution, which is helpful to the convergence of the algorithm. Compared with the traditional genetic algorithm, this not only speeds up the convergence speed but also becomes more stable in searching for the optimal value.

Support Vector Machine (SVM) of RBF Core
SVM is now regarded by many researchers as an excellent binary classification algorithm. It processes data samples with certain predetermined kernel functions and generates a class of samples. A given classification problem is first divided into training and test data sets, and the generated output specifies the class label. SVM can maximize the boundaries of different categories of data and minimize the error of misclassified data. The original form of the soft margin SVM can be expressed in mathematical formula as:

$$\min_{\omega \in R^F} \frac{1}{2}[\omega]_2^2 + C \sum_{i=1}^{n} I(y_i, f_\omega(X_i)) \tag{5}$$

Here, ω is a hyperplane vector used to distinguish two classes. Specify the $I(y, y\wedge)$ loss function. C represents the regularization parameter used to evaluate the smoothness and error of the weight. $fw(xi) = \langle \Phi(xi), w \rangle$ represents the mapping function. The function can map the input data from space d to the corresponding feature space f, where f may be infinite. When the value of f becomes very large, then the kernel function $k(x, y) = \langle \Phi x, \Phi y \rangle$ can be used to calculate the product in the internal space Rf. Among the four basic SVM kernel functions, to solve some originally linear inseparable problems, we choose RBF as the kernel function of SVM. The formula is expressed as:

$$k(x, y) = \frac{exp\|x - y\|^2}{2 \times \sigma^2} \tag{6}$$

Through the parameters in formulas (3) and (4), it can be found that to better realize the classification accuracy of SVM, the parameters of γ and C in the formulas need to be carefully selected. The main parameters in SVM have two C (penalty coefficient) and γ. The penalty coefficient is the tolerance for errors. The higher the penalty coefficient is, the more difficult it is to tolerate errors, and overfitting is prone to occur. The smaller the penalty coefficient is, the more likely underfitting is to occur again. If the penalty coefficient is too large or too small, the generalization ability will deteriorate. The size of γ will affect the range of the Gaussian distribution corresponding to each support vector. If the γ value is too large, the Gaussian distribution will look "long and thin", which will cause the classification curve to only act near the support vector sample. The classification effect for unknown samples will be very poor. In contrast, if the value of γ is selected to be too small, the smoothing effect will be too large, and a particularly high accuracy rate cannot be obtained on the training sets. Therefore, the improved genetic algorithm introduced in the previous section is considered for optimization.

Synthetic Minority Oversampling Technique (SMOTE) Algorithm

The SMOTE algorithm was proposed by Chawla [17] in 2002. The basic idea of the smote algorithm is to expand a few categories of samples to expand the data volume of the original data set. The process of the algorithm is described as follows:

- First, for samples in a category with a small number of samples, x use Euclidean distance as the distance between samples, and obtain its neighbors k.
- Determine sampling magnification N. For each minority sample x From it k neighbors, we randomly select several samples. Here, we assume that the chosen neighbor is xn.
- For each randomly selected neighbor xn, the following formula is used to create a new sample from the original sample.

$$xnew = x + rand(0, 1) \times (\tilde{x} - x) \tag{7}$$

- Add the constructed new sample to the original data set to form a new data set.

3   Speech Recognition and Diagnosis

3.1   Data Preprocessing

Feature selection based on L1-Norm SVM

Since not all features are useful for prediction, even the existence of certain features will improve the accuracy of training in the training set, but it will have a bad effect on the test set. Amin Ul Haq, Jian Ping Li et al. [14] proposed using L1-Norm SVM to extract features of Parkinson's speech data set. Parameters, and at the same time, remove the features whose feature weight is 0 corresponding to the SVM that outputs each parameter, and then merge the feature subsets obtained in each round.

Finally, ten important features were selected: 'PPE', 'spread2', 'D2', 'DFA', 'RPDE', 'MDVP: Fo (Hz)', 'MDVP: Fhi (Hz)', 'MDVP: Flo (Hz)', 'HNR', 'MDVP: Shimmer (dB)'.

The important features are described as follows:

- PPE: pitch parameter.
- Spread2: pitch parameter.
- D2: Nonlinear complexity criterion.
- DFA: Fractal scale power.
- RPDE: Nonlinear analysis.
- MDVP: Fo (Hz): average fundamental frequency.
- MDVP: Fhi (Hz): Maximum fundamental frequency.
- MDVP: Flo (Hz): lowest fundamental frequency.
- HNR: Sounding parameters.
- MDVP: Shimmer (dB): amplitude parameter.

Data Set Expansion Based on the SMOTE Algorithm

Observing the imbalance between the positive and negative samples of the original data set, consider using the SMOTE algorithm to generate data to expand the original data set. Figure 2 is a visual comparison diagram of the original data set and the expanded data set. A total of 294 new data sets were generated through the SMOTE algorithm, which were spliced with the original data set (195 items) into a new data set (489 items in total).

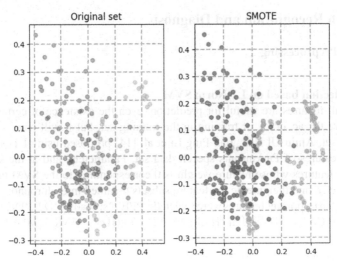

Fig. 2. The sample distribution of the original data set (left), and the sample distribution after the SMOTE algorithm expands the data set (right)

3.2 Improved GA-SVM Model

For the selection of parameters in SVM, it is very important to improve the accuracy of classification. This will also affect the accuracy of the test set. In summary, the selection of relevant parameters will have a great impact on the final classification accuracy of the whole model.

For the parameter selection problem, the first thought of using heuristic search method to select, and the more representative algorithm of heuristic search is genetic algorithm. For the traditional genetic algorithm, there is an obvious defect that the speed of finding the optimal solution is slow, and it is difficult to jump out of the local optimal solution. So think of an improved adaptive genetic algorithm (Improved-GA).

We use the penalty parameter C and γ value as the chromosome and the average accuracy of the 10-fold cross-validation of the SVM as the fitness function. Use SVM to perform 10-fold cross-validation on all data (489 data in total). The 10-fold cross-validation is to divide the entire data set into 10 blocks, use 9 blocks as the training set and the remaining block as the test set, and conduct experiments. Each experiment will obtain the corresponding accuracy rate. Then, loop this step, take a piece of the training set in the previous round as the test set, and reconstitute a new training set from the test set of the previous round and the rest of the training set. Finally, the average of the accuracy of the 10 results is used as the final evaluation index. This can better reflect the generalization performance of the model and make it less sensitive to data division.

3.3 Speech Recognition Algorithm

The model process is described as follows:

- First, the original data set is standardized, and important features are selected through L1-Norm SVM iteration.
- The SMOTE algorithm is used to expand the data set with the selected feature data set.
- The expanded data set is classified using a support vector machine (SVM).
- The parameters C and γ of the SVM are searched using the improved GA, and the average accuracy of the SVM 10-fold cross-validation classification is taken as the fitness function of the improved GA.
- Judge whether the average accuracy rate meets the condition; if it meets the condition, output the result; otherwise, go to step 4.

4 Experiment and Evaluation

4.1 Experiment Setup

Parkinson's Speech Data Set

We used the UCI Parkinson's data set, which is a publicly available data set of the UCI repository created by Max Little of the University of Oxford [13], to classify normal people and PD patients. These data sets contain detailed information about the biomedical speech measurement of 31 patients, of which 8 healthy controls were PD. A total of 195 lines of 6 different vowel sounds corresponding to 31 individuals were recorded, with a duration of 36 s. Among the 195 items, 147 voice measurements were from PD patients, and the rest were from disease-free people. The status column has two values, 0 for disease-free individuals and 1 for PD patients. The patient samples ranged from 0–28 years old and 46–85 years old, with an average of 65.8. Table 1 shows the voice feature information contained in the data set.

Experimental Parameters

For our experiment, we use 10-fold cross-validation to evaluate the performance of the proposed model.

The evaluation criteria used in the evaluation of the final model performance are as follows. Accuracy refers to the ratio of the number of correctly classified examples to the total number of examples. The F1 value represents the harmonic average of precision and recall. Precision represents the accuracy rate, which refers to the proportion of the true positive class (TP) among all judged positive classes (TP+FP). Recall refers to the proportion of all real positive classes (TP+FN) that are judged to be positive classes (TP). Specificity refers to the proportion of all true negative classes (TN+FP) that are judged as negative classes (TN). The calculation formulas of these five evaluation indicators are as follows:

$$Accuracy = \frac{TP + TN}{TP + FP + TN + FN} \tag{8}$$

$$Precision = \frac{TP}{TP + FP} \tag{9}$$

$$Specificity = \frac{TN}{TN + FP} \tag{10}$$

$$Recall = \frac{TP}{TP + FN} \tag{11}$$

$$F_1 = \frac{2 * Precision * Recall}{Precision + Recall} \tag{12}$$

Table 1. Voice feature information

Oxford PD data set Label	Feature
MDVP:Fo(Hz) MDVP:Fhi(Hz) MDVP:Flo(Hz) MDVP:Jitter(%) MDVP:Jitter(Abs) MDVP:RAP MDVP:PPQ Jitter:DDP	Frequency Parameters
MDVP:Shimmer MDVP:Shimmer(dB) Shimmer:APQ3 Shimmer:APQ5 MDVP:APQ Shimmer:DDA	Amplitude Parameters
NHR HNR RPDE	Voicing Parameters
PPE spread1 spread2	Pitch Parameters
D2 DFA	Fractal Scaling

Experimental Environment

The algorithm is deployed on a host with a dual AMD CPU and GTX1650 graphics cards, and the sklearn platform architecture and Python 3.7 are used in the machine learning model to implement the algorithm in this paper.

4.2 Comparison and Analysis of Results

Based on the above model and training, the algorithm of this paper is applied to the public data set UCI Parkinson's speech to test the classification effect. The classic machine learning algorithm SVM, logistic regression, random forest and mainstream integrated learning algorithm gradient boosting decision tree (gradient boosting decision tree, GBDT) are selected. It also compares the experimental results of other researchers in their literature (compare the experimental results of other researchers as shown in Table 3).

Figure 3 shows that traditional machine learning algorithms do not perform well on unbalanced data sets, and at the same time, the parameter selection of specific classifier algorithms is also a major problem. In terms of the classification performance of the original data set, the comprehensive performance of SVM and GBDT is better.

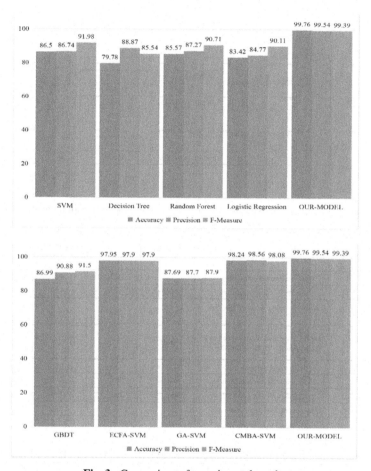

Fig. 3. Comparison of experimental results

Decision Tree performs poorly on unbalanced data sets, which may be due to overfitting during the training phase, which reduces the accuracy of the test phase.

The experimental results of our model show that the accuracy of SVM classification (10-fold CV) was greatly improved after selecting the optimal features and combining the SMOTE algorithm for data enhancement and finding the optimal parameters. The best results show that the accuracy, precision and F-measure can even reach 99.76%, 99.54% and 99.39%, respectively. At the same time, to illustrate the effectiveness of our proposed model more convincingly, we conducted relevant ablation experiments for the method of this paper (the experimental results are shown in Table 2).

Table 2. Comparison of ablation experiments

Dataset	Methodology	Mean Accuracy (%)
Original Dataset	SVM	0.865
	SVM(Grid Search)	0.869
	SVM(GA)	0.876
	SVM(Improve GA)	0.8918
Dataset (10 features)	SVM	0.840
	SVM(Grid Search)	0.867
	SVM(GA)	0.8806
	SVM(Improve GA)	0.9021
Dataset (10 features) +SMOTE	SVM	0.9099
	SVM(Grid Search)	0.9818
	SVM(GA)	0.9921
	SVM(Improve GA)	0.9976

Table 3. Compare the experimental results of other researchers in their literature

Reference	Methodology	Validation Method	Accuracy (%)	Specificity (%)	Sensitivity (%)
[18]	PCA with Random Forest (RF)	~	96.87	99.85	99.75
[9]	Gaussian Process+ 5 features	10-fold CV	96.92	99.29	90
[19]	Ensemble bagging + Genetic Algorithm (GA)	~	98.28	~	~
[20]	Hybrid Relief prior and Bacterial Foraging Optimization SVM (RF-BFO-SVM)	5-fold CV	97.42	91.5	99.29
[21]	Artificial neural networks (ANN)	10-fold CV	96.88	100	95.74
[6]	BiLSTM with Augmentation (Spline)	Holdout	97.1	98.78	95.57
[10]	Random forest + GA + SMOTE	10-fold CV	95.58	97.95	93.19
[22]	Deep Autoencoder Neural Network	~	96.11	89.78	98.15
[23]	Complex-Valued Neural Networks and mRMR Feature Selection Algorithm	10-fold CV	98.12	98.96	99.24
OurModel	SMOTE(10 features) + GA + SVM	10-fold CV	99.76	99.89	99.37

As seen in Table 2, for the original data set, our proposed method of optimizing SVM using GA obtained better classification accuracy, which is sufficient to verify the effectiveness of our proposed GA. However, for the overall data set, the classification accuracy did not reach a good level, so we considered feature filtering and data augmentation for the data set, and then we tested it again and found that it yielded better results for both classifiers.

Such experimental results show that for medical data with a small amount of data and an imbalanced data set, the data enhancement technology combined with the method of classifier parameter optimization can significantly improve the accuracy of classification.

5 Conclusion

This article introduces the application of evolutionary algorithms to machine learning algorithms and verifies the feasibility of using the SMOTE algorithm to generate new data to expand the original imbalanced data set. The experimental results show that the best accuracy of 10-fold cross-validation is 99.76%, precision is 99.54%, and the F1 value is 99.39%. The model can effectively classify Parkinson's speech data sets.

At the same time, the application of data enhancement technology not only improves the accuracy of the classifier but also reduces overfitting and improves the overall performance. When faced with a small, unbalanced and large number of features in Parkinson's speech signal data set, using the model proposed in this article may better solve this type of problem and improve the classification accuracy.

In the future, we will improve the classification algorithm to make this model applicable to larger unbalanced data sets.

Acknowledgment. This work was supported by the Youth Fund Project of the National Natural Fund of China under Grant 62002038.

References

1. Agarwal, A., Chandrayan, S., Sahu, S.: Prediction of Parkinson's disease using speech signal with Extreme Learning Machine. In: International Conference on Electrical, Electronics and Optimization Techniques 2016, pp. 3776–3779. IEEE (2016)
2. Synnott, J., Chen, L., Nugent, C.: WiiPD-objective home assessment of Parkinson's disease using the nintendo Wii remote. IEEE Trans. Inf. Technol. Biomed. **16**(6), 1304–1312 (2012)
3. Synnott, J., Chen, L., Nugent, CD.: The creation of simulated activity data sets using a graphical intelligent environment simulation tool. In: Annual International Conference of the IEEE Engineering in Medicine and Biology Society 2014, EMBC 2014, pp. 4143–4146. IEEE (2014)
4. Saikia, A., Majhi, V., Hussain, M.: A systematic review on application based Parkinson's disease detection systems. Int. J. Emerg. Technol. **10**(3), 166–173 (2019)
5. Chan, M.Y., Chu, S.Y., Ahmad, K.: Voice therapy for Parkinson's disease via smartphone videoconference in Malaysia: a preliminary study. J. Telemed. Telecare **27**(3), 174–182 (2019)
6. Olusola, O., Abayomi, A., Robertas, D.: BiLSTM with data augmentation using interpolation methods to improve early detection of Parkinson disease. In: Conference on Computer Science and Information Systems 2020, pp. 371–380. IEEE (2020)
7. Sahu, B., Mohanty, S.N.: CMBA-SVM: a clinical approach for Parkinson disease diagnosis. Int. J. Inf. Technol. **13**(2), 647–655 (2021). https://doi.org/10.1007/s41870-020-00569-8
8. Kaur, S., Aggarwal, H., Rani, R.: Hyper-parameter optimization of deep learning model for prediction of Parkinson's disease. Mach. Vis. Appl. **31**(5), 31–32 (2020). https://doi.org/10.1007/s00138-020-01078-1
9. Ladimir, D., Tomas, S., Christoph, S.: Speech based estimation of Parkinson's disease using Gaussian processes and automatic relevance determination. Neurocomputing **40**(1), 173–181 (2020)
10. Rohit, L., Hadeel, F.A., Anurag, J.: A hybrid system for Parkinson's disease diagnosis using machine learning techniques. Int. J. Speech Technol. 1–11 (2021). https://doi.org/10.1007/s10772-021-09837-9

11. Sakar, C.O., Serbes, G., Gunduz, A., Tunc, H.C., Nizam, H.: A comparative analysis of speech signal processing algorithms for Parkinson's disease classifcation and the use of the tunable Q-factor wavelet transform. Appl. Soft Comput. **74**, 255–263 (2019)

12. Srinivas, M., Patnaik, L.M.: Adaptive probabilities of crossover and mutation in genetic algorithms. IEEE Trans. Syst. **24**(4), 656–667 (1994)

13. Little, M.A., McSharry, P.E., Roberts, S.J., Costello, D.A.E.: Exploiting nonlinear recurrence and fractal scaling properties for voice disorder detection. Nat. Prec. (2007)

14. Haq, A.U., Li, J.P.: Feature selection based on L1-norm support vector machine and effective recognition system for Parkinsons's disease using voice recordings. IEEE Access **7**, 37718–37734 (2019)

15. Sivaram, M., Batri, K., Amin Salih, M.: Exploiting the local optima in genetic algorithm using tabu search. Indian J. Sci. Technol. **12**(1), 1–13 (2019)

16. Onur, K., Hakan, C., Adi, A.: Robust automated Parkinson disease detection based on voice signals with transfer learning. Expert Syst. Appl. **178**, 115013 (2021)

17. Chawla, N.V., Bowyer, K.W., Hall, L.O., Kegelmeyer, W.P.: SMOTE: synthetic minority oversampling technique. J. Artif. Intell. Res. **16**, 321–357 (2002)

18. Aich, S., Younga, K., Hui, K.L., Al-Absi, A.A.: A nonlinear decision tree based classification approach to predict the Parkinson's disease using different feature sets of voice data. In: International Conference on Advanced Communication Technology 2018, pp. 638–642. IEEE (2018)

19. Fayyazifar, N., Samadiani, N.: Parkinson's disease detection using ensemble techniques and genetic algorithm. In: Artificial Intelligence and Signal Processing Conference 2017, pp. 162–165. IEEE (2017)

20. Cai, Z., Gu, J., Chen, H.L.: A new hybrid intelligent framework for predicting Parkinson's disease. IEEE Access **5**, 17188–17200 (2017)

21. Wang, X.: Data mining analysis of the Parkinson's disease. Masters thesis Submitted to the College of Arts and Sciences, Georgia State University (2014)

22. Kose, U., Deperlioglu, O., Alzubi, J., Patrut, B.: Diagnosing Parkinson by using deep autoencoder neural network. In: Kose, U., Deperlioglu, O., Alzubi, J., Patrut, B. (eds.) Deep Learning for Medical Decision Support Systems. SCI, vol. 909, pp. 73–93. Springer, Singapore (2020). https://doi.org/10.1007/978-981-15-6325-6_5

23. Peker, M., Sen, B., Delen, D.: Computer-aided diagnosis of Parkinson's disease using complex-valued neural networks and mRMR feature selection algorithm. J. Healthc. Eng. **6**(3), 281–302 (2015)

24. Mohamadzadeh, S., Pasban, S., Zeraatkar-Moghadam, J., Shafiei, A.K.: Parkinson's disease detection by using feature selection and sparse representation. J. Med. Biol. Eng. **41**(4), 412–421 (2021). https://doi.org/10.1007/s40846-021-00626-y

A Review of Animal Individual Recognition Based on Computer Vision

Weifang Wang[1,2], Junfeng Wu[1,2(✉)], Hong Yu[1,2], Huanjun Zhang[1], Yizhi Zhou[2], and Yuxuan Zhang[1,2]

[1] School of Information Engineering, Dalian Ocean University, Dalian 116023, China
wujunfeng@dlou.edu.cn
[2] Key Laboratory of Environment Controlled Aquaculture, (Dalian Ocean University) Ministry of Education, Dalian 116023, China

Abstract. With the improvement of modernization, China's animal husbandry and fishery enterprise has ushered in a new vogue of informationization, factorization, and precision farming, and the demand for unique identification of individual animals is growing. Traditional individual animal identification methods, such as footprint identification, molecular biology, and different techniques, have low accuracy, excessive cost, and different risks. RFID technological know-how and implants put on monitoring units and different techniques additionally face invasiveinvasiveness, excessive labor costs, slender application scopes and challenges in promoting a massive place and different issues. Deep learning is enjoying an increasing number of essential positions in the discipline of animal individual identification, which has made it possible for the noninvasive recognition of individual animals. This paper discusses the progress of individual animal recognition using computer vision techniques and its application popularity in different species fields, focuses on the issues and challenges of individual animal recognition, and suggests that future lookup instructions for animal identification are foreseen.

Keywords: Animal recognition · Individual identification · Deep learning · Computer visio · Convolutional neural network

1 Introduction

Accurate knowledge of an individual animal's diet, health, and behavior is a prerequisite for the precise breeding of reared animals and can also provide an important basis for animal protection departments to carry out wildlife protection. Individual animal identification technology can further identify individual animals based on species identification and realize "face recognition" in the animal world. Individual animal identification can provide researchers and breeders with data for decision analysis and thus improve breeding standardization (see Fig. 1). In recent years, with the growing demand for precision breeding in animal husbandry, uncommon animal protection, and preservation of ecological security, the need for the identification of individual animals has become increasingly urgent, and research in this area has gradually received widespread attention from industry and academia [1–3].

© The Author(s), under exclusive license to Springer Nature Singapore Pte Ltd. 2022
Y. Wang et al. (Eds.): ICPCSEE 2022, CCIS 1628, pp. 287–309, 2022.
https://doi.org/10.1007/978-981-19-5194-7_22

(a) Panda individual identification.   (b) Puffer fish individual identification.

Fig. 1. Animal individual identification.

At present, animal individual identification research is gradually developing into one of the hot topics of research and application in industry and academia. According to different practical applications, many experts and scholars have conducted targeted analysis and research on the characteristics of different animals and designed different methods to achieve individual animal identification. Lu [4] researched a set of noncontact individual identification methods based on acoustic feature extraction for individual giant pandas and proposed a giant panda compound Meier frequency cepstrum coefficient to improve the high-frequency resolution of the system according to the acoustic characteristics and frequency distribution of giant pandas as a way to extract the acoustic features of giant pandas. Eric et al. [5] conducted a long-term tracking identification study on individual herd-dwelling domestic animals, analyzed the link between keypoint-based detection and MAP estimation, and proposed a detection-based probabilistic tracking method. The tracking method uses the visible key points of a single animal as input. Guo et al. [6] analyzed animal images to identify species using deep learning techniques to analyze video and still-frame images to develop a system dedicated to automatically detecting animal faces and identifying individuals, addressing the difficulties that limit researchers' ability to quantify data for ecological conservation, analyze animal behavior and enhance conservation while also providing a dataset containing primates and carnivores. Lin et al. [7] summarized the important techniques of preprocessing, target detection, recognition, and tracking of underwater target images with an eye on the basic ideas and methodological features of underwater image enhancement techniques and image recovery techniques, laying the foundation for the identification of individual animals in underwater optical images.

It is extremely challenging to achieve accurate individual identification of animals, as the individual characteristics of different species vary greatly, and the environments in which animals live vary, requiring cross-fertilization of computer technology, artificial intelligence, and zoology. We have conducted a comprehensive review and combined the construction of individual animal identification datasets, individual animal target detection and feature extraction, individual animal identification methods, and applications of individual animal identification, with the aim of providing some reference for relevant experts and scholars in smart agriculture, precision farming, wildlife conservation, and biodiversity censuses.

2 Construction of Animal Datasets

Data acquisition is the basis of individual animal identification, and the target detection, alignment, feature extraction, and identification needed for individual identification all need to be carried out on the basis of a certain scale of datasets. Based on the different living environments and habits of each species of animals, the type of data collected will also have differences, and at the same time, the acquisition methods will be different for different individual animals. As of now, teams studying the field build their small datasets based on their respective research directions. To obtain a large quantity and high-quality data, researchers mostly use color dithering, affine transformation, Gaussian noise, and GAN [8]. Thus, the processed dataset can meet the needs of individual animal identification algorithm research. However, the diversity of data and data size in the dataset are particularly important to construct individual recognition models with high robustness and generalization ability.

2.1 Data Collection Method

Animal data acquisition is generally carried out through cameras or depth cameras, and such methods are camera-based data acquisition methods. There are three commonly used methods: infrared laser camera acquisition, micro bayonet camera acquisition, and depth camera acquisition. The characteristics of the three methods are as follows:

1) Infrared laser camera acquisition. The radiation path of infrared laser cameras is generally 300–5000 m, which is suitable for wildlife data collection over a long distance. Infrared laser cameras cannot be used at short distances due to their concentrated energy and small angles. Infrared cameras can still acquire high-quality data in nighttime environmental conditions.

2) Miniature bayonet camera capture. Miniature bayonet cameras are highly flexible and can be widely used in remote mountainous areas and farming areas for surveillance systems. Integrated video surveillance and individual animal identification with more efficient and intelligent commercial application functions, fusion image processing chip and star algorithm, improved ISP technology, there are 2 DSP processing chips, one is responsible for encoding the H.264 output video stream, and the other is dedicated to JPEG encoding and output image information analysis, two independent chip design can ensure that no matter day or at night can achieve good video and image capture effect.

3) Depth camera acquisition. A depth camera based on a deep learning structured light algorithm with 3D structured light, with the help of structured light devices, can simultaneously capture color images, infrared images, and depth images, identify and analyze individual animals in the scene, generate 3D images, create individual models with depth information, and more effectively defend against attacks on animal body parts with obscurants or patterns that are not photographed. The analysis time is reduced from the previous time. The analysis time is compressed from 1–2 s to milliseconds.

2.2 Datasets Related to Individual Animal Identification

Datasets are the basis of individual animal identification research, and different researchers have constructed datasets with different contents and sizes depending on the objects studied. At present, due to the difficulty of data collection, there are still few datasets applicable to individual animal identification, and they are mainly focused on several species that are more common and have certain application value. Table 1 shows several relevant animal datasets.

1) ImageNet dataset [9]. Created under the direction of Professor Feifei Li at Stanford University, is a dataset on computer vision aspects. The dataset contains more than 14 million images divided into 27 major categories and more than 20,000 minor categories. The ImageNet has been considered an industry benchmark for evaluation in terms of image classification algorithm performance. ImageNet, as an enduring research effort, aims to provide an easily accessible image database to research workers in various countries. Containing more than 2200 natural images of elephants, it facilitates scholars to research the identification of different individuals of the same animal.

Table 1. Animal datasets.

Dataset	Animal category	Data modal	Amount of data	Data chatacteristics
ImageNet	27	Image	14000000	Used for classification, localization and detection tasks
Kaggle Animal-10	10	Image	280000	Used to test different image recognition networks
Animalweb	21	Image	21900	One massive annotated animal face partial layer dataset with 9 key facial feature points
Pascal VOC	20	Image	17125	Mainly doing detection and classification tasks
DLOUFish	1	Image	400	Used to detect and identify the ID of each red fin oriental triggerfish

2) Kaggle Animal-10 dataset (Fig. 2), a dataset on Kaggle contains approximately 10 categories of 28K medium-quality animal images: dogs, cats, horses, spiders, butterflies, chickens, sheep, cows, squirrels, and elephants. The image counts for

each category range from 2K to 5K, satisfying researchers carrying out the task of identifying different individuals between the same animals in.jpeg format. Each category is stored in a separate folder. All images are gathered from Google Images and are examined manually, with some incorrect data simulating the real situation (e.g., images taken by the user's application).

3) AnimalWeb dataset. This dataset [10] was created by Muhammad, Georgios, et al. and contains 21,900 facial images of 334 different animal species, while covering 21 different taxonomic animals, containing approximately 1,000 images of each animal, such as dogs, horses, elephants, butterflies, cats, etc., providing relevant data in the identification of different individuals of the same animal to facilitate scholars to conduct the study. Each facial image has 9 landmarks key facial feature points, with strict manual annotation and alignment operations on the dataset.

4) PASCAL VOC dataset. The dataset was created by Mark Everingham, Luc van Gool and others. The animal category contains cats, dogs, cows, horses, sheep, and birds, with approximately 1000 images of each animal, which can provide data support for researchers to identify individuals among animals of the same species. The 2 main tasks of this dataset are detection and classification, and for target detection, each image corresponds to an annotation file in XML format.

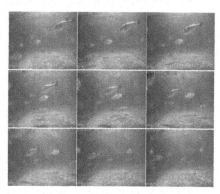

Fig. 2. Dataset of Kaggle Animal-10.   **Fig. 3.** Dataset of DLOUFish.

5) DLOUFish dataset. The dataset was obtained by the Key Laboratory of Environment Controlled Aquaculture Ministry of Education of Dalian Ocean University through placement of depth cameras in aquaculture-related enterprises. The dataset consists of video streams transformed and manually filtered data frames to prevent a large amount of similar data in the dataset from affecting the subsequent use. The dataset has 800 images of redfin oriental puffers captured underwater (see Fig. 3), and each redfin oriental puffer in the image and its unique ID were labeled with an annotation tool to facilitate subsequent research.

2.3 Data Preprocessing

Due to the complex growth environment of animals, the collected data usually have a large noise; for example, the fish images obtained through the underwater camera are usually obtained in the case of unclear water quality, the animals in the breeding area have a high degree of shading, and the animal data images in the field have obstacles blocking them and dim light, making the obtained image data blurred and unclear, which will directly affect the accuracy of individual identification if not processed. Therefore, some preprocessing of the data before image analysis (feature extraction, segmentation, matching, recognition, etc.) is required to improve the algorithm design and recognition accuracy [11].

1) Luminance processing. Generally, methods that do not add image pixels, such as image enhancement and image restoration, as well as superresolution algorithms that do not add additional information to the image, are used. Luminance processing requires the use of many other classical image processing methods, including histogram equalization, grayscale transformation, image smoothing, image restoration, and image superresolution, which mainly solve the problems of image color deviation and blurring. For example, Liu et al. [12] used histogram equalization to image enhance the collected Atlantic salmon data to make the data more usable.

2) Image enhancement. Using deep learning methods for image enhancement, the nucleus of the method relies on a mass of comparable statistics to train a convolutional neural network, which learns relevant information about the image data from the mass of data. It can reap the effect of understanding, judging and predicting the image content. When testing input a blurred image that does not contain detailed information, the neural network automatically adds details to the image to be processed based on what it has learned to achieve image enhancement. For example, Ye et al. [13] proposed a color correction method that is an unsupervised adaptive network for underwater images, and Ma et al. [14] proposed a micro-optical image enhancement algorithm that fuses residual networks and retinex learning.

3) Data expansion. In the absence of massive data, the only way to ensure effective model training is to generate more data from the limited data. The number and diversity of training samples (noisy data) are increased by data expansion to increase noisy data and improve model robustness. Data expansion methods include mirroring, rotation, scale transformation [15], random keying, color dithering, etc.

3   Individual Animal Target Detection

To perform accurate individual recognition of individual animals, it is necessary to first detect individual animals in the image and therefore to perform accurate target detection of individual animals. Target detection application scenarios cover a variety of fields, such as aerospace, robot navigation, intelligent surveillance, and industrial inspection. Individual animal identification was performed on the basis of target detection. Accurately identifying the location of an object from a picture by target detection and accurately classifying that object, and the more accurate the result of target detection, the more accurate the recognition result will be. Target detection has great practical significance and research value.

3.1 Traditional Target Detection Algorithms

Many tasks in traditional target detection algorithms cannot be solved in one step and require multiple steps. In the traditional algorithm, feature extraction mainly adopts the manual selection method. The benefit of this method lies in the selection of more important features based on human subjective experience, but the disadvantage lies in the weak objectivity, easy to ignore some hidden details of features, poor robustness of manually designed features, and low efficiency. Common traditional feature extraction is generally considered in terms of color, gradient, texture, and shape. When traditional methods perform animal target detection tasks, the main feature classifiers usually used are the plain Bayes classifier (Naive Bayes), Rogers Returns, Sofimax classifier, KNN classifier, and SVM. Therefore, traditional animal target detection methods are suitable for relatively small-scale and distinctive animals.

1) HOG+SVM algorithm. This target detection method was proposed in 2006 and is mainly used for pedestrian detection. The extracted features mainly revolve around the underlying features and mid-level features, such as color and texture. The main steps are to first extract the HOG features, train the SVM classifier, extract the target region using sliding windows, perform classification, and finally filter and merge the candidate frames using the NMS algorithm to output the final detection results. HOG+SVM is usually implemented with OpenCV, and the HOG features are usually large in dimensionality, which usually requires PCA dimensionality reduction.

2) DPM algorithm. This target detection algorithm was proposed in 2008, and the DPM algorithm adopts improved HOG features, an SVM classifier and sliding windows detection idea. For any input image, calculate its DPM feature map, perform the calculation of the original DPM feature map with the root filter response map and part filter response map, obtain the final response map, use the latent SVM classifier for training, and finally obtain the detection result.

3.2 Animal Target Detection Algorithm Based on Deep Learning

Deep learning algorithms mostly use an end-to-end scheme, where a graph is an input and the final desired result is output, and the algorithm details and learning process are all left to the neural network. A deep learning algorithm is a method that relies on big data training and automatically extracts features and automatically learns all kinds of features of the task target from a large amount of data, especially the detailed features that can be easily ignored and can be mined out. Therefore, deep learning methods are suitable for relatively large-scale, poorly characterized animals with complex environments. In recent years, more than 90% of the published literature on animal target detection and recognition algorithms uses deep learning algorithms, including Faster-RCNN, ResNet [16], DenseNet, the improved YOLOv3 [17, 18], Inception-V3, etc.

In recent years, new algorithms such as NAS-FPN, EfficientDet and YOLOF have also emerged. Several major deep learning network models are characterized as follows.

1) Faster-RCNN [19]: Ross B. Girshick proposed in 2016 to structurally implement the four-in-one overall network Faster-RCNN (feature extraction (FE), region proposal

network (RPN) bounding box regression (rect refine), classification union), thus greatly improving the overall performance of the system, especially the detection speed. Outperforming R-CNN [20] and Fast-RCNN [21].

2) ResNet: ResNet proposes the idea of residual learning and therefore solves the problem of very deep networks that are difficult or even impossible to train. The network was proposed by Kai-Ming He's team in 2015, and the team also successfully trained a deep network containing 152 layers using residual blocks. It was difficult to train such a deep network before residual learning was proposed because traditional convolutional networks inevitably lose information when passing information layer by layer, which not only causes information loss to a certain extent but also causes problems such as gradient explosion and gradient disappearance the deeper the convolutional layers are. ResNet makes it possible to effectively protect the integrity of the information while simplifying the learning goals and difficulty by continuously passing input information around to the output.

3) Inception-V3: There are four versions of InceptionNet, and each new version is continuously optimized and developed based on the previous version, the first of which was presented by Google in the ILSVRC competition in 2014. Inception-V3, as one of the most representative versions of this series, uses dense components to make local sparse junctions reach near-optimality, uses convolutional kernels of different sizes to obtain perceptual field information of different sizes, and finally fuses these different scales of feature information. To a certain extent, inception solves the problems of overfitting, computational complexity, and gradient disappearance brought by deeper and wider networks.

4) DenseNet [22]: The DenseNet convolutional neural network contains a densely connected part. In this network, any two layers are related to each other, and the input of each layer is the concatenation of the output of all previous layers. Compared with ResNet (deeper) and Inception (wider), this network is designed to be thinner, and the authors start with features to achieve better results and fewer parameters through the extreme use of features.

5) YOLO (You Only Look Once) [23]: Redmon et al. proposed a model in 2016 to address the problems of slow detection and optimization. The algorithm is a one-stage model. YOLO redefines target detection as a regression problem by applying a single CNN to a whole image, dividing the picture into grids and predicting the bounding boxes and class probabilities for the appearance grid, and finally filtering out the bounding boxes with confidence less than a threshold using nonmaximal suppression (NMS). YOLO is very fast, and since the detection problem is a regression problem, it does not require a complex pipeline and can handle real-time video streams [24] with a latency of fewer than 25 ms.

6) NAS-FPN [25]: In 2019, the Google Brain team proposed NAS-FPN (Neural Architecture Search Feature Pyramid Network). The NAS-FPN is a feature pyramid structure using neural network structure search discovery, with top-down and bottom-up connections to perform feature fusion at different scales. The main contribution of this method is to design the search space to cover all possible cross-scale connections and generate multiscale feature representations. NAS-FPN works well with various backbone models, such as MobileNet, ResNet, and AmoebaNet [26].

7) Efficientdet [27]: In a paper published in November 2019, the Google team proposed Efficientdet. Efficientdet is a general term for a series of target detection algorithms, containing 8 algorithms D0-D7. EfficientDet has two main innovations. First, a weighted bidirectional feature pyramid network (BiFPN) that can perform multiscale feature fusion simply and quickly is proposed. Second, a composite scale expansion method is also proposed. This method can scale all backbones, resolutions of feature networks and prediction networks, and depths and widths uniformly. This allows the EfficientDet series to reduce the number of parameters by a factor of 4–9 compared to previous traditional target detection algorithms.

8) YOLOF [28]: YOLOF (You Only Look One-level Feature) is a paper published in CVPR in March 2021. Replace complex feature pyramids with single-layer feature maps to simplify the complexity of partitioning strategies for optimization problems. YOLOF significantly improves the detection performance by designing two core components, a dilated encoder and a uniform matching strategy.

Table 2. Individual animal identification method.

Animal type	Algorithm	Algorithm characteristics	Goal
Golden Monkey [30]	Faster-RCNN, attention, Clustering	Attentional mechanisms make the network focus on the facial features	Improving the accuracy and speed
Chimpanzee [34]	SSD+KLT	SSD detects faces, KLT for facial track tracking, Deep CNN model for recognition	Addressing the limitations of manual data processing
Panda [45]	Faster-RCN, segmentation, Spatial transformation	Detection, segmentation, comparison and prediction stages are intertwined	Conservation and precise management of giant pandas
Brown bear [49]	FaceNet	Bear face detection, redirection and cropping, bear face embedding, and classification	Identification of animals without special markings

(continued)

Table 2. (*continued*)

Animal type	Algorithm	Algorithm characteristics	Goal
Cattle [56]	Inception-V3+LSTM	The former CNN model extracts the features and trains the latter with the extracted features	Individual raw cattle without relying on RF ear tags
Pig [60]	Blinear-CNN+ResNet	Calculate the outer product of different spatial locations and fuse the features of different channels	Realize individual pig identification

4 Methods of Individual Identification of Animals

The identification of individual animals requires capturing small differences in appearance features between similar species. Therefore, it is necessary to design recognition algorithms according to the characteristics of different recognition species. Table 2 summarizes the current more popular algorithms for individual animal identification.

4.1 Typical Animal Individual Recognition Algorithms

Individual Identification of Golden Monkeys Based on Tri-AI Technology. The Sichuan golden monkey is a rare and endangered species endemic to China. To protect and study the Sichuan snub-nosed monkey that inhabits the Qinling mountain system, a research team from Northwestern University has been conducting tracking surveys for many years. Although individual identification of wild golden monkeys, which can be observed close at hand, was achieved as early as 20 years ago, it has been a challenge for researchers to have a better accuracy and faster identification of individual wild golden monkeys. In response to this cutting-edge problem, Professor Guo Songtao of Northwestern University established an animal AI (artificial intelligence) research team, which is deeply cross-fertilized with the field of computer science to protect the Sichuan golden monkey with scientific understanding. Based on the long-term characterization of this species by other teams, Guo Songtao et al. incorporated attention mechanisms into neural networks [29] and designed the first individual recognition system Tri-AI [30] for golden monkeys.

According to Fig. 2 in the literature [30] (see Fig. 4), the Tri-AI system overrides the traditional identification methods that rely on the presence of features on individual animals or purposefully mark individual animals artificially, thus achieving the performance of accurate identification and continuous tracking of individual wild animals for data collection, which provides the possibility of "no observer interference effect" in zoological studies.

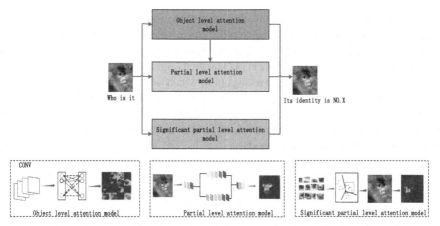

Fig. 4. Deep neural network model with attention mechanism.

The method uses Faster-RCNN as a detector to detect the monkey face region, and this part is added to the object-level attention mechanism, which focuses on detecting the object; the detected monkey face region is truncated and then passed through the attention mechanism network, which focuses on the facial features [31]. Finally, the salient-level attention mechanism with significant facial features is compared with all the individuals in the database by the artificial intelligence clustering algorithm [32] to identify the ID of this golden monkey, and the ID of the individual with the corresponding golden monkey in the database is output. If there is no corresponding individual golden monkey, add it to the database and assign a new ID to this golden monkey.

Wild Video Chimpanzee Face Recognition Based on Deep Learning. Currently, recording and observing animal behavior with video is very common [33], and manual processing of large amounts of data is no longer enough in terms of time and resources to meet the demand, making it extremely difficult and challenging to perform large-scale analysis. Daniel Schofield et al. [34] proposed a deep CNN method to address the limitations due to the human workload [35] required for manual processing and the accuracy of the information. The team obtained 15274 images of 23 chimpanzees to obtain the best overall performance of identification and gender recognition accuracy. Chimpanzee faces were identified to generate a co-occurrence matrix to track demographic changes in aging in social network structures [36].

The workflow of the system is (A) frame extraction from the original video; (B) face detection is implemented using the SSD [37] model; (C) face tracking is implemented using KLT [38], the tracker to group the detected face targets into face trajectories; (D) face recognition and gender recognition [39], which is achieved by training a deep CNN model; (E) the system's input does not require any preprocessing, raw video is sufficient, and the tagged face track data are used as temporal and spatial information; and (F) the temporal and spatial information is then used for social network analysis [40].

A Study on Image Recognition of Giant Pandas Based on Mass Captivity. As a highly endangered animal [41], the conservation and reproduction of giant pandas have attracted widespread attention from scholars, and the identification of individual giant pandas is crucial to achieving more scientific conservation and precise management of giant pandas. Traditional identification methods rely on invasive means, such as identification based on footprint recognition [42] methods, special patterns drawn on the body [43], and RF ear tags with unique markings [44], which lack accuracy and have expensive and easily lost equipment. Chen et al. [45] proposed a noninvasive and fully automated deep learning algorithm that consists of several neural networks that can perform all the tasks of individual recognition of giant pandas, enabling noninvasive image-based individual identification of giant pandas.

The network is divided into four parts: (a) giant panda face detection, using Faster-RCNN as the target detector, including the region suggestion network [46] and classification network, using the already trained ResNet-50 [47], which successfully detected the face region of the giant panda; (b) segmentation network, dividing the giant panda face as a region of interest. Removing the background by generating a binary mask (its black pixel image), localizing the location and extent of the region of interest, again using the already trained ResNet-50, thus achieving an image preprocessing role. (c) Spatial transformation network [48], where the segmented panda face image is affine transformed with the standard panda face image. This module is designed using ResNet-50, and the last layer has 6 neurons corresponding to 6 affine transformation parameters for alignment operations between the segmented panda face image and the standard panda face image. (d) Giant panda identity prediction: the aligned giant panda face image is input to this module, and the giant panda ID prediction is performed by a classification network that still uses the already trained ResNet-50. The face images after the alignment operation are fine-tuned, and each output corresponds to a node indicating the probability value, which indicates how likely the prediction is to be the same panda.

In the experiments, Chen et al. used closed-set recognition and open-set recognition. In the former, manual identity matching is required when an unknown giant panda is predicted; in the latter, no manual matching is required when a panda with unknown identity is predicted, and the output of additional nodes is just the probability value.

Brown Bear Individual Recognition. At present, with the exception of primates, which use visual methods for individual identification, some other species without special markings actually lack more objective scientific methods of identification. Based on this problem, Melanie Clapham et al. [49] applied face recognition technology to propose a method to recognize individual brown bears [50], and the algorithm involves tasks such as target detection, landmark detection, similarity comparison networks [51], and support vector machine-based classifiers. Based on the workflow of FaceNet [52], the pipelined workflow method is divided into four parts: (1) The bear face detection part aims to find the bear face regions and feature point coordinates in the image. This part uses a target detector (OD) [53] and a shape predictor (SP) [54]. od is a convolutional neural network based on sliding windows and dlib for face detection; SP uses dlib and a regression tree ensemble to achieve face alignment and trained using iconic labels from the gold dataset. (2) The purpose of the bear face reorientation and cropping section is to use the detected facial feature point coordinates to reposition the feature points and extract the bear face, use the coordinates of the two eyes for alignment, center and rotate the face to the optimal orientation, and then scale and crop to 150 * 150 pixels. (3) Bear face embedding is essentially a function of learning similarity. The output is a vector of facial features that can be compared with other output vectors. Melanie Clapham et al. trained a deep CNN similarity comparison network with a ResNet-34 architecture to output a 128-dimensional vector of each facial image as a numerical representation of that facial feature in Euclidean space. (4) Bear face classification uses a one-to-one classifier of the support vector machine to train the embedding and labeling of facial features and after training.

The method achieves automatic facial recognition of wild animals, and deep learning is further applied, achieving an average accuracy of 97.7% in the absence of strict facial pose criteria.

Cattle Individual Identification. Precision livestock farming is also necessary for individual cattle identification, and current methods for identifying cattle either require vision [55] or unique RF ear tags. Qiao et al. [56] combined the benefits of Inception-V3 [57] and LSTM [58] networks to introduce a recognition network for individual cattle (see Fig. 5). The former extracts features of cattle in images [59], and the latter network models training features.

Inception-V3 commonly exists in image recognition. In this structure, the most important is the features extracted from the last pooling layer of Inception-V3, which outputs a 2048-dimensional vector representing the cattle body features, and this vector is the input to the LSTM network. LSTM is a temporal recurrent neural network with a robust capability to analyze and have in mind lengthy sequences for input data. For cattle recognition, each and every frame is a 2048-dimensional CNN feature. The output of the LSTM is a confidence value in the range of 0 to 1. The maximum value is used as the predicted cattle identification result, and if it matches the ground truth class label, then it is the individual cattle.

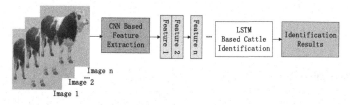

Fig. 5. Cattle identification scheme based on CNN and LSTM.

Pig Face Recognition Algorithm. Hangzhou University of Electronic Science and Technology used VGG-16 [60] as a feature extractor, fused the features extracted from different layers as the final features, and used the final fully connected layer for classification to achieve noninvasive identification of individual pigs [61, 62]. The design makes flexible use of bilinear convolutional neural networks (Blinear-CNN) [63] and multilevel feature fusion structures (see Fig. 6). Thoughtful handling of feature information, in addition to the pig face features extracted from the last convolutional layer

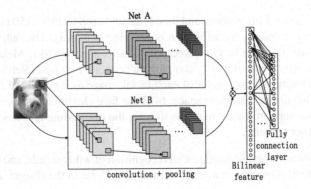

Fig. 6. Improved multilevel bilinear-CNN network structure model.

of the feature extractor itself to perform the outer product, it also performs the outer product with the pig face features extracted from the other previous convolutional layers to achieve the purpose of fusing features of different levels.

Table 3. Recognition accuracy.

Animal type	Number of datasets	Accuracy
Golden Monkey	102399	94.1%
Chimpanzee	15274	92.5%
Panda	6441	96.27%
Brown bear	4674	83.9%
Cattle	5370	91%
Pig	2110	95.73%

4.2 Performance Comparison

The recognition accuracy of the above individual animal recognition algorithms is shown in Table 3. Both have high recognition accuracy and speed. The advantages and disadvantages of these recognition algorithms and their scope of application are shown in Table 4.

Table 4. Advantages, disadvantages and scope of application.

Animal type	Advantages	Disadvantages	Application scope
Golden Monkey	Graded attention mechanism	Longer time to train	Primates, large, medium-sized mammals
Chimpanzee	Reduced data processing time	Model is difficult to understand	Wild animals
Panda	Fusion of multiple neural networks	Training the model is more tedious	Bigger animals
Brown bear	Simple network structure	Manually add new individuals	Animals lacking special markings
Cattle	Combine the advantages of LSTM network	Larger number of parameters	Large range of activities
Pig	Fusing the extracted features at different levels as an outer product	Pig appearance has a large impact on model performance	Farming animals

5 Application of Animal Individual Identification Technology

Individual animal identification technology can be widely applied in livestock production, scientific research, ecological protection, and other fields, allowing people to understand the daily behavior, action trajectory, and living habits of different species in a more scientific way. Accurate identification of each individual animal is necessary for the daily management and analysis of the growth condition of animals.

1) Precision breeding. Precision livestock farming starts with the continuous automatic collection and monitoring of farm animal data, which can thus improve the management of these animals and adjust them accordingly. Smart devices help farmers create optimal living conditions for each animal, given that farms are becoming larger and more complex to manage. It is very important for farmers to track each poultry in real time through individual animal identification technology, to master the growth of each animal, to analyze the growth data and intake of each animal at different times and to feed it rationally to achieve scientific feeding. Using precision livestock farming technology, farmers can make better decisions early in the farming process because their observations are enriched by artificial intelligence technology, and all data can be systematically analyzed, enabling precise farm feeding, which reduces the waste of resources and increases the effectiveness and efficiency of farming. Cui et al. [64] developed a fully automated fish detection algorithm that enables intelligent fish detection by deep learning, the current mainstream technologies in this field. By accurately identifying each fish and feeding it after analyzing different

periods for each fish, Wang [65] proposed a deep learning-based algorithm for real-time detection of caged egg behavior, which provides a basis for accurate breeding by detecting caged hen behavior in real time and changing the breeding strategy at any time according to the actual situation.

2) Animal behavior monitoring. Each species has a unique role, and protecting animals can better maintain biodiversity and protect the biological chain to better maintain the ecological balance. Individual animal identification technology can track the activity trajectory of individual animals in real time to ensure their activities within a safe environment, analyze the popular and cold areas of individual animal activities, conduct individual animal behavior monitoring, and grasp their behavioral habits and characteristics. By tracking and monitoring the behavioral status of different individual animals [66], a reasonable protection and management plan can be formulated to address various abnormal behaviors promptly to enhance the management convenience and protect the safety of animals to a certain extent. Claire et al. [67] proposed an automatic face recognition method for rhesus monkeys based on a nonintrusive identification of individual rhesus monkeys in the video to keep track of their physical condition in real time and to protect them medically in real time. David et al. [68] put forward a face recognition idea to facilitate the identification of individual lemurs. Analysis of factors influencing the size and growth status of natural populations through long-term studies of known individuals; Pranjal et al. [69] designed an algorithm that relies on images for giant panda behavior recognition. Monitoring the behavior of giant pandas through deep learning methods for noninvasive long-term behavioral observation is essential for conservation and understanding their health status.

3) Intelligent classification. Through individual animal identification technology, the length, weight, and maturity of individual animals are accurately predicted. Real-time monitoring of individual animal growth and timely classification of individual animals in the breeding area. It effectively avoids the transmission of epidemic diseases and timely transport of farmed mature animals to the market to avoid overfarming. Huang et al. [70] designed a system for measuring the body length of individual fish and predicated computer vision keypoint detection, which provides the possibility to differentiate eligible individual fish into different regions by detecting the location of key points in the midline of the fish body for fish length measurement.

6 Summary and Outlook

At present, individual recognition of animals is still in its initial stage in terms of the dataset, animal target detection, feature extraction, maturity of application, etc. Many research contents need more in-depth research by experts and scholars in related fields. The relevant elements that have been more studied are mainly focused on the following aspects:

1) The construction of the dataset: The construction of a dataset is the first step in conducting animal individual recognition studies, which require a substantial dataset for support. The data collection methods used for individual animal identification

focus on cameras and ordinary digital cameras that have long prevented the field, including infrared laser cameras, miniature bayonet cameras, and depth cameras, all of which are well adapted to the nighttime environment and can be shot normally under insufficient light conditions. Commonly used data labeling tools include Labelimg, Labelme, RectLabel, and OpenCV/CVAT. Labelimg supports PASCAL VOC format and YOLO format; Labelme supports object detection, data annotation for image semantic segmentation, rectangle, circle, line segment, and point annotation, video annotation, export of VOC and COCO format data; RectLabel supports object detection, image instance segmentation data annotation, export JSON, CSV, and XML format data; OpenCV/CVAT is an online annotation tool that supports image and video data annotation.

2) Data preprocessing: Since animals can move freely and their environment is relatively complex, preprocessing of data is required before individual animal identification algorithms are performed. Removing unnecessary noise from the data further enhances the accuracy of the animal individual recognition algorithm. There are two methods of animal data preprocessing. One is predicated image processing methods. The familiar histogram equalization, grayscale transformation, and so forth all start from the image pixels to deal with the problem of inadequate image quality. The other is a deep learning-based image enhancement method. After the neural network is trained using a substantial amount of similar data, the network learns the relevant information of such data and then automatically completes the missing information based on what it has learned.

3) Animal target detection: The development of animal target detection has been broadly divided into two phases. The first stage is the traditional approach, using image processing and machine learning algorithms [71]. The other phase is the use of emerging deep learning methods for target detection tasks. Traditional methods include the scale-invariance feature transform [72] method and histogram-oriented gradient (HOG) method [73]. Emerging deep learning methods accomplish detection tasks with the help of convolutional neural networks. The method has three features: local connectivity, parameter sharing, and translational invariance. Emerging algorithms come in two forms. One is two-stage detection algorithms, which divide the detection problem into two stages. First, candidate regions (region proposals) are selected, and then the candidate regions are identified. Typical algorithms are RCNN, Fast-RCNN, and Faster-RCNN. The other is one-stage detection algorithms, which directly regress the image to produce the class and coordinate position of the object. More typical algorithms are YOLO and SSD.

4) Feature extraction: Features are the only basis for identity matching, the whole image content is represented by the extracted features, and matching or discrimination is performed according to the features. There are three main forms of feature extraction using traditional machine learning: color-based features, texture-based features and shape-based features. Superior deep learning algorithms, in simple terms, use neural networks to complete the feature description of the target image and extract features automatically without human intervention. There are two kinds of feature extraction network designs commonly used in animal individual recognition algorithms. One is based on ResNet, Inception-V3, VGG, and other basic neural networks to perform

improvement, and the other is combining two and more different network features for optimization modification.

6.1 Major Problems

At present, animal identification technology has been developed to a relatively complete system, including data acquisition, dataset construction, data preprocessing, target detection, feature extraction, target identification, etc., in more than dozens of fruitful animal research studies. These technologies lay the foundation for intelligent farming and effective protection of animals, but there are still problems, and the main problems are three.

1) The dataset is not large enough and lacking in variety. Some animals grow in complex and variable environments, and it is not easy to collect high-quality data. Those who conduct animal identification research usually collect data from the network, perform manual annotation, and use data augmentation to expand the dataset. Expanding the data in this way is prone to the absence of variability in the data sample. As a result, the generalizability of the network is compromised, and optimal results cannot be achieved.

2) It is difficult to detect animal targets with high overlap. Animals generally live in more complex environments, and most of the creatures live in groups and are densely distributed, so the problem of occlusion is serious. This leads to the situation in which some of the acquired data are obscured by individual animals. The overlap between targets in a dense occlusion environment is serious, and the neural network mistakenly considers multiple animals with high overlap as one target individual when performing target detection, leading to a decrease in target detection accuracy and generating missed detections. Both the scale variation and severe occlusion problems encountered in the collected data challenge CNN networks with scale invariance [74, 75].

3) The models have not been fully investigated. Deep learning models have the most advanced performance in individual animal identification tasks. However, when applying it to individual animal identification tasks, pretraining using the human dataset was followed by fine-tuning using the animal dataset. The application of these deep learning models to other individual animal identification tasks did not make any significant innovative modifications to the architecture of the models, and the generalization of the models was poor.

6.2 Outlook

Individual animal identification technology is an effective application of animal precision breeding and protection, which is the development trend of China's agricultural modernization. To improve the modernization of animal breeding and the intelligence of animal protection, future research on individual animal identification can be carried out in the following three areas.

1) Research on new methods to produce high-quality datasets. Substantial and high-quality datasets are the basis for better recognition results. Research on effective animal data collection techniques, data annotation techniques, and so on. We are exploring new methods for producing more scientifically rigorous data, building large-scale, high-quality datasets containing various species of animals, and improving the accuracy of individual animal identification.

2) Actively break through the key technical difficulties of accurate detection and individual identification of animal targets. Promote research on interdisciplinary and in-depth cross-fertilization technologies combining animal morphological analysis, high-speed small target imaging technology, and human visual cognitive reasoning mechanisms. Solving the problem of possible mutual exclusion between speed and accuracy of animal individual recognition algorithms. We will vigorously promote the research of key core technologies for animal target detection and individual identification.

3) Increase investment in R&D and innovation. Encourage the exploration of innovative integration of algorithms in multiple identification fields and increase the investment in funding for individual animal identification technologies. Support the joint development of intelligent monitoring devices for real-time accurate tracking and detection of animals. Solve the problems of quality, clarity, energy, waterproofing and anti-attack of contactless intelligent monitoring devices for animals. Improve the stability and reliability of equipment and develop intelligent monitoring equipment for accurate detection and real-time track tracking of individual animals. The development of individual animal identification technology should be effectively advanced.

Acknowledgments. The National Natural Science Foundation of China (31972846) funded this research, Key Laboratory of Environment Controlled Aquaculture (Dalian Ocean University) Ministry of Education (202205), Major Special Plan for Science and Technology in Liaoning Province (2020JH1/10200002).

References

1. Hou, J., He, Y., Yang, H., et al.: Identification of animal individuals using deep learning: a case study of giant panda. Biol. Conserv. **242**, 1–6 (2020)
2. Sun, Y.K., Huo, P.J., Wang, Y.J., et al.: Automatic monitoring system for individual dairy cows based on a deep learning framework that provides identification via body parts and estimation of body condition score. J. Dairy Sci. **102**(11), 10140–10151 (2019)
3. Kalafut, K.L., Kinley, R.: Using radio frequency identification for behavioral monitoring in little blue penguins. J. Appl. Anim. Welf. Sci. **23**(1), 62–73 (2020)
4. Lu, H.K.: Analysis and Research of Giant Panda Individual Identification System Based on Voiceprint, pp. 1–75. University of Electronic Science and Technology of China (2019)
5. Psota, E.T., Schmidt, T., Mote, B., et al.: Long-term tracking of group-housed livestock using keypoint detection and MAP estimation for individual animal identification. Sensors **20**(13), 3670 (2020)

6. Guo, S.T., Xu, P.F., Miao, Q.G.: Accurate identification of "true and false Monkey King" animal individual identification system is coming. Netinfo Secur. **21**(03), 99 (2021)

7. Lin, S., Zhao, Y.: Review on key technologies of target exploration in underwater optical images. Laser Optoelectron. Prog. **57**(6), 060002 (2020)

8. Terayama, K., Shin, K., Mizuno, K., et al.: Integration of sonar and optical camera images using deep neural network for fish monitoring. Aquacult. Eng. **86**, 102000 (2019)

9. Deng, J., Dong, W., Socher, R., et al.: ImageNet: A large-scale hierarchical image database. In: Proceedings of IEEE Computer Vision & Pattern Recognition, pp. 248–255 (2009)

10. Khan, M.H., Mcdonagh, J., Khan, S., et al.: AnimalWeb: a large-scale hierarchical dataset of annotated animal faces. In: 2020 IEEE/CVF Conference on Computer Vision and Pattern Recognition (CVPR), Seattle, pp. 6939–6948. IEEE (2020)

11. Yuan, H.C., Zhang, S.: Detection of underwater fish based on Faster R-CNN and image enhancement. J. Dalian Ocean Univ. **35**(4), 612–619 (2020)

12. Liu, Z.Y., Li, X., Fan, L.Z., et al.: Measuring feeding activity of fish in RAS using computer vision. Aquacult. Eng. **60**, 20–27 (2014)

13. Ye, X.C., Li, Z., Sun, B.L., et al.: Deep joint depth estimation and color correction from monocular underwater images based on unsupervised adaptation networks. IEEE Trans. Circ. Syst. Video Technol. **30**(11), 3995–4008 (2020)

14. Ma, L., et al.: Learning multi-scale retinex with residual network for low-light image enhancement. In: Peng, Y., et al. (eds.) PRCV 2020. LNCS, vol. 12305, pp. 291–302. Springer, Cham (2020). https://doi.org/10.1007/978-3-030-60633-6_24

15. Chen, G., Song, X.: Quantum color image scaling on QIRHSI model. In: Zeng, J., Qin, P., Jing, W., Song, X., Lu, Z. (eds.) ICPCSEE 2021. CCIS, vol. 1451, pp. 453–467. Springer, Singapore (2021). https://doi.org/10.1007/978-981-16-5940-9_35

16. Jia, Y.X., Fan, S.C., Yi, X.M.: Fish recognition based on significant enhancement and transfer learning. Fish. Modernization **264**(01), 40–48 (2020)

17. Liu, P., Yang, H.B., Song, Y.: Marine biometric algorithm based on improved YOLOv3 network. Appl. Res. Comput. **37**(S1), 394–397 (2020)

18. Wang, Y., Liang, Z., Cheng, X.: Fast target tracking based on improved deep sort and YOLOv3 fusion algorithm. In: Zeng, J., Qin, P., Jing, W., Song, X., Lu, Z. (eds.) ICPCSEE 2021. CCIS, vol. 1451, pp. 360–369. Springer, Singapore (2021). https://doi.org/10.1007/978-981-16-5940-9_27

19. Ren, S.Q., He, K.M., Girshick, R., et al.: Faster R-CNN: toward real-time object detection with region proposal networks. IEEE Trans. Pattern Anal. Mach. Intell. **39**(6), 1137–1149 (2017)

20. Girshick, R., Donahue, J., Darrell, T., et al.: Rich feature hierarchies for accurate object detection and semantic segmentation. In: Proceedings of the IEEE Conference on Computer Vision and Pattern Recognition, Columbus, pp. 580–587. IEEE (2014)

21. Girshick, R.: Fast R-CNN. In: Proceedings of the IEEE International Conference on Computer Vision, pp. 1440–1448 (2015)

22. Huang, G., Liu, Z., Van, D., et al.: Densely connected convolutional networks. In: Proceedings of the IEEE Conference on Computer Vision and Pattern Recognition, Hawaii, pp. 4700–4708. IEEE (2017)

23. Redmon, J., Divvala, S., Girshick, R., et al.: You only look once: unified, real-time object detection, pp. 779–788. IEEE (2016)

24. Shi, J., Zhang, G., Yuan, J., Zhang, Y.: Improved YOLOv3 infrared image pedestrian detection algorithm. In: Zeng, J., Jing, W., Song, X., Lu, Z. (eds.) ICPCSEE 2020. CCIS, vol. 1257, pp. 506–517. Springer, Singapore (2020). https://doi.org/10.1007/978-981-15-7981-3_37

25. Ghiasi, G., Lin, T.Y., Le, Q.: NAS-FPN: learning scalable feature pyramid architecture for object detection. In: Proceedings of the 2019 IEEE/CVF Conference on Computer Vision and Pattern Recognition (CVPR), Long Beach, pp. 7029–7038. IEEE (2019)

26. Shah, S., Wu, W., Lu, Q., et al.: AmoebaNet: an SDNenabled network service for big data cience. J. Netw. Comput. Appl. **119**, 70–82 (2018)

27. Tan, M., Pang, R., Le, Q.V.: EfficientDet: scalable and efficient object detection. In: Proceedings of the 2020 IEEE/CVF Conference on Computer Vision and Pattern Recognition (CVPR), Seattle, pp. 10778–10787. IEEE (2020)

28. Chen, Q., Wang, Y., Yang, T., et al.: You only look one-level feature. arXiv:2103.09460 (2021)

29. Chaudhari, S., Mithal, V., Polatkan, G., et al.: An attentive survey of attention models. ACM Trans. Intell. Syst. Technol. (TIST) **12**(5), 1–32 (2021)

30. Guo, S.T., Xu, P.F., Miao, Q.G., et al.: Automatic identification of individual primates with deep learning techniques. iScience **23**(8), 101412 (2020)

31. Zhang, K.P., Zhang, Z.P., Li, Z.F., et al.: Joint face detection and alignment using multitask cascaded convolutional networks. IEEE Sig. Process. Lett. **23**(10), 1499–1503 (2016)

32. Wang, Z.D., Zheng, L., Li, Y.L., et al.: Linkage based face clustering via graph convolution network. In: 2019 IEEE/CVF Conference on Computer Vision and Pattern Recognition (CVPR), Long Beach, pp. 1117–1125. IEEE (2019)

33. Zhang, J.L., Zeng, G.S., Qin, R.F.: Fish recognition method for submarine observation video based on deep learning. J. Comput. Appl. **39**(2), 376–381 (2019)

34. Schofield, D., Nagrani, A., Zisserman, A., et al.: Chimpanzee face recognition from videos in the wild using deep learning. Sci. Adv. **5**(9), eaaw0736 (2019)

35. Brust, C.A., Burghardt, T., Groenenberg, M., et al.: Toward automated visual monitoring of individual gorillas in the wild. In: 2017 IEEE International Conference on Computer Vision Workshop (ICCVW), Venice, pp. 2820–2830. IEEE (2017)

36. Deng, K., Liu, W., Wang, D.H.: Social network analysis and its application in animal behavior. Acta Theriologica Sin. **39**(03), 87–98 (2019)

37. Liu, W., et al.: SSD: single shot multibox detector. In: Leibe, B., Matas, J., Sebe, N., Welling, M. (eds.) ECCV 2016. LNCS, vol. 9905, pp. 21–37. Springer, Cham (2016). https://doi.org/10.1007/978-3-319-46448-0_2

38. Chen, X., Zou, Q.N., Xie, S.Y., et al.: A key frame automatic selection method for moving object. Comput. Modernization (10), 81–89 (2020)

39. Liu, J.: Individual identification and sex identification based on black muntjac feces, pp. 1–61. Zhejiang Normal University (2012)

40. Van, B.S., Fernandez-Duque, E., Di, F.A.: Demography and life history of wild red titi monkeys (Callicebus discolor) and equatorial sakis (Pithecia aequatorialis) in Amazonian Ecuador: a 12-year study. Am. J. Primatol. **78**, 204–215 (2016)

41. Guan, T.P., Owens, J.R., Gong, M.H., et al.: Role of new nature reserve in assisting endangered species conservation-case study of giant pandas in the Northern Qionglai Mountains, China. PLoS ONE **11**(8), e0159738 (2016)

42. Alli, M.N., Viriri, S.: Animal identification based on footprint recognition. In: 2013 International Conference on Adaptive Science and Technology, Pretoria, pp. 1–4. IEEE (2013)

43. Burghardt, T., Campbell, N.: Individual animal identification using visual biometrics on deformable coat patterns. In: International Conference on Computer Vision Systems: Proceedings, Rio de Janeiro, pp. 1–10. IEEE (2007)

44. Qin, L.: Research and Development of the Information Collection and Management System for Stocking Sheep Based on RFID, pp. 1–48. Inner Mongolia University (2016)

45. Chen, P., Swarup, P., Matkowski, W.M., et al.: A study on giant panda recognition based on images of a large proportion of captive pandas. Ecol. Evol. **10**(7), 3561–3573 (2020)

46. Xu, F.Q., Ding, X.Y., Peng, J.J., et al.: Real-time detecting method of marine small object with underwater robot vision. In: 2018 CEANS-MTS/IEEE Kobe Techno-Oceans (OTO), Kobe, pp. 1–4. IEEE (2018)

47. He, K.M., Zhang, X.Y., Ren, S.Q., et al.: Deep residual learning for image recognition. In: 2016 IEEE Conference on Computer Vision and Pattern Recognition (CVPR), pp. 770–778 (2016)

48. Jaderberg, M., Simonyan, K., Zisserman, A.: Spatial transformer networks. In: Advances in Neural Information Processing Systems, Montreal, pp. 2017–2025. NIPS (2015)

49. Clapham, M., Miller, E., Nguyen, M., et al.: Automated facial recognition for wildlife that lack unique markings: a deep learning approach for brown bears. Ecol. Evol. **10**(23), 12883–12892 (2020)

50. Hilderbrand, G.V., Schwartz, C.C., Robbins, C.T., et al.: The importance of meat, particularly salmon, to body size, population productivity, and conservation of North American brown bears. Can. J. Zool. **77**(1), 132–138 (1999)

51. Chopra, S., Hadsell, R., Lecun, Y.: Learning a similarity metric discriminatively, with application to face verification. IEEE Comput. Soc. **1**, 539–546 (2005)

52. Schroff, F., Kalenichenko, D., Philbin, J.: FaceNet: a unified embedding for face recognition and clustering, pp. 815–823. IEEE (2015)

53. Papageorgiou, C., Poggio, T.: A trainable system for object detection. Int. J. Comput. Vis. **38**(1), 15–33 (2000)

54. Al Arif, S.M.M.R., Knapp, K., Slabaugh, G.: Spnet: shape prediction using a fully convolutional neural network. In: Frangi, A.F., Schnabel, J.A., Davatzikos, C., Alberola-López, C., Fichtinger, G. (eds.) MICCAI 2018. LNCS, vol. 11070, pp. 430–439. Springer, Cham (2018). https://doi.org/10.1007/978-3-030-00928-1_49

55. Wang, K.L., Yuan, H.C.: Aquatic animal image classification method based on transfer learning. J. Comput. Appl. **333**(05), 88–92+110 (2018)

56. Qiao, Y.L., Su, D., Kong, H., et al.: Individual cattle identification using a deep learning based framework. IFAC-PapersOnLine **52**(30), 318–323 (2019)

57. Szegedy, C., Vanhoucke, V., Ioffe, S., et al.: Rethinking the inception architecture for computer vision, pp. 2818–2826. IEEE (2016)

58. Ordóñez, F., Roggen, D.: Deep convolutional and LSTM recurrent neural networks for multimodal wearable activity recognition. Sensors **16**(1), 115 (2016)

59. Kumar, S., Pandey, A., Satwik, K.S., et al.: Deep learning framework for recognition of cattle using muzzle point image pattern. Measurement **116**, 1–17 (2018)

60. Qin, X., Song, G.F.: Pig face recognition algorithm based on bilinear convolution neural network. J. Hangzhou Dianzi Univ. (Nat. Sci). **39**(02), 12–17 (2019)

61. Hansen, M.F., Smith, M.L., Smith, L.N., et al.: Toward on-farm pig face recognition using convolutional neural networks. Comput. Ind. **98**, 145–152 (2018)

62. Gaber, T., Tharwat, A., Hassanien, A.E., et al.: Biometric cattle identification approach based on Webers Local Descriptor and AdaBoost classifier. Comput. Electron. Agric. **122**, 55–66 (2016)

63. Lin, T.Y., Roychowdhury, A., Maji, S.: Bilinear CNN models for fine- grained visual recognition. In: Proceedings of the IEEE International Conference on Computer Vision, Santiago, pp. 1449–1457. IEEE (2015)

64. Cui, S.X., Zhou, Y., Wang, Y.H., et al.: Fish Detection Using Deep Learning. Appl. Comput. Intell. Soft Comput. **2020**, 1–13 (2020)

65. Wang, K.: Research and system development of layer behavior detection method based on deep learning, pp. 1–87. Zhejiang A&F University (2019)

66. Ye, Z.J., Ren, M.: An experimental animal information management system and an individual identity discrimination method for experimental animals. China, CN111523017A[P], 11 August 2020

67. Witham, C.L.: Automated face recognition of rhesus macaques. J. Neurosci. Methods **300**, 157–165 (2018)

68. Crouse, D., Jacobs, R.L., Richardson, Z., et al.: LemurFaceID: a face recognition system to facilitate individual identification of lemurs. BMC Zool. **2**(1), 1–14 (2017). https://doi.org/10.1186/s40850-016-0011-9

69. Swarup, P., Chen, P., Hou, R., et al.: Giant panda behavior recognition using images. Glob. Ecol. Conserv. **26**, e01510 (2021)

70. Huang, I.W., Hwang, J.N., Rose, C.S.: Chute based automated fish length measurement and water drop detection. In: IEEE International Conference on Acoustics, Shanghai, pp. 1906–1910. IEEE (2016)

71. Dong, M.: Object multi-mark recognition algorithm based on machine learning and image processing. Comput. Digit. Eng. **44**(12), 2488–2492 (2016)

72. Lowe, D.G.: Distinctive image features from scale-invariant keypoints. Int. J. Comput. Vis. **2**(60), 91–110 (2004)

73. Dalal, N., Triggs, B.: Histograms of oriented gradients for human detection. In: IEEE Computer Society Conference on Computer Vision & Pattern Recognition, San Diego, pp. 886–893. IEEE (2005)

74. Lin, T.Y., Dollár, P., Girshick, R., et al.: Feature pyramid networks for object detection. In: Proceedings of the IEEE Conference on Computer Vision and Pattern Recognition, Hawaii, pp. 2117–2125. IEEE (2017)

75. Dai, J.F., Qi, H.Z., Xiong, Y.W., et al.: Deformable convolutional networks. In: Proceedings of the IEEE International Conference on Computer Vision, Venice, pp. 764–773. IEEE (2017)

CIRS: A Confidence Interval Radius Slope Method for Time Series Points Based on Unsupervised Learning

Shuya Lei[1], Weiwei Liu[1(✉)], Xudong Zhang[2], Xiaogang Gong[3],
Jianping Huang[2], Yidan Wang[3], Jiansong Zhang[2], Helin Jin[4],
and Shengjian Yu[4]

[1] Artificial Intelligence on Electric Power System State Grid Corporation
Joint Laboratory (State Grid Smart Grid Research Institute Co., Ltd.),
Beijing 102209, China
oweiwlo@163.com
[2] State Grid Zhejiang Electric Power Co., Ltd., Quzhou, China
[3] State Grid Zhejiang Information and Telecommunication Branch, Hangzhou, China
[4] Harbin Institute of Technology, Harbin, China

Abstract. The rise of big data has brought various challenges and revolutions to many fields. Even though its development in many industries has gradually become perfect or even mature, its application and development in complex industrial scenarios is still in its infancy. We run research on single-dimensional time series point anomaly detection based on unsupervised learning: Unlike periodic time series, aperiodic or weakly periodic time series in industrial scenarios are more common. Considering the need for online real-time monitoring, we need to solve the problem of point anomaly detection of oil chromatographic characteristic gases. Thus, we propose a sliding window-based method for the unsupervised single-dimensional time series point anomaly detection problem called the confidence interval radius slope method (CIRS). CIRS is a fusion of knowledge-driven and data-driven methods to realize online real-time monitoring of possible data quality problems. From the experimental results, CIRS has obtained higher PR values than other unsupervised methods by the subject data.

1 Introduction

As a typical representative of the era of intelligent information, big data have a profound impact on the organizational structure of society, economic operation mechanisms, and enterprise decision-making structures [16]. Based on big data-related technologies, it has become particularly important to conduct experiments to discover results or predict needs through events that have occurred and are occurring, thereby improving the decision-making process [8].

Facing the real-time growth of massive data, how to efficiently detect abnormal data patterns is a very important problem, especially real-time online monitoring [20]. In terms of traffic management, real-time abnormal pattern detection

© The Author(s), under exclusive license to Springer Nature Singapore Pte Ltd. 2022
Y. Wang et al. (Eds.): ICPCSEE 2022, CCIS 1628, pp. 310–325, 2022.
https://doi.org/10.1007/978-981-19-5194-7_23

can be performed on road and traffic information, which can quickly respond to certain emergencies and further provide an efficient and scientific decision-making basis for the benign operation of urban traffic [6]. In terms of public opinion monitoring, intelligent analysis of search records and semantics of network keywords can maximize social conditions and public opinion and can effectively crack down on crimes in network emergencies [21]; in security-related fields, through abnormal pattern detection of big data, it can also be timely Detect man-made or natural disasters and terrorist events, and improve emergency response and security capabilities [23].

Industrial big data anomaly detection has the following difficulties:

- Perceptual data sensors have high parameter dimensions and strong data heterogeneity. As the main component of industrial big data, time series data have a high proportion, long time span, and complex time structure information [7]. The characteristics of these industrial big data make traditional anomaly detection methods unable to meet the requirements of efficiency and effectiveness.
- The production process has strong association rules, and the data contain inherent knowledge in the field of manufacturing, but the difficulty of abstraction and application of the domain knowledge seriously affects the availability of detected abnormal points or abnormal fragments [19].
- Abnormal data and normal data appear alternately in the data, and due to the complex causes of abnormality, it is difficult to distinguish the two, which brings great trouble to the design of anomaly labeling and anomaly detection algorithms.

At present, although researchers have carried out research on the above problems of industrial big data anomaly detection difficulties, the existing methods still cannot provide a complete and efficient solution for industrial production, mainly in the following three aspects:

- Most time series data anomaly detection algorithms focus on solving single-dimensional data with periodic or simple patterns, and it is difficult to perform effective real-time abnormal pattern detection on aperiodic single-dimensional time series data.
- The existing anomaly pattern detection algorithms mainly focus on the constraint satisfaction test based on attribute correlation, without integrating domain knowledge, the communication preparation work between the early stage and the industry is insufficient, the later cooperation often has faults, and the evaluation and verification work for actual production is carried out insufficient.
- Most industrial big data anomaly detection algorithms have a high demand for the labeling of abnormal data. Modeling based on high-quality data after removing anomalies or training a classifier for abnormal data and normal data is the mainstream idea. However, due to the increasing complexity of industrial manufacturing systems, industrial database administrators have

gaps in their cognition of abnormal data in industrial systems, and the massive data provided to knowledge workers often have the problem of insufficient data labeling.

Therefore, due to the existence of industrial big data accumulated in the current manufacturing system, abnormal data exist widely and the abnormal sources and data patterns are increasingly complex. To address these challenges, we propose an unsupervised anomaly detection algorithm for industrial time series data to realize single-dimensional real-time monitoring pattern mining in the industrial manufacturing process. The theoretical method studied in this paper can reduce the occurrence rate of abnormal conditions and reduce the loss caused by equipment abnormalities, which has great economic and social significance [11].

To solve the problem of point anomaly detection in industrial time series data, this paper proposes a new unsupervised learning method to realize point anomaly detection for nonperiodic or weakly periodic industrial time series data [10].

2   Related Work

At present, anomaly detection methods for time series data are mainly divided into statistics-based, constraint-based and machine learning-based anomaly detection [5]. Statistics-based methods [24] are a more traditional class of methods that extract time series trends and detect low-quality data points by obtaining statistics from time series, fitting model parameters, or morphological transformation of data. Constraint-based methods aim to use the correlation or statistics of adjacent series to determine whether the value of the series is abnormal. Machine learning-based methods apply traditional classification, clustering and deep learning to time series to detect data exception methods. Since the sequence model is more complex than the vector space model, the application of this type of method to time series is still in its infancy [13]. Among them, research on anomaly detection based on machine learning is the current mainstream research direction, and we focus on the unsupervised learning anomaly detection method here. It can be divided into the following categories:

- Similarity measures: The most popular measures of sequence similarity are sequence similarity based on simple match counts [13] and the normalized length of the longest common subsequence [2]. The former has the advantage of its higher computational efficiency, while the latter can accommodate segments in sequences that contain noise, but is more expensive due to its dynamic programming approach [3].
- Clustering methods: Popular clustering methods include k-means [15], EM algorithms, dynamic clustering [17], stepwise k-means [4], multivariate clustering in principal component space time series and one-SVM [9] etc. The choice of clustering method is application specific, as different clustering methods have different complexities and different adaptations to different numbers, shapes, and sizes of clusters [18].

- Constraint methods: Commonly used anomaly detection algorithms include rule constraint-based methods, such as order dependencies [22], velocity constraints, variance constraints and rejection constraints [14], which can effectively utilize temporal features, to repair highly abnormal data, but this method generally has difficulty meeting the problem of time series anomaly detection with changing patterns [1].

3 Definition

3.1 Multidimensional Time Series

First, the definition and formal expression of the multidimensional timing are given. Multidimensional time series: A series of data points (usually continuous over a period of time) that are sampled and captured by multiple sensors and can be represented as $X \in R^{K \times N}$. Where K represents the number of sensor or attribute$\{a_1, a_2, \ldots, a_K\}$ values, N is the length of the time series. Under different anomaly detection methods, it can be defined as different formal expressions:

- In the single-dimensional time series data point anomaly detection method, sensors or attributes can be used as partitioning units; that is, multidimensional time series can be divided into sets composed of several single-dimensional time series. The single-dimensional time series of each attribute a_i is an ordered set captured by the corresponding sensor at every moment, which can be expressed as $X_i = <x_{i1}, x_{i2}, \ldots, x_{iN}>, \quad i = 1, 2, \ldots, K$.
- In the method of pattern anomaly detection of multidimensional time series data, the time stamp can be used as the partition unit; that is, the multidimensional time series can be divided into an ordered vector group composed of several multidimensional vectors. The multidimensional vector at each moment t is the data value captured by K sensors at the same time, which can be expressed as $X_t = (x_{1t}, x_{2t}, \ldots, x_{Kt}), \quad t = 1, 2, \ldots, N$

3.2 Definition of Research Problem

Second, the definition of the problem studied in this chapter is given. In the abnormal detection of oil chromatography data, the possible data quality problems of one-dimensional time series data fall into the following categories:

- Null, at a certain point in the gas concentration is not monitored or not collected, summarized.
- Zero, the gas concentration value detected at a certain point is zero. However, C2H2 and N2 are normal.
- negative, the monitoring data are negative, and the gas concentration cannot be negative.
- Above the threshold, at a certain point or after a period of time, the concentration of the monitored gas exceeds the value of attention.
- Below the threshold but it is mutated, the gas concentration value detected at a certain time point has a significant mutation compared with the previous time, although it does not exceed the limit.

According to the results caused by the above kinds of point exceptions, they can be classified into three categories, namely, single point big error, single point small error and continuous error. For a one-dimensional time series $X_i = <x_{i1}, x_{i2}, \ldots, x_{iN}>$, there are:

Continuous errors. The so-called continuous error means that in a time series, the error occurs at several consecutive points in time, i.e.,

$$E = <x_{ia}, \ldots, x_{ib}>$$

In the figure on the right, the observed values from the 6th collected data to the 6th collected data are all 0; that is, continuous errors occur here. Serial errors are common in real life. For example, when someone is walking down the road with a smartphone, nearby tall buildings could have a lasting effect on the GPS information collected. In addition, system errors can also lead to continuous errors.

Single large error. A single point error is an error that does not occur continuously s in a time series, but only occurs at a single data point at intervals. A large deviation means that the observed value of a data point deviates significantly from the true value. At the same time, the size of abnormal deviation has a great correlation with the data set. As shown in the 23rd data collected in the figure above, the deviation degree of the data point is large, and the anomaly of this data point is larger. Therefore, the anomaly of this data point is a large mistake.

One small error. Similar to a single point large error, the error does not occur continuously, but only at a single data point on the interval. When the deviation of the data point is small, the 27th acquisition data point in the figure has a small error. The rationale behind single point minor errors is that a person or system is always trying to minimize possible errors. For example, one may make only minor omissions in regulating the valves of a machine (Fig. 1).

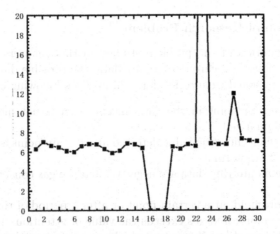

Fig. 1. Single-dimensional time series data exception example

The occurrence of continuous error is usually attributed to the occurrence of null values, negative values or overlimit values, while the occurrence of single point error is usually attributed to the occurrence of zero values and nonoverlimit mutation values. For the out-of-limit judgment standard and nonout-of-limit mutation value detection, the former is generated by domain knowledge (industry standard), that is, knowledge driven; the latter is produced by unsupervised learning, which is data-driven. For the point anomalies of a one-dimensional time series, the threshold value can be determined according to the existing domain knowledge, and the mutation value can be detected by an unsupervised learning method. The existing outlier detection algorithms for time series data include statistical methods (mean square deviation method and box plot method, etc.), clustering methods (DBSCAN clustering, etc.) and specific outlier detection algorithms (One Class SVM, Isolation Forest, etc.).

For the existing statistical methods, the disadvantage of the box plot method is that it needs to extract data from multiple equipment to establish data sets for offline statistical calculation, which is not good for online monitoring. Although the mean square deviation method can be applied to the sliding window method that can be monitored online, the detection results will be affected when the data distribution in the window is too dense or too loose (as described experimentally in Sect. 3.3.2). The clustering and specific outlier detection algorithm only considers the deviation of single point data relative to the whole data but ignores the time order of time series, that is, ignores the difference of mutation criteria in different time stages (local), so it is not suitable to solve the anomaly detection problem. Therefore, this paper proposes a sliding window-based confidence interval radius anomaly detection method to detect local mutation values, which not only takes into account the different local mutation standards caused by the time order but also weakens the influence of overdense local data distributions.

4 Model

For a one-dimensional time series, the size of the sliding window is given as L and the moving step is given as $w(w < L)$. The subsequence in the JTH sliding window can be expressed as:

$$r_{i,j} = \frac{\widehat{\theta}_{i,j} - \check{\theta}_{i,j}}{2}$$

where

$$\hat{\theta}_{i,j} = \mu_{i,j} + \frac{\sigma_{i,j}}{\sqrt{L}} Z_{\frac{\alpha}{2}}$$

$$\check{\theta}_{i,j} = \mu_{i,j} - \frac{\sigma_{i,j}}{\sqrt{L}} Z_{\frac{\alpha}{2}}$$

Represent the upper limit and lower limit of the confidence interval, respectively, $\mu_{i,j}$ and $\sigma_{i,j}$ represent the mean value and mean square deviation of data in the window, the random variable $Z \sim N(0,1)$. α is the confidence level(0.05), and $Z_{\alpha/2} = Z_{0.025} = 1.96$ can be obtained by looking up the table.

This feature reflects the confidence interval radius of the normal distribution of the subsequence when $\alpha = 0.05$. Some researchers think that $r_{(i,j)} > \tau$ indicates that there is a mutation point or abnormal fluctuation in the window(confidence interval radius threshold method). However, if you simply put $r_{(i,j)} > \tau$ as the judgment standard of outlier detection, the judgment result not only relies excessively on the selection of super parameter τ but is also relatively rough, which will result in a large misjudgment rate (the normal data are judged as abnormal data, which will be introduced in detail in Chap. 3).

Therefore, we propose an outlier determination method based on the detection of confidence interval radius change, which not only solves the problem of overdependence on the superparameter τ. ut also reduces the misjudgment rate. The algorithm flow chart is shown in Fig. 2 below. In the actual calculation window of the mean and mean square,

$$\mu_{i,j} = \frac{x_{i,j*w} + x_{i,j*w+2} + \cdots + x_{i,j*w+L-1}}{L}$$

$$\sigma_{i,j} = \sqrt{\frac{\sum_{k=j*w}^{j*w+L-1} (x_{i,k} - \mu_{i,j})^2}{L}}$$

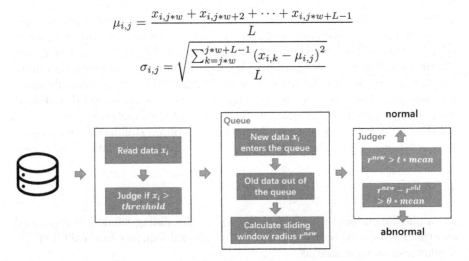

Fig. 2. Confidence interval radius slope method

The calculation time will be wasted if the calculation is carried out point-by-point in each window. Therefore, the following rapid calculation optimization (w unknown) is made:

Fast Calculation of the Mean Value

We know that the mean of the current window and the mean of the next window are shown in Fig. 3.

$$\mu_{i,j} = \frac{x_{i,j*w+1} + \cdots + x_{i,j*w+1+w} + \cdots + x_{i,j*w+L}}{L}$$

$$\mu_{i,j+1} = \frac{x_{i,j*w+1+w} + \cdots + x_{i,j*w+L} + \cdots + x_{i,j*w+L+w}}{L}$$

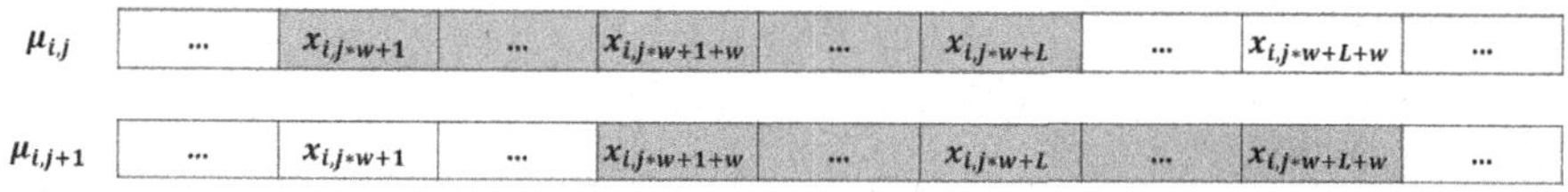

Fig. 3. The mean of the current window and the mean of the next window

Obviously, there are

$$\mu_{i,j+1} = \mu_{i,j} + \frac{(x_{i,j*w+L+1} + \cdots + x_{i,j}jw + L + w) - (x_{i,j*w+1} + \cdots + x_{i,j*w+w})}{L}$$

Fast Calculation of Mean Square Deviation

The mean $\mu = \frac{1}{L}(x_1 + \cdots + x_L)$. For the formula of mean square deviation, we transform it to have

$$\sigma = \sqrt{\frac{1}{L}\left[(x_1 - \mu)^2 + (x_2 - \mu)^2 + \cdots + (x_L - \mu)^2\right]}$$
$$= \sqrt{\frac{1}{L}(x_1^2 + \cdots + x_L^2 - 2x_1\mu - \cdots - 2x_L\mu + \mu^2 + \cdots + \mu^2)}$$
$$= \sqrt{\frac{1}{L}(x_1^2 + \cdots + x_L^2) - \mu^2}$$

Therefore, in the process of fast calculation of mean value and mean square deviation, only the fast calculation of cumulative sum and square cumulative sum is needed.

In the practical application scenario, it is necessary to determine the point anomaly of each real-time data point. Therefore, when the moving step w of the sliding window is 1, the pseudocode of CIRS is given as Algorithm 1.

5   Algorithm Complexity and Feasibility Analysis

This section will give the complexity of the proposed CIRS algorithm. In addition, the necessity and feasibility of using this method are analyzed.

Algorithm complexity Analysis

According to algorithm 2-1, CIRS is a single-dimensional anomaly detection method for timing data points based on a sliding window. In Sect. 2.2, the fast calculation method of the mean value and mean square deviation is given. In each sliding window, it is not necessary to recalcalculate the mean value and mean square deviation of each data point but only to use the data at both

---

Algorithm 1. Confidence interval radius slope method (CIRS)

Require: one-dimensional time series $X = \langle x_1, x_2, \ldots, x_N \rangle$, sliding window size L, superparameters τ and θ

Ensure: ata point anomaly judgment results

  $queue \leftarrow []$

  $sum_x, sum_{x2} \leftarrow 0, 0$

  for $x_i \in X$ **do**

    if x_i is out of bounds or $x_i = 0$ **then**

      x_i will be recorded as abnormal data

      continue

    end if

    if $QueueLength < L$ **then**

      x_i enter queue

      $sum_x, sum_{x2} \leftarrow sum_x + x_i, sum_{x2} + x_i * x_i$

      continue

    end if

    $x_{old} \leftarrow OutOfQueue; x_i EnterQueue$

    $sum_x, sum_{x2} \leftarrow sum_x + x_i - x_{old}, sum_{x2} + x_i * x_i - x_{old} * x_{old}$

    mean and std can be calculated from sum_x and sum_x2

    The upper limit of $\hat{\theta}_i$ and lower limit of $\check{\theta}_i$ can be calculated, and the confidence interval radius distance r_i is calculated

    if $r_i > \tau * mean$ and $r_i - r_{old} > \theta * mean$ **then**

      x_i will be recorded as abnormal data;

    else

      x_i will be recorded as normal data

    end if

    $r_{old} \leftarrow r_i$

  end for

---

ends of the sliding window for calculation. Therefore, the time complexity of the algorithm is reduced from $O(n*L)$ to $O(n)$ by a fast calculation method. Meanwhile, in each sliding window, the algorithm only needs to maintain the confidence interval radius distance r_{old} of the previous window, the sum of data in the window sum_X and the data square and sum_x2, so the spatial complexity of the algorithm is $O(1)$.

The Necessity and Feasibility of Using This Method are Analyzed

This method solves the problem of detecting nonoff-limit mutation values in the anomaly detection of oil chromatography data. As described in Sect. 2.1.2, for the point anomaly of single-dimensional time series data, the off-limit value can be determined according to the existing domain knowledge, while the mutation value can be detected by an unsupervised method. Different from the existing outlier detection algorithms for time series data, the CIRS method overcomes the shortcoming of the box plot method, which needs to establish data sets for off-line statistical calculation but cannot carry out online monitoring. At the same time, the influence of the data distribution in the sliding window bing too dense or too loose on the mean square deviation method is also solved. In addition, other outlier detection algorithms (such as clustering and some specific outlier detection algorithms) ignore the timing of time series and only pay attention to the deviation of single point data relative to the whole data, while CIRS (confidence interval radius slope method) increases the adaptability to different time (local) mutation criteria [12].

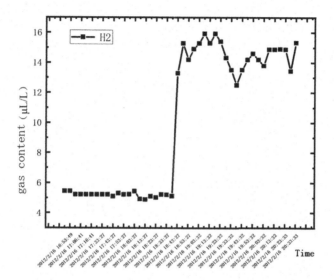

Fig. 4. The distribution of the data changes and enters the next stable period

It can be seen from Sect. 2.1.2 that the point anomaly detection solved in this chapter includes null value, zero value, negative value, overlimit value and nonover-limit mutation value. However, in lines 4–6 of algorithm 2-1 proposed in this paper, the null value, zero value, negative value and overlimit value are not included in the sliding window. It avoids the problem of drastic changes in the normal distribution of data simulation in the sliding window caused by the introduction of large error data into the sliding window. The comparison between Fig. 3-2(g) and Fig. 3-2(h) in the third chapter below shows the seriousness of this problem and the necessity of solving it.

At the same time, the CIRS method includes the nonoff-limit mutation value in the sliding window but does not ignore it, because it takes into account the existence of stationary periods with different distributions of aperiodic or weakly periodic time series data, as shown in Fig. 4. If the detected nonoff-limit mutation value is not included in the sliding window, the normal distribution simulated by the algorithm will stay in the normal distribution followed by the previous period of time series data; that is, the normal distribution limited in the previous stationary period cannot update the change of the distribution in time. Therefore, the CIRS method proposed in this paper has the necessity and feasibility to be researched and used in the actual scenarios studied in this topic.

6   Experiment

6.1   Data Set

The data set used in this paper is the online monitoring data of oil chromatography in transformer equipment in several regions of the State Grid, including province, equipment code, voltage level, monitoring time, H2, CH4, C2H4, C2H2, C2H6, CO, CO2, O2, N2 and TOTALHYDROCARBON (total hydrocarbons), for a total of 14 fields, as shown in Table 1. Among them, total hydrocarbons refer to the sum of the contents of four gases: CH4, C2H4, C2H2 and C2H6.

Table 1. Information about the data sets used

No.	Field name	Type
1	SBBM	Varchar2 (17)
2	DYDJ	Varchar2 (2)
3	LINKEDPROVINCE	Char (180)
4	ACQUISITIONIME	Date
5	H2 (μL/L)	Number (10, 3)
6	CH4 (μL/L)	Number (10, 3)
7	C2H4 (μL/L)	Number (10, 3)
8	C2H2 (μL/L)	Number (10, 3)
9	C2H6 (μL/L)	Number (10, 3)
10	CO (μL/L)	Number (10, 3)
11	CO2 (μL/L)	Number (10, 3)
12	O2 (μL/L)	Number (10, 3)
13	N2 (μL/L)	Number (10, 3)
14	TOTALHYDROCARBON (μL/L)	Number (10, 3)

6.2   Set up

Comparison Method

It has been mentioned that the mean square deviation method can be applied in a sliding window to realize online monitoring of one-dimensional time series data. Therefore, the mean square deviation method based on statistics will be compared with our method. In the previous introduction, the possibility of using the constraint class method for on-line monitoring of one-dimensional time series data is presented. Therefore, the SCREEN algorithm based on the constraint class is compared with our method.

Accuracy Measurement Method

The point anomaly detection problem of one-dimensional time series data is a dichotomous problem, and the commonly used evaluation indexes for dichotomous problems are Precision and Recall. Generally, the concerned class is regarded as the positive class, and the other classes are regarded as the negative class. In this paper, abnormal data are considered, so abnormal data are regarded as the positive class, and normal data are regarded as the negative class. In terms of the test data, the prediction of the model algorithm is correct or incorrect, and the total number of the 4 cases is denoted as:

- TP – Predict the number of positive classes as positive classes
- FN – Predict the number of positive classes as negative classes
- FP – Predict the number of negative classes as positive classes
- TN– Predict the number of negative classes as negative classes.

Moreover, Precision is defined as $P = \frac{TP}{TP+FP}$ and recall is defined as $R = \frac{TP}{TP+FN}$. Among them, the accuracy rate refers to the predicted results, and its practical significance lies in the probability that all the samples predicted to be positive are actually positive, that is, represents the accuracy of the prediction in the positive results. The recall rate is a measure of coverage, which measures the algorithm's ability to identify positive classes.

6.3 Accuracy

Overall Accuracy

In this section, abnormal samples are taken as positive examples. P-r value pairs obtained by online anomaly detection of one-dimensional time series flow using the above methods are shown in Table 2.

Table 2. Comparison of experimental results of various sliding window methods (samples contain 0 value)

Method names	TP	FN	FP	TN	P	R
SCREEN algorithm (constraint class)	43	842	70	6310	0.38053	0.04858
CIRT	73	830	142	6238	0.39037	0.08221
Mean square deviation method	868	20	64	6316	0.93133	0.97747
CIRT-2	874	14	63	6317	0.93276	0.98423
CIRS	876	12	8	6372	0.99095	0.98648

Because the 0 value accounted for 88.06% in abnormal samples, the results of the detection evaluation index of the method susceptible to the 0 value were generally low, while the results of the detection evaluation index of the method unaffected by the 0 value were generally high. Therefore, supplementary experiments were carried out without the value of 0, and the experimental results are shown in Table 3.

Table 3. Comparison of experimental results of various sliding window methods (samples without 0 value)

Method names	TP	FN	FP	TN	P	R
SCREEN algorithm (constraint class)	43	842	70	6310	0.38053	0.04858
CIRT	73	830	142	6238	0.39037	0.08221
Mean square deviation method	868	20	64	6316	0.93133	0.97747
CIRT-2	874	14	63	6317	0.93276	0.98423
CIRS	876	12	8	6372	0.99095	0.98648

As seen from Table 2 and Table 3, the confidence interval radius slope method proposed in this paper has achieved better experimental results than other methods in the abnormal detection of single-dimension timing data points, achieving the expected goal.

Gas Concentration Comparison

Taking the statistical mean square deviation method as an example, this section summarizes the similarities and differences between the statistical mean square deviation method and the proposed confidence interval radius slope method (CIRS) as follows:

- Similarities: Both methods are based on the assumption that data obey normal distribution to detect anomalies.
- Differences: The former is to test whether the data deviates from the three mean square deviations of the current normal distribution as the measurement standard, while the latter is to measure the amplitude of distribution change brought by the data to be tested on the normal distribution.

For the mean square deviation method, when the data distribution is too dense, the allowable range of data is too narrow, resulting in the method bing too sensitive to numerical changes. The experimental comparison is shown in Fig. 5(a) and Fig. 5(b). When the data distribution is too sparse, the allowable range of data is too wide, resulting in poor sensitivity of the method to numerical changes. The experimental comparison is shown in Fig. 5(c) and Fig. 5(d). Where $H2\_normal$ and $H2\_error$ indicate that the detection results are marked as normal data and abnormal data respectively. Lower, upper, mean and STD represent the lower limit, upper limit, data mean and standard deviation of the confidence interval in the current sliding window respectively. R represents the confidence interval radius of the current sliding window.

Obviously, the proposed confidence interval radius slope method (CIRS) can better avoid these two problems.

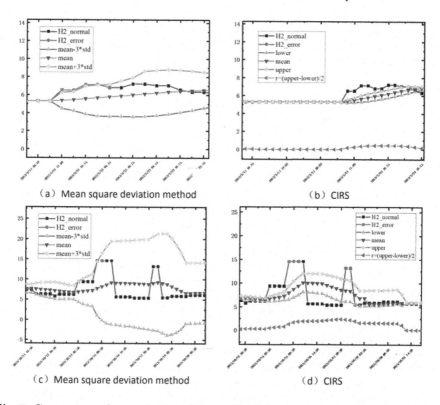

(a) Mean square deviation method   (b) CIRS

(c) Mean square deviation method   (d) CIRS

Fig. 5. Comparison of anomaly detection results of the mean square deviation method and slope method

7   Conclusion

In this paper, the problem of point anomaly detection and the proposed method of point anomaly detection are introduced in detail. At the same time, the optimization method in the process of implementation is given. An unsupervised learning method based on a sliding window, the confidence interval radius slope method (CIRS), is proposed, which not only considers the problem that abnormal points such as exceeding the limit value and 0 value are added to the sliding window and have a great influence on the model data distribution but also takes into account the problem that the model cannot update the simulation distribution in time when the data distribution changes if the abnormal points such as out-of-limit mutation values are not added to the sliding window. At the same time, the abnormal feature measurement index is optimized from a quantitative index to a variable index, which increases the adaptability of the model to changes in data distribution. In the experiments, it is verified that the experimental results of this algorithm are better than those of other algorithms.

Future work may focus on the detection of multidimensional data. Obtaining better anomaly detection results on multidimensional time series with limited labels is also a very worthy problem.

Acknowledgments. The project is supported by State Grid Research Project "Study on Intelligent Analysis Technology of Abnormal Power Data Quality based on Rule Mining" (5700-202119176A-0-0-00).

References

1. Bergman, M., Milo, T., Novgorodov, S., Tan, W.-C.: QOCO: a query oriented data cleaning system with oracles. Proc. VLDB Endow. **8**(12), 1900–1903 (2015)
2. Budalakoti, S., Srivastava, A.N., Akella, R., Turkov, E.: Anomaly detection in large sets of high-dimensional symbol sequences (2006)
3. Cai, R., Zhang, H., Liu, W., Gao, S., Hao, Z.: Appearance-motion memory consistency network for video anomaly detection. In: Proceedings of the AAAI Conference on Artificial Intelligence, vol. 35, no. 2, pp. 938–946 (2021)
4. Chakrabarti, K., Keogh, E.J., Mehrotra, S., Pazzani, M.J.: Locally adaptive dimensionality reduction for indexing large time series databases. ACM Trans. Database Syst. **27**(2), 188–228 (2002)
5. Chandola, V., Banerjee, A., Kumar, V.: Anomaly detection: a survey. ACM Comput. Surv. (CSUR) **41**(3), 1–58 (2009)
6. Chen, X., Wang, H.: Data-driven prediction of foodborne disease pathogens. In: Zeng, J., Qin, P., Jing, W., Song, X., Lu, Z. (eds.) ICPCSEE 2021, Part I. CCIS, vol. 1451, pp. 106–116. Springer, Singapore (2021). https://doi.org/10.1007/978-981-16-5940-9_8
7. Corain, M., Garza, P., Asudeh, A.: DBSCOUT: a density-based method for scalable outlier detection in very large datasets. In: 2021 IEEE 37th International Conference on Data Engineering (ICDE), pp. 37–48 (2021)
8. Gogacz, T., Torunczyk, S.: Entropy bounds for conjunctive queries with functional dependencies. In: Benedikt, M., Orsi, G. (eds.) 20th International Conference on Database Theory, ICDT 2017, 21–24 March 2017, Venice, Italy. LIPIcs, vol. 68, pp. 15:1–15:17. Schloss Dagstuhl - Leibniz-Zentrum für Informatik (2017)
9. Kumar, B., Sinha, A., Chakrabarti, S., Vyas, O.P.: A fast learning algorithm for one-class slab support vector machines. Knowl. Based Syst. **228**, 107267 (2021)
10. Lee, D., Shin, K.: Robust factorization of real-world tensor streams with patterns, missing values, and outliers. In: 2021 IEEE 37th International Conference on Data Engineering (ICDE), pp. 840–851 (2021)
11. Li, D., Shi, S., Zhang, Y., Wang, H., Luo, J.: an anomaly detection method based on learning of "scores sequence". In: Zhou, Q., Miao, Q., Wang, H., Xie, W., Wang, Y., Lu, Z. (eds.) ICPCSEE 2018. CCIS, vol. 902, pp. 296–311. Springer, Singapore (2018). https://doi.org/10.1007/978-981-13-2206-8_25
12. Liu, F., et al.: Anomaly detection in quasi-periodic time series based on automatic data segmentation and attentional LSTM-CNN. IEEE Trans. Knowl. Data Eng. **34**(6), 2626–2640 (2022)
13. Liu, H., Li, X., Li, J., Zhang, S.: Efficient outlier detection for high-dimensional data. IEEE Trans. Syst. Man Cybern.: Syst. **48**(12), 2451–2461 (2017)

14. Lopatenko, A., Bravo, L.: Efficient approximation algorithms for repairing inconsistent databases. In: Chirkova, R., Dogac, A., Tamer Özsu, M., Sellis, T.K. (eds.) Proceedings of the 23rd International Conference on Data Engineering, ICDE 2007, The Marmara Hotel, Istanbul, Turkey, 15–20 April 2007, pp. 216–225. IEEE Computer Society (2007)

15. Nairac, A., Townsend, N.W., Carr, R., King, S., Cowley, P., Tarassenko, L.: A system for the analysis of jet engine vibration data. Integr. Comput. Aided Eng. 6(1), 53–66 (1999)

16. Peng, Z., Luo, M., Li, J., Xue, L., Zheng, Q.: A deep multi-view framework for anomaly detection on attributed networks. IEEE Trans. Knowl. Data Eng. 34(6), 2539–2552 (2022)

17. Sequeira, K., Zaki, M.J.: ADMIT: anomaly based data mining for intrusions. In: Proceedings of the Eighth ACM SIGKDD International Conference on Knowledge Discovery and Data Mining, 23–26 July 2002, Edmonton, Alberta, Canada, pp. 386–395. ACM (2002)

18. Song, S., Zhang, A., Wang, J., Yu, P.S.: SCREEN: stream data cleaning under speed constraints. In: Sellis, T.K., Davidson, S.B., Ives, Z.G. (eds.) Proceedings of the 2015 ACM SIGMOD International Conference on Management of Data, Melbourne, Victoria, Australia, 31 May–4 June 2015, pp. 827–841. ACM (2015)

19. Wang, S., Wu, L., Cui, L., Shen, Y.: Glancing at the patch: anomaly localization with global and local feature comparison. In: 2021 IEEE/CVF Conference on Computer Vision and Pattern Recognition (CVPR), pp. 254–263 (2021)

20. Wang, X., Dong, X.L., Meliou, A.: Data X-ray: a diagnostic tool for data errors. In: Sellis, T.K., Davidson, S.B., Ives, Z.G. (eds.) Proceedings of the 2015 ACM SIGMOD International Conference on Management of Data, Melbourne, Victoria, Australia, 31 May–4 June 2015, pp. 1231–1245. ACM (2015)

21. Yin, S., Yi, Y., Wang, H.: Data cleaning about student information based on massive open online course system. In: Zeng, J., Jing, W., Song, X., Lu, Z. (eds.) ICPCSEE 2020, Part I. CCIS, vol. 1257, pp. 33–43. Springer, Singapore (2020). https://doi.org/10.1007/978-981-15-7981-3_3

22. Yin, W., Yue, T., Wang, H., Huang, Y., Li, Y.: Time series cleaning under variance constraints. In: Liu, C., Zou, L., Li, J. (eds.) DASFAA 2018. LNCS, vol. 10829, pp. 108–113. Springer, Cham (2018). https://doi.org/10.1007/978-3-319-91455-8_10

23. Zhang, A., Song, S., Wang, J.: Sequential data cleaning: a statistical approach. In: Özcan, F., Koutrika, G., Madden, S. (eds.) Proceedings of the 2016 International Conference on Management of Data, SIGMOD Conference 2016, San Francisco, CA, USA, 26 June–01 July 2016, pp. 909–924. ACM (2016)

24. Zhang, K., Hutter, M., Jin, H.: A new local distance-based outlier detection approach for scattered real-world data. In: Theeramunkong, T., Kijsirikul, B., Cercone, N., Ho, T.-B. (eds.) PAKDD 2009. LNCS (LNAI), vol. 5476, pp. 813–822. Springer, Heidelberg (2009). https://doi.org/10.1007/978-3-642-01307-2_84

LSN: Long-Term Spatio-Temporal Network for Video Recognition

Zhenwei Wang[1,2,3], Wei Dong[1,2,3], Bingbing Zhang[4(✉)], and Jianxin Zhang[1,2,3(✉)]

[1] School of Computer Science and Engineering, Dalian Minzu University, Dalian, China
jxzhang0411@163.com
[2] SEAC Key Laboratory of Big Data Applied Technology, Dalian Minzu University, Dalian, China
[3] Institute of Machine Intelligence and Bio-Computing, Dalian Minzu University, Dalian, China
[4] School of Information and Communication Engineering, Dalian University of Technology, Dalian, China
icyzhang@mail.dlut.edu.cn

Abstract. Although recurrent neural networks (RNNs) are widely leveraged to process temporal or sequential data, they have attracted too little attention in current video action recognition applications. Therefore, this work attempts to model the long-term spatio-temporal information of the video based on a variant of RNN, i.e., higher-order RNN. Moreover, we propose a novel long-term spatio-temporal network (LSN) for solving this video task, the core of which integrates the newly constructed high-order ConvLSTM (HO-ConvLSTM) modules with traditional 2D convolutional blocks. Specifically, each HO-ConvLSTM module consists of an accumulated temporary state (ATS) module as well as a standard ConvLSTM module, and several previous hidden states in the ATS module are accumulated to one temporary state that will enter the standard ConvLSTM to determine the output together with the current input. The HO-ConvLSTM module can be inserted into different stages of the 2D convolutional neural network (CNN) in a plug-and-play manner, thus well characterizing the long-term temporal evolution at various spatial resolutions. Experiment results on three commonly used video benchmarks demonstrate that the proposed LSN model can achieve competitive performance with the representative models.

Keywords: Video action recognition · High-order RNN · Long-term spatio-temporal · ConvLSTM · HO-ConvLSTM

1 Introduction

Action recognition based on video is a basic problem in the field of computer vision. It has a wide range of applications, such as video surveillance, human-computer interaction and social video recommendation. Video action recognition is a challenging problem because it needs to model two key and complementary cues at the same time, namely, spatial and temporal information.

© The Author(s), under exclusive license to Springer Nature Singapore Pte Ltd. 2022
Y. Wang et al. (Eds.): ICPCSEE 2022, CCIS 1628, pp. 326–338, 2022.
https://doi.org/10.1007/978-981-19-5194-7_24

In the past decade, due to the rapid development of deep learning in the field of computer vision, the models based on deep learning have also increased rapidly to solve the task of video action recognition. Generally, the current action recognition methods can be divided into the following four branches. The first branch is mainly two-stream network, first proposed by Simonyan et al. [1]. It processes the video into RGB frames and optical flow and then inputs the two into a separate convolutional neural network (CNN) for computation, where the outputs of the two CNNs are fused at the end of the network. A variant of it is the temporal segment network (TSN) [2]. TSN represents video segments by sampling from uniformly segmented segments, while the two-stream fusion network [3] focuses on fusing two streams CNN features. These models rely on data in optical flow format to obtain temporal information because when video understanding needs to infer temporal context, the role of a single RGB frame of spatial information is limited. The second branch of video recognition employs 3D CNNs [4, 5] that stack 3D convolution (Conv. for brief) [6] to model temporal and spatial semantics jointly. However, 3D CNN models have a large number of parameters, resulting in huge computational costs. In addition, after several local convolution operations, the valuable information of long-term ranges has already been weakened. The final branch mainly pays attention on computation efficiency, and the integration of separate temporal modeling modules into 2D CNNs for lightweight video recognition models has attracted increasing attention [2, 7, 8]. Finally, some researchers characterize temporal context in video information by using recurrent neural networks (RNNs) and their variants, long short-term memory (LSTM), as they are competitive in processing sequential data. One typical work is to place the LSTM module at the end of the CNNs. As a counterparty, Ballas et al. [9] stacked several LSTM modules to the backbone, aiming at modeling the temporal information from different layers.

As mentioned above, although RNN and LSTM [10] are competitive in processing sequential data, they have attracted little attention in modeling the temporal information of videos. RNN uses the hidden state of the previous moment and the input of the current moment to determine the hidden state of the current moment. Only when the gap between the previous frame and the place where it is needed is small can RNNs use the previous information. As that gap grows, RNNs become unable to learn long-term dependence. Higher-order RNNs (HO-RNNs) [11, 12] can solve the problem that RNNs only capture the information contained in all previous inputs from the hidden state of the previous moment to update the hidden state. The HO-RNN updates the current hidden state using hidden states from multiple historical moments, which are generalized forms of higher-order Markovian [13], which explicitly characterize long-term temporal dependencies. Su et al. further proposed a Convolutional Tensor-Train-based LSTM (Conv-TT-LSTM) [14], extending the high-order LSTM to a high-order ConvLSTM for spatiotemporal prediction tasks. In conclusion, HO-RNNs are very suitable for long-term spatio-temporal modeling. Our method is a novel and effective network that uses nonmainstream higher-order RNNs [11, 12] to model long-term temporal information. To the best of our knowledge, no literature on the use of higher-order RNNs in video action recognition has been proposed.

In this paper, we design the High-Order ConvLSTM (HO-ConvLSTM) as a new sptio-temporal modeling module, which can be inserted into different stages of a 2D

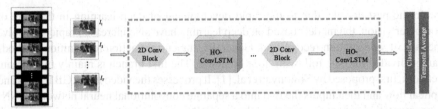

Fig. 1. The overview of our LSN network model for video recognition. LSN effectively integrates the newly constructed high-order ConvLSTM (HO-ConvLSTM) modules with traditional 2D convolutional blocks to generate more powerful video representation. HO-ConvLSTM modules are inserted into different stages of the 2D deep recognition architecture in a plug-and-play manner, thus well characterizing the long-term temporal evolution of video at various spatial resolutions. See Sect. 2.2 for details.

CNN to model spatiotemporal relationships in videos. Specifically, we propose a long-term spatio-temporal network (LSN), and its overall architecture is illustrated in Fig. 1. For each output of different convolution stages, the dimensionality is reduced by one 2D Conv. Then, we enter the HO-ConvLSTM module to update hidden states in each time step, whose core includes the accumulation temporal state (ATS) module and a standard ConvLSTM module. Then, another 2D Conv. is performed to restore the feature dimension. The HO-ConvLSTM module can be inserted into the different stages of 2D CNNs for modeling long-term temporal correlation. Unlike the Conv-TT-LSTM [14] method, they employ stacked LSTM as the network architecture, but we insert the HO-ConvLSTM module into 2D CNNs. Our method can make full use of 2D Conv. to construct very deep networks. Since 2D CNNs have a lower computational cost than 3D CNNs and two-stream framework, after combining 2D CNNs with the HO-ConvLSTM module, 2D Conv. can be utilized to learn spatial features, and the HO-ConvLSTM module can capture long-term temporal features at varying spatial resolutions. The whole network will possess good spatio-temporal modeling ability. We perform extensive experiments on three challenging video dataset benchmarks, Something-Something V1 (S-S V1 for brief) [15], Something-Something V2 (S-S V2 for brief) [15] and Diving48 [16]. The results demonstrate the effectiveness of the given LSN model for this task. The contributions of our work can be summarized as follows:

(1) We propose a novel LSN model for solving the video action recognition task, which inserts the newly constructed HO-ConvLSTM module into different stages of the 2D CNN architecture, focusing on capturing the long-term spatio-temporal correlation of spatio-temporal features at different spatial resolutions. To the best of our knowledge, this is the first task of utilizing HO-ConvLSTM for video action recognition.

(2) Our HO-ConvLSTM module consists of an accumulated temporary state module as well as a standard ConvLSTM module. It utilizes the ATS module to accumulate several previous hidden states to one temporary state, which will get into the standard ConvLSTM to determine the output together with the current input, thus effectively modeling the long-term dependencies of video features.

(3) Extensive experiments on three commonly used benchmarks for video action recognition demonstrate the effectiveness of the proposed LSN model. It achieves competitive performance with representative methods and exhibits good computational efficiency.

2 Method

First, we briefly revisit LSTM [10] and its variant ConvLSTM [17] for long-term modeling. Then, we introduce the details of the long-term spatio-temporal network proposed in this work, including the overall framework of LSN as well as the accumulate temporary state module.

2.1 LSTM and ConvLSTM

LSTM [10] is a variant of RNN that has proven useful in many sequence tasks. It is widely used in one-dimensional sequence learning [18]. The difference between LSTM and RNN mainly lies in its storage unit ($C(t)$), which can learn long-term dependency information. Only the information that conforms to the algorithm authentication will be left, and the past cell status $C(t-1)$ will be forgotten through the forget gate. Whether the final unit output will be propagated to the final state $H(t)$ is further controlled by the output gate.

The key equations are as follows:

$$[I(t); F(t); \tilde{C}(t); O(t)] = \sigma(Wx(t) + Uh(t-1)) \tag{1}$$

$$C(t) = C(t-1) \circ F(t) + \tilde{C}(t) \circ I(t) \tag{2}$$

$$H(t) = O(t) \circ \sigma(C(t)) \tag{3}$$

where $\sigma(.)$ denotes a sigmoid applied to the input gate $I(t)$, forget gate $F(t)$ and output gate $O(t)$, and a $\tanh(.)$ applied to the memory cell $\tilde{C}(t)$ and cell state $C(t)$. $\circ$ denotes the elementwise product. LSTM can only model 1D sequences, however, it cannot model spatio-temporal data such as video.

ConvLSTM [17] addresses the limitation by extending LSTM to model spatio-temporal structures within each cell, i.e., the states, cell memory, gates and parameters are all encoded as high-dimensional tensors. Then, (1) can be replaced by (4):

$$[I'(t); F'(t); C'(t); O'(t)] = \sigma(W * x(t) + U * H(t-1)) \tag{4}$$

where $*$ defines the convolution between states and parameters as in convolutional neural networks.

2.2   Long-Term Spatio-Temporal Network

Overview of the Long-Term Spatio-Temporal Network.   An overview of the structure of the long-term spatio-temporal network (LSN) proposed in this paper is shown in Fig. 1. The network utilizes sparse sampling [2] to model video-level temporal information as the input of the network. It employs 2D convolution [19, 20] to construct a very deep network as a backbone model, which can be utilized to learn spatial features. The high-order ConvLSTM (HO-ConvLSTM) module proposed in this work is inserted between the different stages of the 2D CNNs to extract long-term temporal information. A classifier is defined at the end of the network, and a temporal average pooling operation is performed.

Our goal is to build an effective and efficient network to capture long-term temporal and spatial information for video action recognition. Therefore, the network proposed in this work should meet two requirements: (1) the ability to capture long-term temporal correlation and (2) reducing the number of parameters as much as possible. Inspired by [14], we design the HO-ConvLSTM model with a residual connection for video action recognition. The specific method is to insert HO-ConvLSTM into each stage of 2D CNNs to model the long-term spatiotemporal information of video spatiotemporal features at different spatial resolutions.

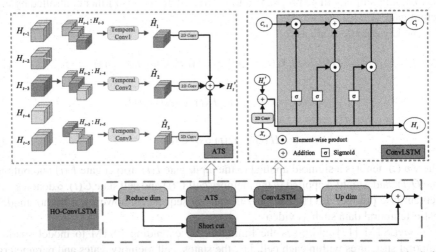

Fig. 2.   Overall architecture of our Ho-ConvLSTM module. The core of Ho-ConvLSTM includes an ATS module and a standard ConvLSTM module. It utilizes the ATS module to accumulate several previous hidden states to one temporary state, which will enter the standard ConvLSTM to determine the output together with the current input, thus effectively modeling the long-term dependencies of video features.

The HO-ConvLSTM Module.   As shown in Fig. 2, the core of our HO-ConvLSTM mainly consists of an accumulate temporary state (ATS) module and a ConvLSTM module. In addition, there is a dimensionality reduction operator before the ATS module and a dimensionality upgrade operation after the ConvLSTM module. The dimensionality reduction operation can significantly reduce the amount of parameters required to compute the HO-ConvLSTM with little change in the final performance of the network. This module is designed to be a plug-and-play form that can be easily inserted into any existing 2D CNN model.

Next, we will specifically describe the details of the ATS Module proposed in this work. In (4), the original hidden state $H(t)$ is used as the input of the module, but this causes the gradient to explode in the calculation process. To solve the problem, we use a sliding window to group adjacent steps together. As shown in Fig. 2, $H_{t-1}, ..., H_{t-3}$ in the ATS model consists of a group. The state of the hidden layer at the first N moments is $H_{t-1}, ..., H_{t-N}$, and the order is defined as M. These N moments pass through a sliding window of size $N - M + 1$, and the step length of each sliding is 1. The sliding window divides the states into overlapping M groups, and each group has $N - M + 1$ hidden layer states. These M groups are subjected to temporal convolution with a convolution kernel size of $3 \times 1 \times 1$. The number of output channels of the temporal convolution is set to ranks, and the ranks are less than the number of input channels. Use such a convolutional layer to reduce the number of channels in each group. The purpose of this is to learn the temporal information between hidden layer states at different moments. Not only can the parameters be reduced, but the previous information can also be merged for better temporal modeling. It can be expressed in the following form:

$$H_i' = Sequeeze(Conv1D_i(Stack(Slide(H_{(t-1)}, ..., H_{(t-i)})))) \qquad (5)$$

where H_i' represents the result after preprocessing. For each output of the preprocessing module, we use a 2D convolution with a kernel size of 3×3 and then add and operate it after the convolution. Then, fusion hidden states representing high-order information are fed to the standard ConvLSTM module in (5).

3 Experiments

In this section, we will perform experiments on three video benchmarks to evaluate the effectiveness of our LSN model. We first briefly introduce three video datasets utilized for the model evaluation and then describe the details of the experiment implementation. Finally, we report and discuss experiment results of LSN, including a comprehensive ablation study on the S-S V1 [15] dataset as well as compared experiments on the S-S V2 [15] and Diving48 datasets [16].

3.1 Datasets

S-S V1 and S-S V2 [15] are two crowdsourced datasets constructed by performing the same action with different objects. There are 108K videos in V1 and 221K videos in V2 with 174 categories in each dataset. Diving48 contains about 18K videos of 48 fine-grained dive classes. In terms of dataset preparation, for S-S V1, we directly employ RGB frames on the official website. In regard to S-S V2 and Diving48, we follow the video processing method of the TSM [21] and process the video data into RGB frames.

3.2 Implementation Details

We take the 2D ResNet-50 model pre-trained on ImageNet as the backbone and embed the proposed HO-ConvLSTM model to construct the LSN. For the data preprocessing, we follow TSN [2] to perform LSN. In addition, for LSN + SS (snippet sampling) mentioned in Table 4 and Table 5, we follow [8]. To keep sparse sampling still efficient, we duplicate the weights of the first convolutional layer of the network to fuse the segments without affecting the computation changes in the rest of the network afterwards. In this work, we choose $T = 8$ or $T = 16$ and fix $K = 3$ after the ablation study. We adopt the SGD optimizer with an initial learning rate of 0.01, the weight decay is set to 5e-4, and the momentum is 0.9. On the S-S V1 and S-S V2 datasets, the networks are trained within 40 epochs with an initial learning rate of 0.01, dropped by 0.1 after 20 and 30 epochs, respectively. On the Diving48 dataset, the networks are trained within 20 epochs with an initial learning rate of 0.01, decreased by 0.1 after 10 and 15 epochs, respectively. The initial learning rate of HO-ConvLSTM is set to 0.05 with a weight decay of 5e-4. To avoid overfitting, we also add dropout to the last fully connected layer, setting it equal to 0.5. For inference, we employ the same strategy as described in [8] for LSN + SS.

3.3 Ablation Study

For the ablation experiment phase, we use the classic 2D CNN ResNet-50 as the backbone network and conduct experiments on the S-S V1 dataset to evaluate the effect of different components of the LSN, including the kernel size of the input matrix (W), the size of the convolution kernel (U) of the output matrix at the previous moment, the number of ranks, the number of orders, and the number and position of HO-ConvLSTM. To reduce the computational cost, we utilize resized images of 112×112 pixel size with $T = 8$ to test LSN and report Top-1 accuracy on the validation set of the S-S V1 dataset.

Comparison of the ConvLSTM and HO-ConvLSTM Modules. To evaluate the temporal modeling ability of the given HO-ConvLSTM module, we compare the baseline TSN [2], ConvLSTM [17] and the proposed LSN model with HO-ConvLSTM. Both ConvLSTM and HO-ConvLSTM modules are inserted between the penultimate residual stage (i.e., $res4$) and the last one (i.e., $res5$) of ResNet-50. Table 1 compares the effects of input frames between ConvLSTM and HO-ConvLSTM on Top-1 accuracy.

As shown in Table 1, both ConvLSTM and HO-ConvLSTM obviously gain performance improvements over the baseline of TSN [2]. This is mainly due to the strong temporal modeling capabilities of ConvLSTM and HO-ConvLSTM. When the input is 8 frames, the two models gain similar classification performance. However, when the input comes to 16 frames, ConvLSTM and HO-ConvLSTM obtain recognition accuracy values of 41.63% and 42.27%, respectively. HO-ConvLSTM gains 0.64% accuracy improvement over ConvLSTM, showing the effectiveness of HO-ConvLSTM for the video recognition task. This may be because HO-ConvLSTM can fuse more hidden layer states to obtain more helpful information, proving that HO-ConvLSTM has better modeling long-term correlation capabilities. At the same time, it can be found that a higher Top-1 accuracy can be obtained by using a more significant number of input frames,

Table 1. Comparison of different modules using ResNet-50 as backbone with input size of 112 × 112 pixels achieved on the validation set of S-S V1 dataset.

Method	Frames	Top-1 Acc. (%)
TSN [2]	8	19.70
ConvLSTM [17]	8	40.46
ConvLSTM [17]	16	41.63
HO-ConvLSTM	8	40.16
HO-ConvLSTM	16	**42.27**

which also proves the importance of long-term modeling for video action recognition application.

Parameter Effects of HO-ConvLSTM Modules. Then, we perform experiments to evaluate the effects of using various parameters for HO-ConvLSTM. HO-ConvLSTM is inserted into different positions of the 2D CNN backbone, aiming at learning the state of the hidden layer at different times with LSTM. Meanwhile, the input dimensionality of 2D convolution is heavily determined by ranks. Therefore, we first conduct experiments to select an appropriate rank, as shown in Fig. 3, and an optimal accuracy of 42.27% is achieved with a rank of 128. Meanwhile, it achieves the second highest recognition accuracy of 42.15% by using ranks of 64, and the smaller ranks decrease the recognition performance of the LSN model.

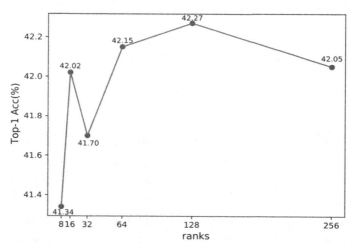

Fig. 3. Top-1 accuracy (%) results of LSN under various ranks achieved on the validation set of S-S V1 dataset (fixing the order and steps).

In addition, the HO-ConvLSTM module in LSN also includes two other important parameters, i.e., order and steps. Order is the number of hidden layer states at the previous historical moment, and steps is the total number of past steps adopted to calculate the next step. For different settings, the current layer will have different hidden layer states. Table 2 lists the compared Top-1 accuracy results of LSN under various order and steps on validation set of S-S V1 dataset when fixing ranks to 128. This table shows that LSN achieves its optimal performance of 42.27% based on both order and steps set to 5. However, there is no obvious rule to choose the better order and steps values.

Table 2. Top-1 accuracy (%) results of LSN under various order and steps achieved on the validation set of S-S V1 dataset (fixing ranks to 128).

Order	Steps				
	1	2	3	4	5
1	41.77	-	-	-	-
2	41.78	42.01	-	-	-
3	42.42	42.46	42.27	-	-
4	42.12	42.19	41.88	41.89	-
5	42.37	41.77	41.92	42.17	**42.47**

Location and Number of HO-ConvLSTM. Finally, we evaluate the performance impact of inserting different numbers of HO-ConvLSTMs at different locations in the backbone network. Unless otherwise specified, we insert HO-ConvLSTM right after the residual stage. Specifically, we insert one or multiple HO-ConvLSTM modules in different positions. Table 3 summarizes the results of different configurations.

Table 3. Top-1 accuracy (%) results of LSN under different quantities and locations on the validation set of S-S V1 dataset.

Configuration	Top-1	Params	GFLOPs
after *res2*	42.57	24.7 M	28.26
after *res3*	42.12	24.8 M	20.29
after *res4*	42.47	26.1 M	19.08
after *res5*	37.69	29.8 M	18.86
after *res2&res3*	42.47	25.7 M	31.19
after *res2&res4*	**45.78**	**26.9 M**	**29.98**
after *res3&res4*	44.57	27.0 M	22.01
after *res2&res3&res4*	42.68	27.9 M	32.92

We can see that only one HO-ConvLSTM module insertion can significantly improve the baseline with some computational burden. In the case of inserting only one module, the module after *res*5 at the lowest spatial resolution performs the worst, while modules at higher resolution perform much better and the one after *res*4 achieves the highest accuracy. These results clearly show that it is more critical to capture fine, local motion at higher resolution feature maps. For the case of inserting only two modules, it can be observed that the two modules further improve the performance. After inserting *res*2 and *res*4, the best result is obtained, which is 3.21% higher than the best result of inserting a single module. We also tried to insert three modules. According to Table 3, we can see that inserting more modules can bring performance gains. However, it also introduces additional computational costs.

3.4 Comparison with Representative Methods

Finally, we compare several representative video recognition methods on the S-S V1, S-S V2 and Diving48 datasets. For experiments on S-S V2 and Diving48 datasets, we also insert HO-ConvLSTM into the position after *res*2 and *res*4. To make a fair comparison, we also utilize 224 × 224 images as the input of LSN. Table 4 lists the comparative results on S-S V1 and S-S V2 datasets.

Table 4. Comparison of LSN with the representative methods on validation sets of S-S V1 and V2 datasets (LSN + SS means using the snippet sampling method on the basis of LSN).

Method	backbone	Frames × Clips × Crops	S-S V1(top1)	S-S V2(top1)	GFLOPs
TSN [2]	2D BN-Inc	8 × 1 × 1	19.5	30.0	16
Non-local [22]	3D R50	32 × 2 × 1	44.4	-	168 × 2
bLVNet-TAM [23]	2D R50	16 × 2 × 1	48.4	61.7	48 × 2
CPNet [24]	2D R34	24 × 16 × 6	-	57.7	-
SmallBig [25]	2D R50	8 × 2 × 3 + 16 × 2 × 3	51.4	64.5	942
TEA [26]	2D R50	16 × 2 × 3	52.3	-	65 × 6
CorrNet [27]	3D R50	32 × 10 × 1	49.3	-	115 × 10
MoViNet-A1 [28]	MoViNet	50 × 1 × 1	-	62.7	-
VidTr-M [29]	ViT-B	16 × 4 × 1	-	61.9	-
VidTr-L [29]	ViT-B	32 × 2 × 1	-	63.0	-
LSN (ours)	2D R50	16 × 1 × 1	**45.7**	**56.1**	**123**
LSN + SS (ours)	2D R50	16 × 2 × 1	**50.4**	**62.3**	**127 × 2**

As shown in the table, our LSN + SS model respectively achieves its optimal Top-1 accuracy values of 50.4% and 62.3% on the S-S V1 and S-S V2 datasets, respectively, with GFLOPs of 127 × 2. Compared with the classic 2D CNN baseline TSN, the LSN outperforms the TSN by a large margin on both datasets. When we compare the LSN model with more 2D and 3D CNN-related methods, such as Non-local [22], bLVNet-TAM [23] and CPNet [24], LSN and LSN + SS achieve better performance with these models with similar computation costs. Then, compared with more recent video recognition networks, including SmallBig [25] TEA [26], CorrNet [27], MoViNet-A1 [28] and VidTr [29], LSN + SS also shows competitive performance. Although LSN + SS is inferior to three representative models, it occupies less computing resource. Besides, the compared results on the Diving48 dataset are also provided in **Table 5**, in which our LSN + SS* achieves the Top-1 accuracy of 84.0% and ranks second place among these models, demonstrating its competitive performance with the representative models.

Table 5. Comparison of LSN with the representative methods on the validation set of the Diving48 dataset. * means training from imagenet.

Method	Backbone	Top-1 Acc. (%)
TimeSformer-L [30]	ViT-B	81.0
TimeSformer-HR [30]	ViT-B	78.0
SlowFast [31]	3D R101	77.6
TimeSformer [30]	ViT-B	75.0
RSANet-50 [32]	2D R50	84.2
LSN + SS (ours)	2D R50	82.3
LSN + SS* (ours)	2D R50	84.0

4 Conclusion

In this work, we mainly exploited HO-ConvLSTM as the temporal modeling module in 2D CNNs for video action recognition and proposed the long-term spatio-temporal network (LSN). The core of HO-ConvLSTM includes an ATS module and a standard ConvLSTM module. In the ATS module, previous hidden states could be employed to fuse one temporary state, and then the update of ConvLSTM in each time step could employ more historical information covering a long temporal range. Through our exquisite design, HO-ConvLSTM could be arbitrarily inserted into all stages of 2D CNNs. In this way, HO-ConvLSTM can model the long-term correlation of features with different spatial resolutions. Extensive experiments on three video benchmarks demonstrate that LSN performs on par with representative methods. Thus, LSN provides an effective option for video recognition that effectively integrates temporal modeling into 2D CNN backbones. In the future, we will integrate HO-ConvLSTM with 3D CNNs to model global long-term spatial and temporal dependencies for video recognition task.

Acknowledgment. This work was partially supported by the National Natural Science Foundation of China (61972062, 61902220), the Young and Middle-aged Talents Program of the National Civil Affairs Commission, and the University-Industry Collaborative Education Program (201902029013).

References

1. Simonyan, K., Zisserman, A.: Two-stream convolutional networks for action recognition in videos. arXiv preprint arXiv:1406.2199 (2014)
2. Wang, L., Xiong, Y., Wang, Z., Qiao, Y., Lin, D., Tang, X., Van Gool, L.: Temporal segment networks: towards good practices for deep action recognition. In: Leibe, B., Matas, J., Sebe, N., Welling, M. (eds.) ECCV 2016. LNCS, vol. 9912, pp. 20–36. Springer, Cham (2016). https://doi.org/10.1007/978-3-319-46484-8_2
3. Feichtenhofer, C., Pinz, A., Zisserman, A.: Convolutional two-stream network fusion for video action recognition. In: Proceedings of the IEEE Conference on Computer Vision and Pattern Recognition, pp. 1933–1941 (2016)
4. Feichtenhofer, C.: X3D: expanding architectures for efficient video recognition. In: Proceedings of the IEEE/CVF Conference on Computer Vision and Pattern Recognition, pp. 203–213 (2020)
5. Tran, D., Bourdev, L., Fergus, R., Torresani, L., Paluri, M.: Learning spatiotemporal features with 3D convolutional networks. In: Proceedings of the IEEE International Conference on Computer Vision, pp. 4489–4497 (2015)
6. Ji, S., Xu, W., Yang, M., Yu, K.: 3d convolutional neural networks for human action recognition. IEEE Trans. Pattern Anal. Mach. Intell. **35**, 221–231 (2012)
7. Zhou, B., Andonian, A., Oliva, A., Torralba, A.: Temporal relational reasoning in videos. In: Ferrari, V., Hebert, M., Sminchisescu, C., Weiss, Y. (eds.) ECCV 2018. LNCS, vol. 11205, pp. 831–846. Springer, Cham (2018). https://doi.org/10.1007/978-3-030-01246-5_49
8. Zhu, X., Xu, C., Hui, L., Lu, C., Tao, D.: Approximated bilinear modules for temporal modeling. In: Proceedings of the IEEE/CVF International Conference on Computer Vision, pp. 3494–3503 (2019)
9. Ballas, N., Yao, L., Pal, C., Courville, A.: Delving deeper into convolutional networks for learning video representations. arXiv preprint arXiv:1511.06432 (2015)
10. Hochreiter, S., Schmidhuber, J.: Long short-term memory. Neural Comput. **9**, 1735–1780 (1997)
11. Soltani, R., Jiang, H.: Higher order recurrent neural networks. arXiv preprint arXiv:1605.00064 (2016)
12. Yu, R., Zheng, S., Anandkumar, A., Yue, Y.: Long-term forecasting using tensortrain RNNs. arXiv preprint arXiv:1711.00073 (2017)
13. Sato, M.: Learning chaotic dynamics by recurrent neural networks. In: Proceedings of International Conference on Fuzzy Logic and Neural Networks (1990)
14. Su, J., Byeon, W., Kossaifi, J., Huang, F., Kautz, J., Anandkumar, A.: Convolutional tensor-train LSTM for spatio-temporal learning. arXiv preprint arXiv:2002.09131 (2020)
15. Goyal, R., et al.: The "something something" video database for learning and evaluating visual common sense. In: Proceedings of the IEEE International Conference on Computer Vision, pp. 5842–5850 (2017)
16. Li, Y., Li, Y., Vasconcelos, N.: RESOUND: towards action recognition without representation bias. In: Ferrari, V., Hebert, M., Sminchisescu, C., Weiss, Y. (eds.) ECCV 2018. LNCS, vol. 11210, pp. 520–535. Springer, Cham (2018). https://doi.org/10.1007/978-3-030-01231-1_32
17. Xingjian, S., Chen, Z., Wang, H., Yeung, D.-Y., Wong, W.-K., Woo, W.-C.: Convolutional LSTM network: a machine learning approach for precipitation nowcasting. In: Advances in Neural Information Processing Systems, pp. 802–810 (2015)
18. Pascanu, R., Mikolov, T., Bengio, Y.: On the difficulty of training recurrent neural networks. In: International Conference on Machine Learning, pp. 1310–1318 (2013)
19. Gao, S., Cheng, M.-M., Zhao, K., Zhang, X.-Y., Yang, M.-H., Torr, P.H.: Res2net: a new multi-scale backbone architecture. IEEE Trans. Pattern Anal. Mach. Intell. **43**, 652–662 (2019)

20. He, K., Zhang, X., Ren, S., Sun, J.: Deep residual learning for image recognition. In: Proceedings of the IEEE Conference on Computer Vision and Pattern Recognition, pp. 770–778 (2016)
21. Lin, J., Gan, C., Han, S.: TSM: temporal shift module for efficient video understanding. In: Proceedings of the IEEE/CVF International Conference on Computer Vision, pp. 7083–7093 (2019)
22. Wang, X., Girshick, R., Gupta, A., He, K.: Non-local neural networks. In: Proceedings of the IEEE Conference on Computer Vision and Pattern Recognition, pp. 7794–7803 (2018)
23. Fan, Q., Chen, C.-F., Kuehne, H., Pistoia, M., Cox, D.: More is less: learning efficient video representations by big-little network and depthwise temporal aggregation. arXiv preprint arXiv:1912.00869 (2019)
24. Liu, X., Lee, J.-Y., Jin, H.: Learning video representations from correspondence proposals. In: Proceedings of the IEEE/CVF Conference on Computer Vision and Pattern Recognition, pp. 4273–4281 (2019)
25. Li, X., Wang, Y., Zhou, Z., Qiao, Y.: Smallbignet: integrating core and contextual views for video classification. In: Proceedings of the IEEE/CVF Conference on Computer Vision and Pattern Recognition, pp. 1092–1101 (2020)
26. Li, Y., Ji, B., Shi, X., Zhang, J., Kang, B., Wang, L.: Tea: temporal excitation and aggregation for action recognition. In: Proceedings of the IEEE/CVF Conference on Computer Vision and Pattern Recognition, pp. 909–918 (2020)
27. Wang, H., Tran, D., Torresani, L., Feiszli, M.: Video modeling with correlation networks. In: Proceedings of the IEEE/CVF Conference on Computer Vision and Pattern Recognition, pp. 352–361 (2020)
28. Kondratyuk, D., et al.: MoViNets: mobile video networks for efficient video recognition. In: Proceedings of the IEEE/CVF Conference on Computer Vision and Pattern Recognition, pp. 16020–16030 (2021)
29. Zhang, Y., et al.: VidTr: video transformer without convolutions. In: Proceedings of the IEEE/CVF International Conference on Computer Vision, pp. 13577–13587 (2021)
30. Bertasius, G., Wang, H., Torresani, L.: Is space-time attention all you need for video understanding? arXiv preprint arXiv:2102.05095 (2021)
31. Feichtenhofer, C., Fan, H., Malik, J., He, K.: Slowfast networks for video recognition. In: Proceedings of the IEEE/CVF International Conference on Computer Vision, pp. 6202–6211 (2019)
32. Kim, M., Kwon, H., Wang, C., Kwak, S., Cho, M.: Relational selfattention: what's missing in attention for video understanding. arXiv preprint arXiv:2111.01673 (2021)

A Novel Fish Counting Method Based on Multiscale and Multicolumn Convolution Group Network

Yuxuan Zhang[1,2], Junfeng Wu[1,2(✉)], Hong Yu[1,2], Shihao Guo[1,2], Yizhi Zhou[2], and Jing Li[1,2]

[1] School of Information Engineering, Dalian Ocean University, Dalian 116023, China
wujunfeng@dlou.edu.cn
[2] Key Laboratory of Environment Controlled Aquaculture (Dalian Ocean University) Ministry of Education, Dalian 116023, China

Abstract. An accurate grasp of the number of fish in the breeding pond or fixed waters can provide an important basis for bait placement and reasonable fishing, and these data can also provide the necessary data support for accurate breeding. Due to the high density of fish in the real underwater environment, the strong occlusion and the large amount of adhesion, it is difficult to count fish, and the accuracy is low. Considering the above issues, we present a new approach to a fish counting method based on a multiscale multicolumn convolution group network. To enhance the counting accuracy and reduce the complexity of the network, this method uses an asymmetric convolution kernel to change the traditional convolution kernel, which increases our network depth and appreciably reduces the size of the network. In the backbone network, a convolutional group is used to replace a single convolutional layer to enhance the learning capacity of the network. The back of the net introduces the spatial structure of the pyramid and the multicolumn dilated convolution, which preserves the different scaling properties of fish data and improves the capabilities of the fish counting algorithm. To check the performance of the algorithm, this work collects and labels the DLOU3 fish dataset suitable for counting fish and conducts simulation experiments on the DLOU3 fish dataset using our algorithm. The experiments are compared with other popular fish counting algorithms in terms of the mean absolute error (MAE) and mean square error (MSE). The MAE and MSE of the final experimental results of our method are 5.36 and 6.56 and 23.67 and 32.52 in the two test sets, respectively, and the best performance among the five groups of algorithms is obtained.

Keywords: Fish counting · Neural networks · Asymmetric convolution · Dilated convolution

1 Introduction

In the aquaculture industry, in the process of continuous development, a series of problems have caused some harm, such as poor control of breeding density and improper

Y. Wang et al. (Eds.): ICPCSEE 2022, CCIS 1628, pp. 339–353, 2022.
https://doi.org/10.1007/978-981-19-5194-7_25

treatment of bait. These problems can lead to large-scale fish kills, resulting in huge economic losses for aquaculture companies and farmers. Therefore, it is very important to appropriately respond to risks and ensure the safety and health of the breeding environment. It is necessary to monitor the number, density and distribution of fish in real time to provide data for the safety of the breeding environment. On the other hand, when allocating the density and quantity of fish populations, it is beneficial to control and adjust fishing volume, adjust the production structure of aquaculture, optimize the spatial distribution of fisheries, and promote the sustainable development of fisheries.

1.1 Related Work

In the early stages of development of computer technology. Congrong Zhu [1] used the machine vision method to process and analyze fry images and build a model through a CCD camera and image processing software. The method is more accurate at 45 to 50 fish. Ling Huang et al. [2] used image processing methods, and specific values were calculated by comparing the pixel points of the fish in the image with the pixel points in the whole image. When there is a gap in the individual size of the fish group, the precise value decreases. Shuo Wang et al. [3] used computer methods for image processing, which requires collecting video images of fish flowing through a fixed channel and is not suitable for high-density underwater fish counting. Song Fan et al. [4] used an endpoint-based refinement method to achieve counting by counting endpoints in the connection domain, but the limited space of the connectivity domain does not accommodate large-scale counting. The above methods are only suitable for a small number of fish and are not suitable for high-density fish school pictures. There is no suitable way to obtain these key data quickly and efficiently for large-scale fish. In recent years, fish density statistics based on deep learning methods have certain research value.

The emergence of deep learning methods has brought a great deal of research and applications to the academic community. Researchers use deep learning for object detection [5], image classification [6], image processing [7], attitude estimation [8] and other research directions. In the field of counting, Zhang et al. [8] proposed a more accurate and efficient implementation of density statistics using multicolumn neural networks with convolution kernels of different sizes. Sindagi VA et al. [9] proposed a branched network structure that fuses global semantics and partial semantics of images to enhance feature learning. Li et al. [10] abandoned a branch network and proposed a deep and narrow network, and the back-end network used dilated convolution. Compared with ordinary convolution, dilated convolution can obtain global information more easily. Liu et al. [11] use a network where 3 images of different sizes can be entered at the same time. Branching networks convolve images of different sizes to obtain richer features.

In the field of fish counting, there are also some proposed methods of their own. Jing Li et al. [16] introduced an attention mechanism. The front end obtains basic detailed features through a single-column network and then processes images through a convolutional neural network. The attention module integrates different scales. However, the network learning ability is relatively weak. The current research on fish is very extensive. Related studies include fish localization based on deep learning [12], fish detection [13], fish individual identification [14], fish weight measurement [15], etc. Deep learning technology will bring more changes to the fishing industry in the future. Deep learning technology will bring more changes to the fishing industry in the future.

2   The Proposed Method

2.1   Fish Counting Network

We conduct research based on current problems and propose our solutions. Figure 1 is our network structure.

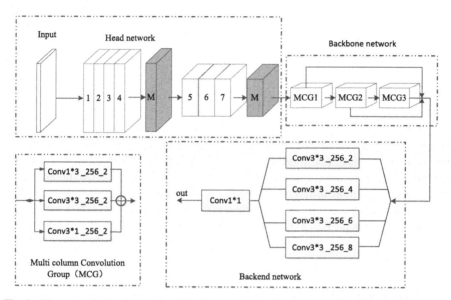

Fig. 1. The network has 3 modules: Head network, Backbone network and Backend network. MCG1, MCG2, and MCG3 in the backbone network are composed of multicolumn convolution groups.

The counting network consists of three parts, including 12 layers of convolutional layers. Table 1 lists the specific parameters of each layer of the network. Among them, Conv3*3 is a common convolution operation, Conv3*1 and Conv1*3 are asymmetric convolution operations, and 64 represents the number of convolution channels. 2, 4, 6, and 8 in the convolutional layers of the back-end network represent the convolution hole rate.

The picture enters the network through the input end, and the convolution operation is first performed on the head network. The shallow network learns the underlying information through convolution, such as the pattern of the fish body, mouth, fins, etc. After the pooling operation, the size of the parameter matrix is reduced, the complexity of the entire network is simplified, and the corresponding calculation amount is reduced. In the backbone network, three sets of multicolumn convolution groups are used to perform convolution operations to enhance feature extraction, and by adding channels in

Table 1. Fish counting network structure.

Head network		Backbone network		Backend network	
Convolution layer	Convolution kernel	Convolution layer	Convolution kernel	Convolution layer	Convolution kernel
1	Conv_3*3(64)	MCG1	Conv_1*3(256)	11	Conv_3*3(256)2
2	Conv_3*1(64)		Conv_3*3(256)		Conv_3*3(256)4
3	Conv_1*3(64)		Conv_3*1(256)		Conv_3*3(256)6
4	Conv_3*3(64)	MCG2	Conv_1*3(256)		Conv_3*3(256)8
M	Maxpooing		Conv_3*3(256)	12	Conv1*1_1
5	Conv_3*1(64)		Conv_3*1(256)		
6	Conv_1*3(64)	MCG3	Conv_1*3(256)		
7	Conv_3*3(64)		Conv_3*3(256)		
M	Maxpooing		Conv_3*1(256)		

the convolutional layer by using more convolutional kernels, the channels can hold more information. The three convolution kernel specifications in each multicolumn convolution group are set to 3*3, 3*1, and 1*3. Compared with a single convolutional layer, this method learns more differentiated feature information and improves the convolution effect of the network. In the backend network, we use convolutional layers with different expansion rates (2, 4, 6, 8) in 4 branching network layers, which allows for targeted learning of large and small targets. Finally, the predicted density map is generated by a 1*1 convolution.

2.2 Asymmetric Convolution

We use asymmetric convolution in the front-end network to improve the accuracy of the network's counts. Layers 1 to 7 in the head network are composed of a mixture of asymmetric convolution and ordinary convolution. The asymmetric convolution is a decomposition process of a common convolution structure. Szegedy C et al. [17] split the traditional n*n of the same size to obtain a combination of n*1 and 1*n. This has the following advantages: (1) Reduce computational difficulty to speed up computation. (2) A nonlinear structure is added to the network by this splitting structure. (3) Smaller convolution kernels participate in the operation to reduce the risk of network overfitting.

2.3 Spatial Pyramid Structure

We use multicolumn convolutional groups in the designed backbone network to enhance the feature extraction capability, and such convolutional groups improve the accuracy of counting. A training method based on the ACNet [18] network uses a set of convolutional layers instead of a single convolutional layer of n*n convolutional kernels. We use a multicolumn structure to convolve the input information using 3 sets of convolution kernels with different contents. After the convolution is completed individually, these channels are superimposed (see Fig. 2).

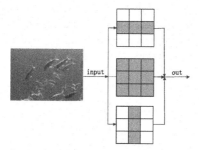

Fig. 2. The multicolumn convolution group contains 3 columns of convolutional layers. Each layer uses a nonstandard convolution kernel. This structure has a better feature extraction effect for flipped and rotated objects, and this method can be used in the training phase to enhance feature extraction.

2.4   Multicolumn Convolution Group

The spatial pyramid structure used by the back-end network consists of multicolumn dilated convolution (see Fig. 3). This structure we propose can enhance the network's ability to learn fish population characteristics at different scales and improve the accuracy of network prediction.

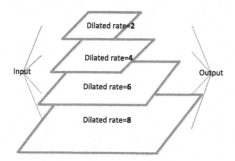

Fig. 3. Use a feature pyramid structure composed of a hole rate of (2, 4, 6, 8) hole convolution, and output after channel fusion at the end, expand network acceptance, and improve our network ability to extract global feature information.

2.5   Predicted Density Plot for Count

Our method generates a corresponding density map based on the original map and implements fish counting based on the density map data, so the quality of the density map affects the accuracy of the prediction result.

Due to the distortion of the fish in the picture due to the shooting angle problem, to truly reflect the relative size, each fish is marked, and the size is estimated by the distance between each point that has been marked and the neighboring points around itself. Through the geometric adaptive Gaussian kernel algorithm, the marked point is converted into a corresponding area, and the size covered by this area reflects the size of the fish in the picture. If there is a point in pixel xi, we denote this as an incremental function $\delta(x - xi)$. We use the following function to represent an image with N marked headers.

$$H(x) = \sum_{i=1}^{N} \delta(x - xi) \tag{1}$$

Achieving the density conversion requires the calculation of $F(x) = H(x) * G\sigma(x)$. This process requires the use of a Gaussian kernel. Each x_i represents a simplified pattern of marked fish. x_i is also the relative position of this fish.

Since the problem of near big and far small appears on almost every picture of the fish, we make the density map closer to the original map and match the true distribution (the size of the fish is different in different locations) and therefore use a density mapping method based on an adaptive Gaussian kernel:

$$F(x) = \sum_{i=1}^{N} \delta(x - xi) * G\sigma i(x), with \sigma i = \beta \overline{d}^{i} \tag{2}$$

This formula is used to calculate the average distance between the m fish heads closest to the fish head x_i. The value of $\overline{d}^{i}$ here indicates the relative size of the fish. To calculate the density of the fish around the pixel, it is necessary to multiply $\delta(x - xi)$ by the Gaussian kernel $G\sigma i(x)$. In the experiment, $\beta = 0.3$.

2.6 Loss Function

This paper uses Euclidean distance to measure the difference between the estimated density map and the ground truth. The loss function is defined as follows:

$$L(\theta) = \frac{1}{2N} \sum_{i=1}^{N} \| F(xi, \theta) - Fi \|_{2}^{2} \tag{3}$$

where L is the loss function, N is the number of input images for each iteration of our training, x_i denotes a specific input image, θ represents our network parameters, $F(xi, \theta)$ is the density map generated by the network after training based on the input images, F_i represents the labeled in the true density map, and L is the total loss value.

2.7 Dataset Construction

In this paper, our goal is to accurately count fish images from any camera view. To train and test the algorithm, this paper collects and organizes the fish density dataset and manually labels the data in the dataset. In this paper, our dataset consists of three parts: a training set, test set and validation set. Images in the training set help train the convolutional neural network, and the obtained network weights are obtained. The test set is used to test with the trained weight files and generate predicted density maps for each image. This paper visualizes it by means of a heatmap to display the number of fish more intuitively.

Since there is currently no public high-density fish image dataset, this paper collects and organizes the DLOU3 dataset by itself and manually annotates it. Part of the data comes from the production workshop of aquaculture enterprises, and part of the data comes from random capture outdoors. The shooting angle of this type of image is from the top-down perspective. Some of the data come from the Internet, and some of these data are obtained from underwater real scenes. The DLOU3 dataset finally integrates more than 700 kinds of high-density fish pictures, which has good extensiveness and diversity (see Fig. 4).

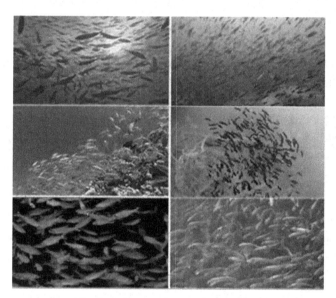

Fig. 4. Partial DLOU3 dataset pictures.

We reduce the training difficulty by further processing the dataset. The images in the training set are scaled down by cropping the original images. The traditional cropping method is random cropping, that is, a part of the image is intercepted from a random position of the picture. However, because the randomly cropped picture does not have a fixed cropping position, the loss of important content in the image will seriously affect the feature learning of the neural network. Therefore, in this paper, the data are processed using the sliding cropping method [16]. A fixed step size sliding crop is selected to perform the cropping work. After reaching the edge, another starting position is picked, the process is repeated, and all the data are finally obtained (see Fig. 5).

In the training network phase, we process these cropped images to generate new annotation files with annotation information corresponding to the location and number of fish in the cropped images, using these processed data to reduce the difficulty of training the network. In the testing phase, we directly use the original images and the original annotation files for validation.

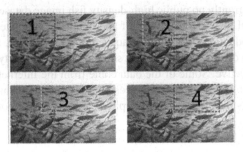

Fig. 5. Slide crop process.

The results of cropping the same image using two different cropping methods are shown in Fig. 6.

(a)   (b)

Fig. 6. A is an adjacent image obtained by sliding cropping, which contains more overlapping parts and has fewer missing parts, and the data preserve more feature details. b is the image obtained by random cropping, and the image of the fish in the edge area is more severe.

The DLOU3 dataset has 700 fish images, Table 2 shows the specific number distribution, the training set contains 400 images, the test set is two groups A and B is a total of 300 images, 100 of the images in Test Set A are from the same video, with the number of fish ranging from 80–105, and 200 images in Test Set B consist of 8 different species of fish. The number of fish per image ranged from 80–550.

Table 2. Data distribution of the dataset.

Number of fish range	Number of pictures
80–90	48
90–120	180
120–150	116
150–180	164
180–210	108
210–350	40
350–450	24
450–550	18

3 Experiments

3.1 Evaluation Metric

We evaluated the performance of the algorithm mainly by MAE (mean absolute error) and MSE (mean squared error), which are the price indicators of the currently available mainstream counting methods, according to the evaluation metrics proposed in the existing works [16]. Smaller results of MAE and MSE indicate lower prediction error and better performance of the algorithm, which are defined as follows:

$$MAE = \frac{1}{N} \sum_{1}^{N} |Zi - \widehat{Z}i| \tag{4}$$

$$MSE = \sqrt{\frac{1}{N} \sum_{1}^{N} (Zi - \widehat{Z}i)^2} \tag{5}$$

N is the sum of all input images, and Zi is the sum of the tagged fish in the i-th picture.

And $\widehat{Z}i$ is the prediction result corresponding to the i-th image.

3.2 Experimental Design

The test set is divided into two groups. Group A is 100 pictures of a single species of fish, and the number of fish in each picture ranges from 80 to 110. Group B contains 200 pictures of 8 species of fish, and the number ranges from 90 to 550. MAE and MSE take the best value of all the results, and the prediction time takes the average of 3 test results.

3.3 Results and Analysis

We have designed experiments to test the practical effects of our approach. We find other related methods and perform multiple training and testing by reproducing these algorithms on our DLOU3 dataset and record the optimal results. The results of our predictions on DLOU3 are given in Table 3, which contains the predictions of the five methods.

Table 3. Prediction results of 5 methods on the DLOU3 dataset.

Algorithms	MAE	MSE	Test sets
MCNN [8]	10.02	12.23	A
MCNN [8]	29.02	40.63	B
Zhou et al. [19]	8.48	9.57	A
Zhou et al. [19]	27.68	38.53	B
FishCount [16]	7.56	8.83	A
FishCount [16]	26.95	36.87	B
CSRnet [20]	7.03	8.49	A
CSRnet [20]	24.97	34.89	B
Ours	**5.36**	**6.56**	A
Ours	**23.67**	**32.52**	B

The MAE and MSE results of the MCNN in test set A reach 10.02 and 12.23, respectively, and the MAE and MSE results in test set B reach 29.02 and 40.63, respectively. Due to the overall use of larger convolution kernels (5*5, 7*7, 9*9), the MCNN network does not have a narrow and deep 3*3 convolution kernel. The MAE and MSE of the network proposed by Zhou et al. are 8.48 and 9.57 in test set A and 27.68 and 38.53 in test set B. The network uses a smaller 3*3 convolution kernel to improve the learning ability of the network, and they use a three-column structure design to learn different sizes of features. Low correlation between branch structures and the feature extraction of different scales is insufficient. The MAE and MSE results of FishCount network in test set A reach 7.56 and 8.83, respectively, and in test set B, the MAE and MSE results reach 26.95 and 36.87, respectively. The network as a whole lacks a larger receptive field, and the global feature extraction is insufficient. The MAE and MSE results of the CSRnet network in test set A reach 7.03 and 8.49, respectively, and the MAE and MSE results in test set B reach 24.97 and 34.89, respectively. The network employs a convolutional layer with a rate = 2 expansion rate to extend the receptive domain of the network and enhance the network's acquisition of global information. However, the use of the same continuous void rate causes the "grid effect&quot, and the features contained in some pixels cannot participate in the convolution and cannot be effectively extracted. The MAE and MSE results of the network proposed in this paper reach 5.36 and 6.56 in test set A, respectively, and the MAE and MSE results in test set B reach 23.67 and 32.52, respectively. The error is the smallest among the five sets of test results.

To see the results more intuitively, the density map is output by means of a heatmap (see Fig. 7). From the figure, it can be found that the prediction maps generated by our method are closer to the fish distribution of the test input images.

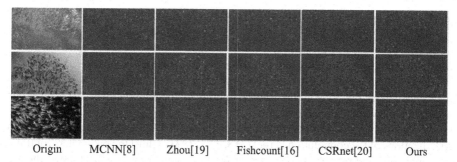

| Origin | MCNN[8] | Zhou[19] | Fishcount[16] | CSRnet[20] | Ours |

Fig. 7. The original and predicted heatmaps are shown in blue. The bright dots in the picture indicate where the predicted fish will appear.

3.4 Ablation Experiment

Asymmetric Convolution Module Experiment and Analysis
We designed experiments to test the effect of our proposed asymmetric convolution. In the front-end structure, one set of ordinary 3*3 convolutional layers is set, keeping other structures unchanged, and the other set of networks sets asymmetric convolutional layers (3*1, 1*3) and ordinary convolutional layers. Adjust the parameters for training without changing other network structures. We give the prediction results of mixed convolution and normal convolution in the test set in Table 4.

Table 4. Prediction results using hybrid convolution and ordinary convolution.

Convolution method	MAE	MSE	Total forecast speed/(s)	Test sets	Weight size (KB)
Ordinary convolution	5.89	7.10	58.03 s	A	1352
Hybrid convolution	**5.36**	**6.56**	57.84 s	A	776
Ordinary convolution	24.17	33.87	116.74 s	B	1352
Hybrid asymmetric	**23.67**	**32.52**	114.23 s	B	776

The prediction time in the table is the total prediction time for predicting 100 pictures and 200 pictures, and the weight is the size of the first 7 layers of the front-end network. It can be seen that the overall weight of asymmetric convolution is only 776 KB when the number of layers is the same. With fewer parameters. Ordinary convolution weights reach 1352 KB. The 7-layer hybrid method using asymmetric convolution has total prediction times of 57.84 s and 114.23 s in the tests of test sets A and B, respectively. Asymmetric convolution reduces parameters by splitting ordinary convolutions and speeds up the overall running speed of the network. After splitting, the network can analyze more diverse spatial features and extract more spatial feature information. The MAE and MSE errors of the 7-layer hybrid structure in test set A are 5.44 and 6.56, respectively, and the MAE and MSE errors in test set B are 23.67 and 32.52, respectively. In general, choosing a 7-layer hybrid asymmetric convolution structure is an optimal choice.

Multicolumn Convolution Group Module Experiment and Analysis

We design experiments to test a multicolumn convolutional group (MCG), and experiments are carried out. One group uses the traditional convolution layer, the other group uses the multicolumn convolution group, the Chinese convolution kernel scale uses the same 3*3 structure, and the third group uses the multicolumn structure with different specifications of the convolution kernel (3*3, 3 *1, 1*3). Adjust the parameters for training separately, and perform two groups of A and B predictions. Table 5 shows the test results of asymmetric convolution and ordinary convolution.

Table 5. Prediction results of the multicolumn convolution convolution group and ordinary volume.

Convolution method	MAE	MSE	Total forecast speed/(s)	Test sets	Weight size (KB)
Ordinary convolution	8.42	10.22	56.14	A	106584
Same convolution kernels	6.67	8.10	62.24	A	144672
Different convolution kernels	**5.38**	**6.59**	58.23	A	119328
Ordinary convolution	28.80	39.16	108.82	B	106584
Same convolution kernels	26.12	35.78	119.22	B	144672
Different convolution kernels	**23.70**	**32.57**	113.91	B	119328

The experimental results show that our different convolutional kernel schemes predict time on the ordinary convolutional layer lags behind the fish, but the errors are smaller. In test set A, the MAE and MSE of different convolutional kernel schemes are reduced by 3.04 and 3.63, respectively, compared to the normal scheme, and in tester B, the MAE and MSE of different convolutional kernel schemes are reduced by 5.10 and 6.59, respectively, compared to the normal scheme. Second, compared with the two groups of multicolumn convolution groups, the prediction time of a group of convolution kernels with different specifications is shorter by 4.01 s and 7.67 s. At the same time, the prediction accuracy is higher, and the MAE and MSE in test sets A and B are reduced by 1.29 and 1.51 and 2.42 and 3.21, respectively. Different sizes of convolution kernels are better than using a multicolumn structure with all 3*3 sizes for feature processing. This asymmetric structure has better feature extraction capability for flipped and rotated objects. The 1*3 convolution can correctly extract the upside-down object features, and the 3*1 convolution can correctly extract the horizontally flipped object features. The postures of the underwater fish images are changeable, and most of the crowds are standing or sitting. This is the largest difference between fish and crowd images. The columnar structure can better extract fish features of different poses and achieve better prediction results.

Spatial Pyramid Module Experiment and Analysis

We design experiments to test the effectiveness of the spatial pyramid module in back-end network experiments. We tested using a set of networks with this structure removed and networks that maintain the spatial pyramid structure. The other position structures of the networks were left unchanged and retrained and tested on the DLOU3 dataset. Finally, we count the best results in each of the two test sets. The prediction results of the two groups of networks in the test set are shown in Table 6.

Table 6. Prediction results after and before removing the spatial pyramid module.

Convolution module	MAE	MSE	Total forecast speed/(s)	Test sets	WeightSize (KB)
Not Spatial Pyramid Module	6.47	8.42	54.71	A	81688
Spatial Pyramid Module	5.36	6.56	58.11	A	119328
Not Spatial Pyramid Module	25.85	35.65	105.34	B	81688
Spatial Pyramid Module	**23.67**	**32.52**	114.05	B	119328

The table shows that the prediction error of using the spatial pyramid module network is lower, and the MAE and MSE results in test set A are reduced by 1.11 and 1.86 to 6.47 and 8.42, respectively. In test set B, the MAE and MSE results decreased by 2.18 and 3.13 to 25.85 and 35.65, respectively. The weight of the back-end spatial pyramid module is 37640 KB, accounting for 32% of the total weight, but after using this structure, the total prediction time of test sets A and B increases by 3.4 s and 8.71 s, respectively, and the total prediction time increases by 5.85% and 7.63%. The prediction time increases by 0.034 s and 0.043 s for each picture in test sets A and B, respectively. The spatial pyramid structure has less influence on the actual prediction time.

In addition, we are also conducting multiple experiments on the DLOU3 dataset to find the best combination of void ratios. After many tests, two groups of the current optimal combinations are given. The prediction results for the two optimal combinations on the DLOU3 dataset are given in Table 7.

The combination of (2, 3, 4, 5) achieves minimum errors of 5.33 and 6.47 for MAE and MSE in test set A, respectively. Test set A contains 100 images of a single fish, and the scale variation of the fish in the whole picture is small. In this case, the dense combination of (2, 3, 4, 5) is better. The combination of (2, 4, 6, 8) achieves minimum errors of 23.67 and 32.52 for MAE and MSE in test set B, respectively. Test set b contains 200 multitype fish pictures, and the scale variation of the fish in the whole picture is larger. In this case, it is better to use (2, 4, 6, 8) this scattered combination.

Table 7. Results of the two optimal combinations on the DLOU3 datasets.

Dilated rate combination	MAE	MSE	Test sets
Dilated rate(2,3,4,5)	5.33	6.47	A
Dilated rate(2,4,6,8)	5.36	6.56	A
Dilated rate(2,3,4,5)	23.80	32.79	B
Dilated rate(2.4.6.8)	23.67	32.52	B

4 Conclusion

Because fish in the real underwater environment have high density, serious occlusion, and large adhesion, they are difficult to count, and the counting accuracy is low. This paper proposes a fish counting method based on a multiscale multicolumn convolution group network. We organize and label the DlOU3 dataset and conduct simulation experiments on this dataset. The dataset has high-quality fish pictures from various angles, and the fish poses are abundant and close to the real situation. The dataset contains approximately 110,000 fish. The algorithm proposed in this paper achieves MAE and MSE results of 5.36 and 6.56 in test set A of the DLOU3 fish dataset, respectively, and 23.67 and 32.52 in test set B, respectively. The error is smaller than that of the mainstream MCNN, FishCount, CSRNet and other frameworks. In the future, our proposed algorithm can adapt to different scales of data and improve the transferability of the method.

Acknowledgment. This is a project funded by the National Natural Science Foundation of China (31972846), Key Laboratory of Environment Controlled Aquaculture (Dalian Ocean University) Ministry of Education (202205), and Major Special Plan for Science and Technology in Liaoning Province (2020JH1/10200002).

References

1. Zhu, C.: A novel fries counting method based on machine vision technique. Fishery Modernization **36**(2), 25–28 (2009)
2. Huang, L., Hu, B., Cao, N.: The novel fries counting method based on image processing. Hubei Agric. Sci. **51**(9), 1880–1882 (2012)
3. Wang, S., Fan, L., Liu, Y.: Theresearch of turbot fry counting method based on computer vision. Fishery Modernization **43**(03), 34–38+73 (2016)
4. Fan, S., Liu, J., Yang, Y.: Research and realization of fry counting based on image recognition technogy. Fisheries Sci. **27**(04), 210–212 (2008)
5. Cai, Z., Vasconcelos, N.: CascadeR-CNN: delving into high quality object detection. In: 2018 IEEE/CVF Conference on Computer Vision and Pattern Recognition (CVPR), Salt Lake City, pp. 6154–6162. IEEE (2018)
6. Zoran, D., Chrzanowski, M., Huang, P., et al.: Towards robust image classification using sequential attention models. In: 2020 IEEE/CVF Conference on Computer Vision and Pattern Recognition (CVPR),Seattle, pp. 9480–9489. IEEE (2020)

7. Cong, W., Zhang, J., Niu, L., et al.: DoveNet: deep image harmonization via domain verification. In: 2020 IEEE/CVF Conference on Computer Vision and Pattern Recognition (CVPR), Seattle, pp. 8391–8499. IEEE (2020)
8. Gkioxari, G., Toshev, A., Jaitly, N.: Chained predictions using convolutional neural networks. In: Leibe, B., Matas, J., Sebe, N., Welling, M. (eds.) ECCV 2016. LNCS, vol. 9908, pp. 728–743. Springer, Cham (2016). https://doi.org/10.1007/978-3-319-46493-0_44
9. Zhang, Y., Zhou, D., Chen, S., et al.: Single-image crowd counting via multi-column convolutional neural network. In: 2016 IEEE Conference on Computer Vision and Pattern Recognition (CVPR), Las Vegas, pp. 589–597. IEEE (2016)
10. Sindagi, V.A., Patel, V.: Generating high-quality crowd density maps using contextual pyramid CNNs. In: 2017 IEEE International Conference on Computer Vision (ICCV), Venice, pp. 1879–1888. IEEE (2017)
11. Liu, L., Qiu, Z., Li, G., et al.: Crowd counting with deep structured scale integration network. In: 2019 IEEE/CVF International Conference on Computer Vision (ICCV), Seoul, pp. 1774–1783. IEEE (2019)
12. Christensen, J., Galeazzi, R., Mogensen, L., et al.: Detection, localization and classification of fish and fish species in poor conditions using convolutional neural networks. In: 2018 IEEE OES Autonomous Underwater Vehicle Symposium, Portugal, pp. 1–6. IEEE (2018)
13. Fan, L., Liu, Y., Yu, X., et al.: Fish motion detecting algorithms based on computer vision technologies. Trans. Chinese Soc. Agric. Eng. (Trans. CSAE) 27(07), 226–230 (2011)
14. Zhang, J., Zeng, G., Qin, R.: Fish recognition method for submarine observation video based on deep learning. J. Comput. Appl. 39(02), 376–381 (2019)
15. Fernandes, A., Turra, E., Alvarenga, R., et al.: PSII-6 deep learning image segmentation for extraction of body measurements and prediction of body weight in Nile tilapia. J. Anim. Sci. 97(Suppl. 3), 236–237 (2019)
16. Li, J., Wu, J., Yu, H., et al.: Fish density estimation algorithm based on redundancy cutting. Comput. Digital Eng. 48(12), 2864–2868+2911 (2020)
17. Szegedy, C., Vanhoucke, V., Ioffe, S., et al.: Rethinking the inception architecture for computer vision. In: IEEE Conference on Computer Vision and Pattern Recognition, Las Vegas, pp. 2818–2826. IEEE (2016)
18. Ding, X., Guo, Y., Ding, G., et al.: ACNet: strengthening the kernel skeletons for powerful CNN via asymmetric convolution blocks. In: 2019 IEEE/CVF International Conference on Computer Vision, Seoul, pp. 1911–1920. IEEE (2019)
19. Zhou, Y., Yu, H., Wu, J., et al.: Fish density estimation with multiscale context enhanced convolutional neural network. J. Commun. Inf. Netw. 004(003), 80–88 (2019)
20. Li, Y., Zhang, X., Chen, D.: CSRNet: dilated convolutional neural networks for understanding the highly congested scenes. In: Proceedings of 2018 IEEE/CVF Conference on Computer Vision and Pattern Recognition, Salt Lake City, pp. 1091–1100. IEEE (2018)

A Low Spectral Bias Generative Adversarial Model for Image Generation

Lei Xu[1]([⊠]), Zhentao Liu[1], Peng Liu[1], and Liyan Cai[2]

[1] Peng Cheng Laboratory, Shenzhen, China
{xul06,liuzht,liup01}@pcl.ac.cn
[2] Sun Yat-Sen University, Guangzhou, China
caily3@mail2.sysu.edu.cn

Abstract. We propose a systematic analysis of the neglected spectral bias in the frequency domain in this paper. Traditional generative adversarial networks (GANs) try to fulfill the details of images by designing specific network architectures or losses, focusing on generating visually qualitative images. The convolution theorem shows that image processing in the frequency domain is parallelizable and performs better and faster than that in the spatial domain. However, there is little work about discussing the bias of frequency features between the generated images and the real ones. In this paper, we first empirically demonstrate the general distribution bias across datasets and GANs with different sampling methods. Then, we explain the causes of the spectral bias through the deduction that reconsiders the sampling process of the GAN generator. Based on these studies, we provide a low-spectral-bias hybrid generative model to reduce the spectral bias and improve the quality of the generated images.

Keywords: Deep learning applications · Image generation models · Generative adversarial network

1 Introduction

The extraordinary generative adversarial networks (GANs) in synthesizing realistic images have been proven, yet there are still some long-standing issues for the community to address, such as upsampling methods [1, 2], avoiding mode collapse [3, 4] and reducing performance biases [5–7].

Previous work typically evaluates image generation methods in a visual way. Traditional GANs such as WGAN-GP [8] and StarGAN [9] try to conquer these problems and improve the visual quality of the generated images by designing specific and sophisticated network architectures or losses. In particular, both the inception score [4] and FID [10] are designed to assess the quality of the generated images and thus lack frequency-domain analysis. Meanwhile, current analytical tools are mainly focused on studying these reasons from spatial features, with few observations in the frequency features.

This work is supported in part by the National Key Research and Development Program of China under Grant no. 2020YFB1806403.

Recently, many studies [11, 12] have shown that despite the low accuracy FID of the generated images, their frequency distributions may be strange. Furthermore, many people [13] have found that high-frequency features, even imperceptible to humans, still greatly contribute to the generalization ability of convolutional neural networks.

The bias in the frequency domain mentioned above is collectively referred to as the spectral bias. In this paper, we introduce a specific spectral tool to demonstrate spectral bias among the generated images of GANs and those real ones. With this tool and mathematical analysis, we systematically investigate the spectral deviation variation of generated images by GANs and demonstrate the reason for spectral deviations to see how these long-standing problems behave in the frequency features. Finally, we conduct in-depth experiments and provide a low-spectral-bias hybrid generative model to reduce spectral bias and improve the quality of the generated images.

Our paper is developed according to the logical order of the questions. In Sect. 2, we first introduce the related works, and in Sect. 3, we introduce our developed tool to show the spectral bias of real images and generated images. In Sect. 4, we will discuss the reason for the spectral bias by reconsidering the upsampling process of the GAN generator. Next, in Sect. 5, we develop a novel hybrid generative model to suppress difference. We then conclude this paper in Sect. 6.

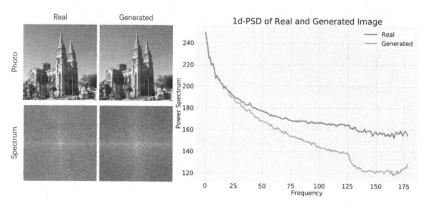

Fig. 1. Illustration of the spectral bias. **Left:** The missing details in the generated image show that there are not enough high-frequency components, and the dark corners in the spectra show that the high frequencies match worse. **Right:** The concave part at high frequency illustrates the disappearance of different frequency components.

2   Related Work

GANs [1] have recently been widely used in computer vision and graphics applications. To improve the training stability and the quality of the generated images, many variants have been proposed [8, 9]. However, many problems remain. The first is how to evaluate the quality of the generated image and evaluate the efficiency of different models. Commonly, we can use many different evaluation metrics, e.g., Inception Score [4],

Frechet Inception Distance [10] (FID), and Earth Mover's Distance [14]. Our aim is to demonstrate spectral bias. There are different tools and models that will be presented in our work to complement the current visual measure and show different insights into the limitations of the current GANs. Currently, many measures can be used to quantify the performance of different models; however, they ignore bias in the frequency features and only consider similarity in the spatial features.

Upsampling is used in GANs to transform the latent code from low-dimensional space to high-dimensional space. However, it is proven that checkerboard artifacts will be introduced by deconvolution in high-frequency features [12], while [11, 15] proposed matching the frequency by the spectral regularization method. We reconsider upsampling in this process to investigate the spectral bias between different frequency distributions of real images and generated images. Furthermore, we also evaluate the performance of GANs on different real photo datasets, such as LSUN Church Outdoor [16] and CAT [17].

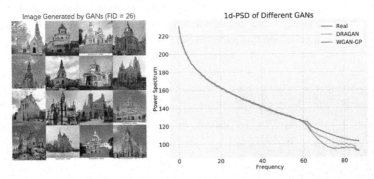

Fig. 2. General observation of the spectral bias. **Left:** Despite the high-quality images generated by GAN, the spectral bias in the frequency domain is obvious. **Right:** The concave parts of the high frequencies of the models.

3 The Spectral Bias of Generated Images

In this section, we first show the observations of spectral bias of different generated images. To understand the spectral bias, we adopt the one-dimensional power spectrum density (1D-PSD) as our first tool to visualize this spectral bias. This tool is defined by the radial distribution of the power spectrum. We then analyze the spectral bias in the resize-convolution generator and deconvolution denerator. In this process, we manually normalized each spectrum to [0, 255].

$$PSD_{1d}(w_k) = \int_0^{2\pi} \|\mathcal{F}(w_k \cos(\phi)), (w_k \sin(\phi))\|^2 \, d\phi, \tag{1}$$

where $\mathcal{F}O$ is the two-dimensional discrete Fourier transform (2D-DFT) of any two-dimensional image, and let $k = 0, 1, 2, 3, \ldots, M/2 - 1$ denote an $M \times M$ image. For

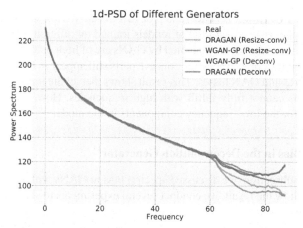

Fig. 3. General observation of the spectral bias. We pay attention to the distortions of both de convolutions at high frequencies.

multichannel images, we convert them to a single-channel grayscale and then perform 2D-DFT.

3.1 Spectral Bias in the Resize-Convolution Generator

In this section, we first adopt resizing convolution for the generator, as previous work shows that we should be careful with deconvolution, as it brings high frequency components such as checkerboard patterns. We use the upsampling layer with nearest padding, followed by a convolutional layer.

Instance Level Case. We first choose WGAN-GP as the base generative model. To obtain a pair of real and generated images, we implement a tool for reconstructing GANs. The reconstructed image is still generated by the same pretrained WGAN-GP, which shows that the model tends to make the input image close to the generated image in the latent space.

The removed patterns imply a lack of high frequencies. Comparing the real and generated images in Fig. 1, for the spectrum, four dark corners indicate the absence of high frequencies, which can be confirmed by the 1d-PSD curve. The mismatch of the 1d-PSD curves between the real image and the generated image indicates that the generated image cannot keep the original high-frequency components, which is exactly the spectral bias.

Distribution Level Case. The instance-level bias seems unconvincing, so we consider the distribution difference. Using pretrained WGAN-GP and DRAGAN models, we generate $N = 8000$ images to obtain an average 1D-PSD, which allows us to analyze the spectral bias between the generated images and the real images in the distribution of low- and high-frequency components.

We calculate FID (lower is better) and plot 1d-PSD curves to evaluate image quality. Figure 2 shows the average 1D-PSD of models trained on different data. Although we can visually see that the images generated by GANs are of high quality, which is proven by the low FID scores, high-frequency concave parts still appear in GANs with different network structures and GAN losses. This result shows that all the images generated with the GAN models cannot fully fulfill with high frequencies. However, the situation is relatively good for the low-frequency components in the model.

3.2 Spectral Bias in the Deconvolution Generator

In the previous subsection, we discussed spectral bias in GANs with resize-convolution generators. To show the result, we conduct several experiments to show that GANs with deconvolutional generators suffer from different spectral biases. In Fig. 3, the spectral bias shown at high frequencies is opposite to the case of adjusted convolution. The deconvolution generator introduces more high-frequency artifacts compared to resize-convolution generators, which lose information in the high-frequency part. We will determine why in the next section.

4   The Reason of Different Spectral Bias

Previous works [11, 12] remind us to be careful with deconvolution, as it brings high-frequency components such as checkerboard patterns. We consider the upsampling process in the GAN models as an energy reconstruction process. In this process, the original continuous signal is a sampled signal with certain sampling rates. Taking a one-dimensional continuous signal, we further study the discrete signal of this process. Other researchers [5] have proven that DNNs prefer low-frequency features to high-frequency components, but the latter still make efforts to promote the effectiveness of the network [13]. In this section, our goal is to determine the cause of spectral bias.

The sampled process is as follows:

$$\bar{f}(t) = \sum_{n=-\infty}^{\infty} f(nT_S)\delta(t - nT_S) \tag{2}$$

where $f(n)$ is the sampled signal, and the sampling period is denoted by T_S. A low-pass filter denotes K, and the impulse response denotes $h(t)$. We can obtain that the signal $f(t)$ in the spatial domains can be constructed as follows:

$$f(t) = (\bar{f} * h)(t) = \sum_{n=-\infty}^{\infty} f(nT_S)h(t - nT_S) \tag{3}$$

which has the Fourier Transform $\mathcal{F}(\mu)$ [18]:

$$F(\mu) = (\bar{F} \cdot H)(\mu) = \sum_{n=-\infty}^{\infty} f(nT_S)e^{-j \cdot 2\pi \mu nT_S} \cdot \mathcal{H}(\mu), \tag{4}$$

where $*$ is the convolution operator, $\cdot$ denotes pointwise multiplication, and $\mathcal{H}(\mu)$ is the FT of parameter $h(t)$.

The discrete former process can be described as follows. Suppose that we want to separate out of M samples of $\overline{\mathcal{F}}(\mu)$ of period $\mu = 0$ to $\mu = 1/T_s$.

$$\mu = \frac{m}{KT_S}, \quad m = 0, 1, 2, 3, \ldots, K-1 \tag{5}$$

Make substituting for μ into Eq. (4), and $\mathcal{F}(m)$ denotes the transformation result:

$$\mathcal{F}(m) = \sum_{n=0}^{K-1} f(nT_S) e^{-j-2\pi \mu mn/K} \mathcal{H}\left(\frac{m}{KT_S}\right) \tag{6}$$

In the following, we discuss three common interpolation methods separately, i.e., the zero padding method, nearest padding method, and linear interpolation method.

When interpolating, we set the sampling interval T_S twice the original and change the spatial resolution of the space to the original double. Since $\mu = 1/T_S$,, the coordinates on the frequency axis will become half of the original. Literature [11] provides a more detailed proof and shows that all frequencies above $M/2$ observed in the spatial domain possibly generate noise during upsampling. Therefore, we can infer that deconvolution will generate high-frequency noise, which explains why the red and purple curves corresponding to the images generated by the two deconvolution generators in Fig. 3 are convex at the ends. This is a manifestation of the chess-card effect in high frequency noise, since the Dirac δ function obviously does not introduce any noise. For nearest padding and linear interpolation, we explain the recesses at high frequencies, as shown in Fig. 2 and Fig. 3.

Table 1. Table captions should be placed above the tables.

DRAGAN	Hybrid-0	Hybrid-1	Hybrid-5	Hybrid-6
FID	24.99	24.03	26.68	25.27

5 Solutions

According to Sect. 4, generative adversarial networks cause spectral bias. Therefore, we propose using a hybrid generative model instead of just applying adjusted convolution or deconvolution. More specifically, a hybrid generative model is constructed by adjusting the convolution block size of the classical deconvolution generator by replacing the last k deconvolution blocks.

We think that we could use a deconvolution block to help preserve the distribution during the training stages, while the resize-convolution block is used to remove high-frequency artifacts caused by deconvolution in the output image.

Experimental Results of DRAGAN. We train DRAGAN models by using LSUN Church Outdoor for fifty epochs with the structure of a total of 6 upsampling blocks. We show the results of the 1D PSD curve and FID in Fig. 4 and Table 1. "Hybrid-k" here means that the last k block is set as a resize-convolution block instead of a deconvolutional block.

Our experimental result of the 1d-PSD of DRAGAN with parameter $k = 1$ is very similar to the real 1D-PSD line and maintains a low FID, which indicates that when $k = 1$, high-frequency features are removed. Here, 'Hybrid-0' and 'Hybrid-6' represent that we have adopted different kinds of structures. We strongly recommend applying our Hybrid-1 to GANs to suppress the spectral bias introduced by classical generators.

Experimental Result on the Model of Relativistic GAN. We also test our hybrid generative model to Relativistic GAN [19], which adds relativism to GANs. We also train Relativistic Standard GAN (RSGAN) with the CAT dataset by using 3 different generators for 40k iterations [19]. We further train the Relativistic Average Standard GAN (RaSGAN) using the same setting. Since the patterns in the CAT dataset [17] are simpler, we set a total of 4 upsampling blocks with hybrid hyperparameter $k = 1$, which represents three same deconvolutional blocks with a resized convolutional block. The result is shown in Fig. 5.

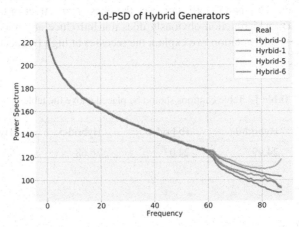

Fig. 4. DRAGAN trained on the LSUN Church Outdoor with different hybrid generators.

In Fig. 5, the experimental results show that the original RSGAN and RaSGAN with a deconvolution Hybrid-0 generator cannot fit the real 1d-PSD at the ends of the curve while training in the CAT dataset. In fact, the images generated by different kinds of generator models are very similar, but the FID of GANs using the Hybrid-1 generator drops slightly, showing that the images of the Hybrid-1 generator are better visually. When the proposed Hybrid-1 generator is used, the bias can be effectively removed. It is clear that the Hybrid-1 generator has the properties of classical generators, which

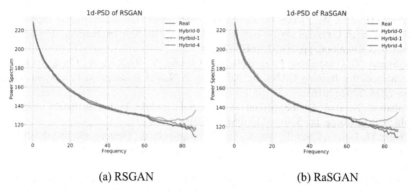

(a) RSGAN   (b) RaSGAN

Fig. 5. 1d-PSD of RSGAN and RaSGAN with Hybrid-0, Hybrid-1 and Hybrid-4 trained on CAT datasets.

means they can generate higher-quality images and perform better than the classical deconvolution generators and resize-convolution generators in the frequency features.

To suppress spectral bias, we need to reconstruct the generator as a hybrid generator simply by replacing the last deconvolutional block with a resize-convolutional block or adding a different resize-convolution block to the generator.

6   Conclusion

In this paper, we explore the limitations of generative adversarial networks in the frequency domain by real images and generated images. We also analyze experiments using the lens of our spectral tools, yielding multiple interesting explanations for spectral bias. We briefly reiterate our conclusion:

- As FID metrics that do not consider frequency distribution may not be sufficient to reflect the true spectral bias between real and generated images, we introduce the 1d-PSD. The specific performance of the spectral bias is that the low frequency fits better, and the high frequency part produces noise or information loss.
- The cause of the frequency difference is the upsampling layer, which corresponds to various kernel functions, such as the deconvolution layer and resize-convolution layer.
- The hybrid generator is chosen instead of the classical deconvolution generator or the resize-convolution generator to suppress the spectral bias.

References

1. Goodfellow, I., et al.: Generative adversarial nets. In: Advances in Neural Information Processing Systems, pp. 2672–2680 (2014)
2. Khayatkhoei, M., Singh, M.K., Elgammal, A.: Disconnected manifold learning for generative adversarial networks. In: Advances in Neural Information Processing Systems, pp. 7343–7353 (2018)

3. Zhang, D., Zuo, W., Zhang, D., Zhang, H.: Time series classification using support vector machine with Gaussian elastic metric kernel. In: 2010 20th International Conference on Pattern Recognition, pp. 29–32. IEEE (2010)
4. Salimans, T., Goodfellow, I., Zaremba, W., Cheung, V., Radford, A., Chen, X.: Improved techniques for training GANs. In: Advances in Neural Information Processing Systems, pp. 2234–2242 (2016)
5. Rahaman, N., et al.: On the spectral bias of neural networks. In: International Conference on Machine Learning, pp. 5301–5310. PMLR (2019)
6. Zhang, D., Lin, L., Chen, T., Wu, X., Tan, W., Izquierdo, E.: Content-adaptive sketch portrait generation by decompositional representation learning. IEEE Trans. Image Process. 26(1), 328–339 (2016)
7. Zhang, K., Zuo, W., Chen, Y., Meng, D., Zhang, L.: Beyond a Gaussian denoiser: residual learning of deep CNN for image denoising. IEEE Trans. Image Process. 26(7), 3142–3155 (2017)
8. Gulrajani, I., Ahmed, F., Arjovsky, M., Dumoulin, V., Courville, A.C.: Improved training of wasserstein GANs. In: Advances in Neural Information Processing Systems, pp. 5767–5777 (2017)
9. Choi, Y., Choi, M., Kim, M., Ha, J.-W., Kim, S., Choo, J.: StarGAN: Unified generative adversarial networks for multidomain image-to-image translation. In: Proceedings of the IEEE Conference on Computer Vision and Pattern Recognition, pp. 8789–8797 (2018)
10. Heusel, M., Ramsauer, H., Unterthiner, T., Nessler, B., Hochreiter, S.: GANs trained by a two time-scale update rule converge to a local nash equilibrium. In: Advances in Neural Information Processing Systems, pp. 6626–6637 (2017)
11. Durall, R., Keuper, M., Keuper, J.: Watch your upconvolution: CNN based generative deep neural networks are failing to reproduce spectral distributions. In: Proceedings of the IEEE/CVF Conference on Computer Vision and Pattern Recognition, pp. 7890–7899 (2020)
12. Odena, A., Dumoulin, V., Olah, C.: Deconvolution and checkerboard artifacts. Distill (2016). http://distill.pub/2016/deconv-checkerboard
13. Wang, H., Wu, X., Huang, Z., Xing, E.P.: High-frequency component helps explain the generalization of convolutional neural networks. In: Proceedings of the IEEE/CVF Conference on Computer Vision and Pattern Recognition, pp. 8684–8694 (2020)
14. Karras, T., Aila, T., Laine, S., Lehtinen, J.: Progressive growing of GANs for improved quality, stability, and variation. arXiv preprint arXiv:1710.10196 (2017)
15. Chen, X., Gupta, A.: An implementation of faster RCNN with study for region sampling. arXiv preprint arXiv:1702.02138 (2017)
16. Yu, F., Seff, A., Zhang, Y., Song, S., Funkhouser, T., Xiao, J.: LSUN: construction of a large-scale image dataset using deep learning with humans in the loop. arXiv preprint arXiv:1506.03365 (2015)
17. Zhang, W., Sun, J., Tang, X.: Cat head detection - how to effectively exploit shape and texture features. In: Forsyth, D., Torr, P., Zisserman, A. (eds.) ECCV 2008. LNCS, vol. 5305, pp. 802–816. Springer, Heidelberg (2008). https://doi.org/10.1007/978-3-540-88693-8_59
18. Baddeley, R.: The correlational structure of natural images and the calibration of spatial representations. Cogn. Sci. 21(3), 351–372 (1997)
19. Jolicoeur-Martineau, A.: The relativistic discriminator: a key element missing from standard GAN. arXiv preprint arXiv:1807.00734 (2018)

Research on Video Object Detection Methods Based on YOLO with Motion Features

Chuanxin Xiao, Peng Liu[✉], Yalei Zhou, Weiping Liu, Ruitong Hu, Chunguang Liu, and Chao Wu

Yangzhong Intelligent Electrical Research Center, North China Electric Power University, Yangzhong, China
liupeng@ncepu.edu.cn

Abstract. Aiming at the fixed-view video surveillance scene, this paper proposes a video object detection method that combines motion features and YOLO. The method uses the method of filtering video frames without motion features and segmenting video frames with motion features to reduce the reasoning pressure of the YOLO algorithm model. In this process, video frames containing moving objects are first obtained by the moving object detection module. Second, the moving target will be recognized by the object of interest recognition module. Finally, the background decision module records and analyzes the detection results to obtain background model updates or result output. It detects moving objects without using traditional background modeling methods. Experiments based on the CDnet2014 dataset show that our method improves the missed detection rate by 0.098% and the average inference speed per frame by 45.62% compared with the YOLO-based humanoid detection method. Furthermore, the method has superior performance in scenarios where target objects appear less frequently (substations, transmission lines, and hazardous areas).

Keywords: Moving object detection · Intelligent video surveillance · Background difference · YOLOv4

1 Introduction

Video surveillance is widely used in many situations because of its intuitive, simple and accurate features, such as home security, urban security, national road traffic, national substations and various banking institutions and other key infrastructure [1, 2]. To meet the needs of office automation, unattended intelligent video surveillance continues to develop in the field of security. Most of the time video surveillance images are without abnormal conditions or display a static background that no one cares about for a long time. Therefore, we detect and filter most of the nonmotion or static background pictures in the surveillance video, which can improve the accuracy and speed of target detection.

At present, the most widely used image recognition detection technology is in the field of video surveillance. Many researchers have performed much research in the field of video surveillance, including moving object detection algorithms and deep learning target detection algorithms. Common moving object extraction algorithms include

Y. Wang et al. (Eds.): ICPCSEE 2022, CCIS 1628, pp. 363–375, 2022.
https://doi.org/10.1007/978-981-19-5194-7_27

the optical flow method [3], frame difference method [4], and background difference method [5]. In terms of moving object detection, L. Wu et al. [6] proposed a moving object detection scheme based on embedded video surveillance. The algorithm removes the influence of complex backgrounds in nonmotion areas. Cao Lijing [7] et al. proposed a new method for moving target recognition under a static background using real-time background updating and replacement to improve the adaptability of background subtraction to the environment. In addition, deep learning detection algorithms are commonly used in industry, such as the RCNN series, YOLO [8] series and SSD [9]. W. Boyuan et al. [10] proposed a pedestrian detection model based on the improved YOLOv4 algorithm, which combines the new SPP (Spatial Pyramid Pooling) network and K-means clustering algorithm with the YOLOv4 model to facilitate feature extraction, which improves the detection performance. R. Fei and L. Ou [11] proposed a video object detector consisting of a deep reinforcement learning-based correction module for image detector object tracking. It has better detection performance than the single-frame baseline Faster-RCNN. Peng Qiwei [12] et al. used GMM to model the background in substation pedestrian detection to initially detect moving pedestrians. At the same time, the YOLO algorithm also applies topedestrian recognition. This method only obtains a better detection effect by comparing and analyzing the detection results of the two. Although this method can improve the detection accuracy, it also increases the detection delay. Yi Lu [13] et al. used the current object position and its motion information to propose a motion prediction model combined with a deep learning target detection algorithm. Compared with the detection method based on deep learning, the proposed method achieves better performance with fast detection.

After research and analysis, we summarize the advantages of moving object detection and deep learning target detection and propose a YOLOv4 video target detection method based on motion feature detection. First, the motion feature detection algorithm is used to detect and filter the video sequence images. Second, the ROI of motion features in the image is calculated, and motion feature segmentation is performed. Then, the detection result of the moving target is obtained according to the YOLO algorithm detection. The experimental results verify the performance superiority of the proposed method.

2   Method Construction

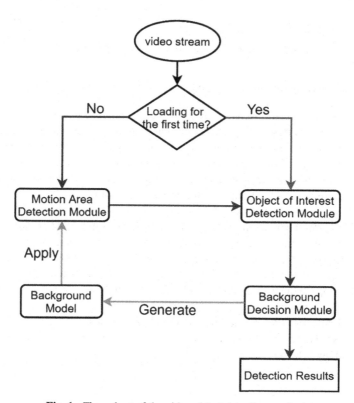

Fig. 1. Flow chart of the video object detection method

The flowchart of the video object detection method is shown in Fig. 1. The method in this paper consists of three execution modules: moving object detection, object of interest recognition and background decision. First, we continuously detect N frames of video using the YOLOv4 object detection algorithm. If no object of interest is detected, the background model is initialized, as indicated by the red arrow in the flowchart. After the background model is initialized, the moving object detection module starts to segment the foreground area containing the moving object and simultaneously records the coordinate position of the area containing the moving object. Next, the moving object detection module transmits the cropped video frames to the object-of-interest detection module for object type identification and recording, as indicated by the blue arrow in the flowchart. Finally, the background decision module performs the update decision of the background model and the output of the detection result. As shown in the flowchart, the orange arrow represents the generation and application of the background model.

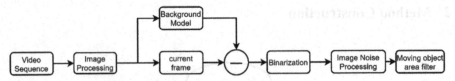

Fig. 2. Algorithm flow chart of moving object detection

2.1 Moving Object Detection Module

The algorithm flow of the moving object detection module is shown in Fig. 2. The main function of this module is to detect motion regions in video sequences and extract foreground images containing moving objects. In this paper, it is difficult to accurately extract moving objects due to noises such as lighting, leaves, and weather. Subsequently, this paper proposes a background modeling method for YOLO. In the process, to further eliminate the influence of external conditions such as lighting and leaves in the picture, this module performs noise processing operations such as grayscale processing, Gaussian filtering, binarization and closing operations on the video image. After that, we use the frame difference method to quickly extract the regions of moving objects. When the background model exists, the operation steps of the motion detection module are as follows:

Step 1: Suppose the background model is $B_r(x, y)$, and the current frame in the video sequence is $F_k(x, y)$.

Step 2: First, the video frames are smoothed using Gaussian filtering to reduce the effect of noise. Then, the difference between $F_k(x, y)$ and $B_r(x, y)$ is made to obtain the difference image $I_r(x, y)$.

$$I_r(x, y) = F_k(x, y) - B_r(x, y) \tag{1}$$

Step 3: Perform the binarization operation on $I_r(x, y)$ to obtain the binarized image $I_{br}(x, y)$.

Step 4: $I_{br}(x, y)$ uses morphological operations to further remove isolated and discrete noise points. After that, the foreground target image $D(x, y)$ is obtained.

Step 5: First, obtain the height and width of $D(x, y)$, and set the threshold T used to calculate the minimum rectangular box area MinArea. Then, all rectangles in $D(x, y)$ whose area is less than MinAra are filtered.

$$MinArea = height \cdot width \cdot T \tag{2}$$

Step 6: The X set and the Y set are composed of the valid rectangular frame coordinate points (x, y) obtained in Step 5. After that, the largest foreground area containing moving objects is calculated. Record the two coordinate points (x_{min}, y_{min}) and (x_{max}, y_{max}) used for foreground area segmentation. That is, x_{min} and y_{min} are taken from the minimum values in the sets X and Y; x_{max} and y_{max} are taken from the maximum values in the sets X and Y.

The effect of moving object region extraction is shown in Fig. 3. (a) background model; (b) current video frame containing moving objects; (c) binary image; (d) extracted moving object regions.

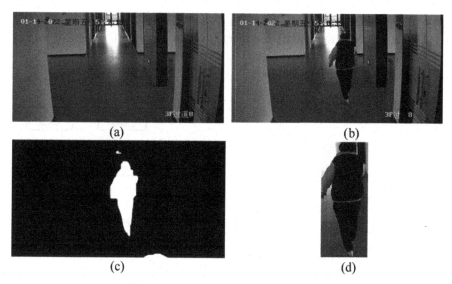

(a)   (b)

(c)   (d)

Fig. 3. Moving object detection renderings

2.2 Object of Interest Detection Module

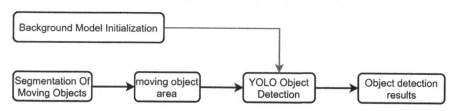

Fig. 4. Algorithm flow chart of object detection of interest

This module selects the YOLOv4 deep learning target detection algorithm of the YOLO series to detect and identify the type of objects of interest. In this module, we use a deep learning object detection algorithm for background modeling and object of interest identification with the following advantages: 1) It ensures that unnecessary noise is generated in background modeling by detecting objects of interest in video frames. 2) It has high target detection accuracy and can meet the real-time requirements of intelligent video surveillance.

The algorithm flow of the object detection module of interest is shown in Fig. 4. The working tasks of the object detection module of interest are as follows: (1) The red arrow in the flowchart shows that when the method is initialized and loaded, the module detects whether there is an object of interest in the continuous video frames to select the initialization of the background model. (2) Receive the coordinate value of the area including the moving foreground sent by the motion detection module, and use the coordinate value to segment the moving area including the foreground target on the original image. Next, the segmented moving foreground region images are sent to the YOLOv4 object detection algorithm to identify objects of interest.

2.3 Background Decision Module

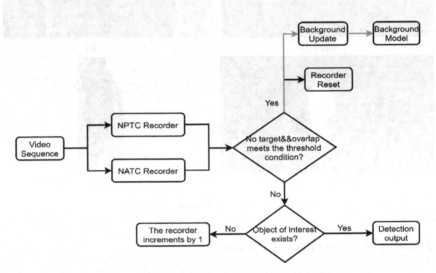

Fig. 5. Algorithm flow chart of background decision

The background decision module updates the background model and outputs the detection results. The algorithm flow of this module is shown in Fig. 5. The two event recorders of the background decision module are as follows: (1) NPTC (NoProspectTargetCount): when there is no moving object foreground in motion detection, the recorder adds 1; (2) NATC (NoAnomalousTargetCount): when the object recognition module of interest does not detect when the object of interest is reached, the recorder increments by 1. First, the decision module calculates the detection results of the current motion detection module or the object recognition module of interest and selects the relevant event recorder to record the detection results. Second, the decision-making module compares whether the NPTC and NATC loggers meet the threshold conditions for background updating. To exclude the influence of noise on the results in complex environments, the two event recorders only record events when they are triggered continuously; otherwise, the recorders are reset. The threshold value of the NATC recorder can be appropriately selected according to the frequency of events in the application scenario. The threshold size of the NATC recorder can be determined by the processing speed of the deep learning object detection algorithm.

3 Experiment and Evaluation

3.1 Implementation Details

The core devices used in the experiments for this method are a 2.80 GHz Intel i5 CPU, 7.7 GB RAM, and a GeForce GTX 1060 6 GB GPU. The method uses the Python language and the PyTorch deep learning framework. The YOLO V4 algorithm is used to detect objects of interest in it. OpenCV is used to extract moving object regions in video sequences.

3.2 Dataset Preparation and YOLOv4 Model Validation

The method in this paper adopts the YOLOv4 model to support the research of the object detection module of interest. We train a YOLOv4 model using three categories of cat, dog and human on the PASCAL VOC 2007 dataset [14]. After that, the inference performance of the YOLOv4 model under different input scales is verified on the PASCAL VOC 2007 test set.

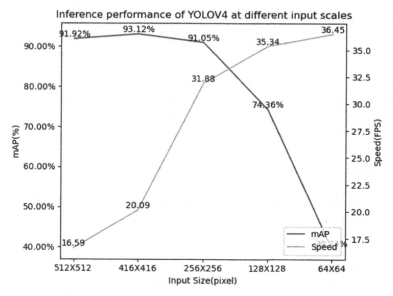

Fig. 6. Performance analysis under different input scales

Figure 6 shows the inference performance of the YOLOv4 model under different input scales. The mAP value represents the average value of all detected class APs. As the scale of the image decreases, the value of mAP decreases continuously, and the processing speed of the image keeps increasing. As the size of the input image decreases, the feature pixels used for target detection are seriously lost, which leads to a decrease in detection accuracy.

3.3 Motion Video Object Detection and Recognition

This paper uses the CDnet2014 [15] dataset to test the method performance. We mainly use the pedestrian detection dataset in this dataset, which contains 10 different pedestrian detection video sequences. For this, we choose people as the object of interest to identify. Finally, the robustness of the method to foliage noise is tested by overpassing video sequences. See Table 1 for reports of video sequences. This is a set of 10 videos, mostly consisting of pedestrians. In this experiment, the YOLOv4 model is used to detect an image input size of 128 × 128.

Table 1. Pedestrian test dataset from CDW-2014

Name	Resolution	Total frames	Ground-truth
backdoor	320 × 240	2000	800
busStation	360 × 240	1250	844
copyMachine	720 × 480	3400	3127
cubicle	352 × 240	7400	2806
office	360 × 240	2050	1440
pedestrians	360 × 240	1099	591
peopleInShade	380 × 244	1199	531
PETS2006	720 × 576	1200	1171
skating	540 × 360	3900	1599
sofa	320 × 240	2750	2027

Table 2. The experimental situation of MDR and APTPF for each video sequence

Video name	YOLOv4-MDR	YOLOv4-APTPF	OursMethod-MDR	OursMethod-APTPF
backkdoor	1.37%	34 ms	1.25%	18 ms
busStation	6.75%	33 ms	9.71%	27 ms
copyMachine	10.87%	38 ms	10.20%	37 ms
cubicle	13.58%	32 ms	22.24%	12 ms
office	0%	34 ms	0%	24 ms
pedestrians	9.31%	33 ms	4.06%	21 ms
peopleInShade	5.46%	34 ms	2.82%	21 ms
PETS2006	0.51%	47 ms	7.94%	42 ms
skating	9.32%	37 ms	9.69%	21 ms
sofa	19.43%	33 ms	9.67%	26 ms

[a]MDR stands for missing detection rate.
[b]APTPF stands for average processing time per frame.

The method in this paper and the YOLOv4 algorithm detect the dataset and record the video frame numbers and detection results where there are objects of interest. After that, the recorded results are compared with the ground truth in the dataset, and the MDR value is calculated. The APTPF value is obtained by averaging the processing time of all video frames in each video sequence. Table 2 shows the experimental results of our method and YOLOv4 under different video sequences, showing the experimental situation of MDR and APTPF for each video sequence. Figure 7 shows a comparison of our method and YOLOv4 on the missed detection rate and inference speed on different video sequences. Figure 8 represents the ratio of MDR and inference speed between our method and the YOLO v4 method. Experiments show that the average missed detection rate of our method on 10 video sequences is increased by 0.098%, and the average inference speed of each frame is increased by 45.62%.

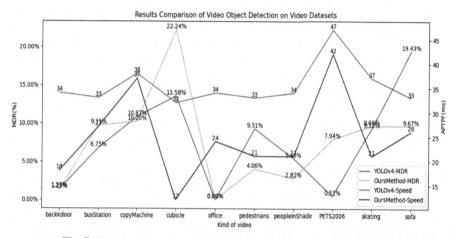

Fig. 7. Results comparison of video object detection on video datasets.

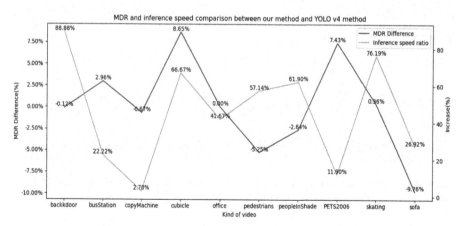

Fig. 8. MDR and inference speed comparison between our method and the YOLO v4 method

Except for copyMachine and PETS2006 video sequences, our method performs well on object detection on the CDW-2014 dataset. There are always scattered moving objects in the video frames of the two video sequences, causing the still picture filtering function to not work. Therefore, this method is not suitable for monitoring scenes where the detected objects of interest are scattered or appear frequently. The reason for the above situation is that the relatively scattered moving objects and high-frequency objects will cause the motion area segmentation effect to be insignificant. When detecting moving objects in a cubicular video sequence, the minimum rectangular box area threshold used for moving region screening is set too large so that smaller moving objects are filtered out. Therefore, the method in this paper has a large missed detection rate. In this case, an excessively large missed detection rate can be avoided by selecting an appropriate area threshold.

By reducing the input size of the image to 64x64, the detection effect of our method is better than that of YOLOv4, as shown in Fig. 9. By comparing the detection results of the two, the method in this paper has a better detection effect under the smaller image input size. This is because the image with segmentation has less loss of feature pixels of moving objects than without segmentation. Figure 10 shows the motion region detection results on the overpass dataset with leaf shaking. The results show that the method has good detection robustness under the influence of leaf noise. Figure 11 and Fig. 12 show the detection results of our method in a real surveillance environment. The results show that the method in this paper also has a good detection effect in a real monitoring environment with less noise.

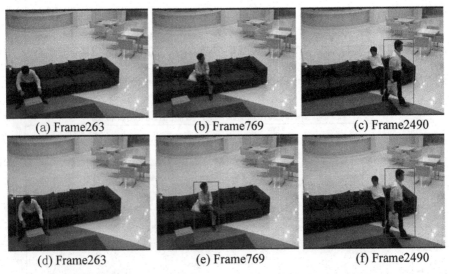

(a) Frame263   (b) Frame769   (c) Frame2490

(d) Frame263   (e) Frame769   (f) Frame2490

Fig. 9. Detection results on video sequence sofa. (a)–(c) are the detection images of the YOLOv4 model. (d)–(f) are the detection images of our method.

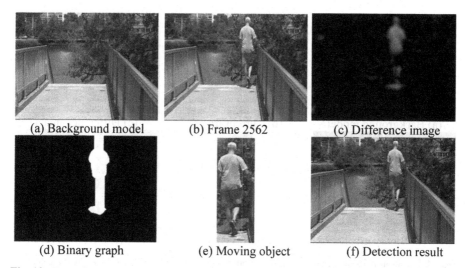

(a) Background model   (b) Frame 2562   (c) Difference image

(d) Binary graph   (e) Moving object   (f) Detection result

Fig. 10. Detection results in the dataset with leaf interference. (a)–(e) are the detection and segmentation of motion regions; (f) is the final detection result of our method

4 Conclusion

This paper combines the characteristics of moving target detection and deep learning target detection and implements a YOLOv4 video target detection method based on motion features. The proposed method can filter video frames without moving objects and only detect video frames that contain moving objects after being segmented. Therefore, this method can not only effectively reduce the false positive rate of target detection but also reduce the reasoning pressure of the YOLOv4 algorithm model. Experiments show that the average per-frame inference speed of our method on the experimental dataset is increased by 45.62%, while the missed detection rate is increased by 0.098%. The method can also handle complex scenes such as lighting, foliage and weather. Afterwards, the effectiveness of our method is verified by real-time video monitoring, which can achieve real-time monitoring of 25 frames per second. This method is not suitable for monitoring scenarios where target objects are too scattered and appear frequently.

In addition, our method can freely select suitable deep learning target detection algorithms and objects of interest with motion effects in practical application scenarios. Our method relies on the accuracy of deep learning object detection algorithms, and the higher the accuracy is, the better the background modeling. Therefore, selecting an appropriate deep learning target detection algorithm can greatly improve the performance advantage of this method.

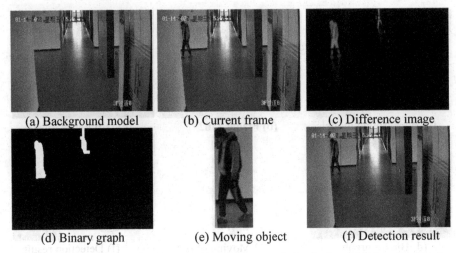

(a) Background model　　(b) Current frame　　(c) Difference image

(d) Binary graph　　(e) Moving object　　(f) Detection result

Fig. 11. Detection results in a real monitoring environment during the day. (a)–(e) are the detection and segmentation of motion regions; (f) is the final detection result of our method

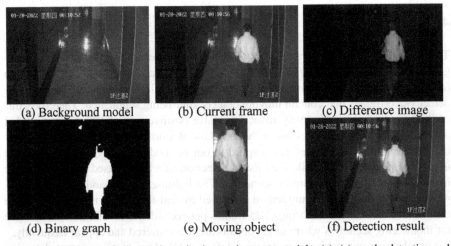

(a) Background model　　(b) Current frame　　(c) Difference image

(d) Binary graph　　(e) Moving object　　(f) Detection result

Fig. 12. Detection results in a real monitoring environment at night. (a)–(e) are the detection and segmentation of motion regions; (f) is the final detection result of our method

References

1. Shifa, A., et al.: MuLViS: multi-level encryption based security system for surveillance videos. IEEE Access **8**, 177131–177155 (2020). https://doi.org/10.1109/ACCESS.2020.3024926
2. Hou, H., Li, N., Liu, D.: Research on moving object detection in intelligent video surveillance. Comput. Technol. Dev. **22**(2), 49–52 (2012)
3. Wan, Y., Han, Y., Lu, H.: The methods for moving object detection. Comput. Simul. **23**(10), 221–226 (2006)

4. Lipton, A.J., Fujiyoshi, H., Patil, R.S.: Moving target classification and tracking from real-time video. In: Proceedings Fourth IEEE Workshop on Applications of Computer Vision, WACV 1998 (Cat. No. 98EX201), pp. 8–14 (1998). https://doi.org/10.1109/ACV.1998.732851

5. Alzughaibi, A.D., Hakami, H.A., Chaczko, Z.: Review of human motion detection based on background subtraction techniques. Int. J. Comput. Appl. **122**, 1–5 (2015)

6. Wu, L., Liu, Z., Li, Y.: Moving objects detection based on embedded video surveillance. In: 2012 International Conference on Systems and Informatics (ICSAI 2012), pp. 2042–2045 (2012). https://doi.org/10.1109/IC-SAI.2012.6223453

7. Lijing, C., Ke, T., Zhonglin, F.: Moving target recognition under static background. In: 2014 Fourth International Conference on Instrumentation and Measurement, Computer, Communication and Control, pp. 117–120 (2014). https://doi.org/10.1109/IMCCC.2014.32

8. Redmon, J., Divvala, S., Girshick, R., Farhadi, A.: You Only Look Once: Unified, Real-Time Object Detection, pp. 779–788 (2016)

9. Liu, W., Anguelov, D., Erhan, D., Szegedy, C., Reed, S., Fu, C.-Y., Berg, A.C.: SSD: single shot multibox detector. In: Leibe, B., Matas, J., Sebe, N., Welling, M. (eds.) ECCV 2016. LNCS, vol. 9905, pp. 21–37. Springer, Cham (2016). https://doi.org/10.1007/978-3-319-464 48-0_2

10. Boyuan, W., Muqing, W.: Study on pedestrian detection based on an improved YOLOv4 algorithm. In: 2020 IEEE 6th International Conference on Computer and Communications (ICCC), pp. 1198–1202 (2020). https://doi.org/10.1109/ICCC51575.2020.9344983

11. Fei, R., Ou, L.: A video object detector based on deep reinforcement learning and correction module. In: 2020 IEEE 3rd International Conference on Electronics and Communication Engineering (ICECE), pp. 1–5 (2020). https://doi.org/10.1109/ICECE51594.2020.9353034

12. Peng, Q., et al.: Pedestrian detection for transformer substation based on Gaussian mixture model and YOLO. In: 2016 8th International Conference on Intelligent Human–Machine Systems and Cybernetics (IHMSC), pp. 562–565 (2016). https://doi.org/10.1109/IHMSC.2016.130

13. Lu, Y., Chen, Y., Zhao, D., Li, H.: Hybrid deep learning based moving object detection via motion prediction. In: 2018 Chinese Automation Congress (CAC), pp. 1442–1447 (2018). https://doi.org/10.1109/CAC.2018.8623038

14. Everingham, M., Van Gool, L., Williams, C.K.I., Winn, J., Zisserman, A.: The pascal visual object classes (VOC) challenge. Int. J. Comput. Vis. **88**(2), 303–338 (2010)

15. Goyette, N., Jodoin, P., Porikli, F., Konrad, J., Ishwar, P.: Changedetection.net: a new change detection benchmark dataset. In: 2012 IEEE Computer Society Conference on Computer Vision and Pattern Recognition Workshops, pp. 1–8 (2012). https://doi.org/10.1109/CVPRW.2012.6238919

TU-Net: U-shaped Structure Based on Transformers for Medical Image Segmentation

Jiamei Zhao, Dikang Wu, and Zhifang Wang[✉]

Department of Electronic Engineering, Heilongjiang University, Harbin 150080, China
wangzhifang@hlju.edu.cn

Abstract. Recently, the development of deep learning technology in medical image segmentation has become increasingly mature, and the symmetric U-Net has made breakthrough progress. However, because of the inherent limitations of convolution operations, U-Net has some shortcomings in the interaction of global context information. For this reason, this paper proposes TU-Net based on transformers. TU-Net can strengthen the modeling of global context information, enhance the extraction of detailed information and reduce the computational complexity of the algorithm. In patch embedding, successive convolutional layers with small convolutional kernels are proposed to extract features. Cross Attention-Skip is proposed to complete the fusion of shallow and deep features during the skip connection process. TU-Net is performed on the Synapse dataset to segment eight abdominal organs. The experimental results show that TU-Net is superior to ViT, V-Net, U-Net and Swin-Unet.

Keywords: Medical image segmentation · TU-Net · Cross attention-skip

1 Introduction

In recent years, deep learning has penetrated various fields of medical image analysis. Among them, medical image segmentation is an essential part. It can accurately obtain the shape and spatial information of targets of interest, which plays a vital auxiliary role in the process of doctors' diagnosis.

Among the existing segmentation methods, U-Net [1] proposed by Ronneberger et al. in 2015 has achieved great progress. The proposal of U-Net is the real meaning of the application of fully convolutional networks to the field of medical image segmentation, which has become the mainstream method in many segmentation tasks. U-Net is completely symmetric and adopts the topology of encoder, decoder and skip connection. The encoder is a contracted path for capturing context information. The decoder is an expanded path for precise localization. Then, the concatenation operation is used to fuse different resolution features of the encoder and decoder by skip connection, which achieves the purpose of reducing information loss. Due to the excellent segmentation performance of U-Net, a series of improved models have been proposed to meet the

© The Author(s), under exclusive license to Springer Nature Singapore Pte Ltd. 2022
Y. Wang et al. (Eds.): ICPCSEE 2022, CCIS 1628, pp. 376–386, 2022.
https://doi.org/10.1007/978-981-19-5194-7_28

actual segmentation requirements. For instance, classic network modules such as the Residual module [2], Dense module [3], Inception module [4], and Attention module [5] are added to better the precision of network segmentation. In addition, there are some improvements that can directly process 3D volumes instead of just processing 3D volumes into 2D slices to complete segmentation [6, 7]. 3D U-Net [8] and V-Net [9] are proposed to realize end-to-end processing of 3D volumes. These U-Net-based methods are widely used in multi-organ segmentation, heart segmentation, brain segmentation, etc., and play a good auxiliary role in doctors' diagnosis.

At present, although U-Net-based methods have achieved better results, there are still some shortcomings. Due to the limited and fixed receptive field of convolution operations, U-Net has limitations in modeling long-range relations [10]. Generally, the information at diagonal positions in the same medical image cannot be directly convolved together, so global information cannot be effectively extracted. As a result, patients who exhibit larger differences in texture, shape and size of structures generally produce weaker performance. Most of the existing solutions improve the extraction of contextual information and enhance the grasp of low-level information by introducing self-attention mechanism [11], image pyramid [12] and deeper convolutional layers [13]. However, the above methods lead to more computation. Recently, numerous researchers have applied Transformer to deal with computer visual tasks [14]. It was initially used in the domain of natural language processing [5–15], and it has shown superior capabilities in machine translation, document classification [16], etc. Compared with CNN-based methods, Transformer has limitations in the extraction of low-level details and requires larger datasets for pre-training. However, the Transformer itself has positional embedding and global self-attention mechanism, which are powerful in capturing long-range dependencies. For example, Vision Transformer (ViT) [17] is the first image processing network based entirely on Transformer. It takes 2D images with positional embeddings as input and achieves comparable performance to CNN-based methods but requires pre-training on large datasets. TransUNet [10] is a simple combination of CNN and Transformer. First, the CNN is applied to obtain detailed information, and then the information processed by the CNN is sent to the Transformer to extract global information. Finally, the information with different resolutions is concatenated through skip connections. Algorithms with similar combinations include TransFuse [18], MedT [19], TransAttUnet [20], etc., all of which reflect that the combination not only strengthens the network's long-distance modeling capabilities but can also better restore local spatial details. In addition, the structure of Swin-Unet [21] is similar to U-Net, which is also composed of an encoder-decoder with skip connection. Specifically, the swin transformer block [22] is used as the visual center to replace the convolutional layers. Patch merging and patch expanding are proposed to complete the up and down sampling operations. The successful practice of ViT, TransUNet, Swin-Unet and other networks has fully proven the great potential of Transformer in computer vision tasks.

Based on Swin-Unet, this paper proposes TU-Net to complete the segmentation of medical images. The main differences are that, on the one hand, Swin-Unet adopts large convolutional kernels in patch embedding, while TU-Net adopts successive convolutional layers with small convolutional kernels to complete the extraction of different features. The advantage is that it is more accurate than block-by-block positional encoding and can effectively reduce the computational complexity. On the other hand, in the process of encoder-decoder information transmission, Swin-Unet performs a simple concatenation operation on the feature maps of the encoder L layer and the decoder L-1 layer. However, TU-Net explores a new skip connection mechanism named Cross Attention-Skip, which applies the self-attention mechanism in Transformer to represent the feature maps. Under limited computational conditions, TU-Net can achieve better segmentation results. Specifically, it can be summarized as follows:

1) The patch embedding block with less computational complexity is proposed, which used the successive convolutional layers with small convolutional kernels to replace the large ones;

2) A more effective skip connection mechanism named Cross Attention-Skip is explored, which is used to transmit the feature information of the encoder and decoder.

2   Method

2.1   Algorithm Framework

The symmetric structure of TU-Net is shown in Fig. 1. Concretely, the encoder consists of blocks such as patch embedding, swin transformer, and patch merging. Symmetrically, the decoder is composed of blocks such as linear projection and expanding, swin transformer, and patch expanding. The feature information of the two parts is transmitted through skip connections. In particular, patch embedding reduces the convolutional kernel size and uses successive convolutional layers. In the skip connection, Cross Attention-Skip is utilized for feature fusion of the encoder and decoder. These make positional encoding more precise and help in the recovery of low-level details.

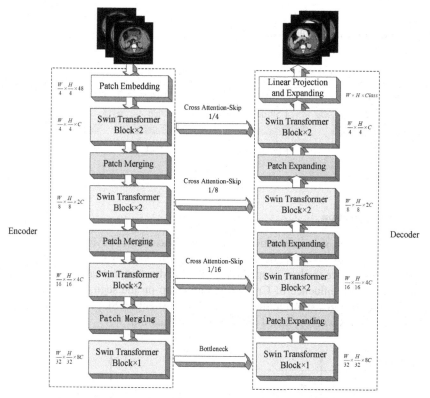

Fig. 1. Framework of TU-Net

2.2 Patch Embedding

In this paper, the patch embedding block is used to change the dimension of the patches and encode the relative position of its spatial information to achieve accurate positioning, which is shown in Fig. 2. The patch embedding block first crops the input feature map $x \in R^{H \times W \times 3}$ into $N = \frac{H}{S} \times \frac{W}{S}$ small patches. Then, all the patches are flattened and converted into a high-dimensional vector $x \in R^{\frac{H}{S} \times \frac{W}{S} \times C}$ by linear projection, where H is the height, W is the width, S is set to 4 and C is an arbitrary dimension. Different from ViT applying large convolutional kernels such as 16×16 in the patch embedding block, this paper changes them into successive convolutional layers with small kernels after several experiments, which can achieve better results. First, the computational budget of a convolutional layer is proportional to $K^2 D_1 D_2$, where K represents the kernel size and D_1 and D_2 represent the input and output dimensions, respectively. Thus, successive convolutional layers with small convolutional kernels can lessen the use of many parameters and effectively reduce the computational complexity without diminishing the receptive field. In addition, the patch embedding block encodes pixel-level spatial information by using smaller ones, which can increase the number of channels of the feature map. Its extraction is more comprehensive and accurate than block-by-block position encoding. As shown in Fig. 2, four convolutional layers are formed by four small convolutional kernels, with strides of [1, 2]. Similarly, there is a GeLU layer [23] and a normalization layer [24] after each convolutional layer.

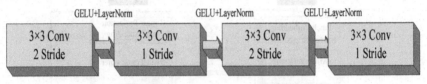

Fig. 2. Patch embedding block

2.3 Skip Connection

This paper uses the skip connection method to concatenate the feature maps of the encoder and the corresponding decoder and merges the shallow features with the deep features. It can reduce the loss of spatial information due to down sampling. This paper mainly studies the influence of Cross Attention-Skip on the segmentation results. In Fig. 3, (a) and (b) represent Attention-Skip and Cross Attention-Skip, respectively, where W-MSA represents window-based multi-head self-attention.

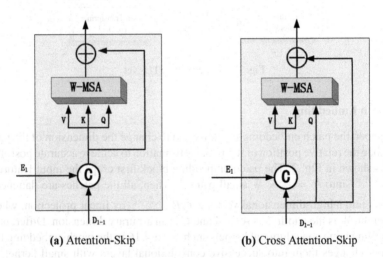

(a) Attention-Skip   (b) Cross Attention-Skip

Fig. 3. Skip connection block

First, E_L is the output of the L layer feature map in the encoder, where $E_L \in \frac{H}{2^{L-1}} \times \frac{W}{2^{L-1}} \times 2^{L-1}C$. D_{L-1} is the output of the $L - 1$ layer feature map in the decoder, where $D_{L-1} \in \frac{H}{2^{L-1}} \times \frac{W}{2^{L-1}} \times 2^{L-1}C$. Attention-Skip directly concatenates the two feature maps on the dimension of the channel and transfers the obtained feature map $\frac{H}{2^{L-1}} \times \frac{W}{2^{L-1}} \times 2^L C$ to the swin transformer block through the self-attention mechanism. Cross Attention-Skip in Fig. 3(b) concatenates E_L and D_{L-1} on the dimension of the channel to serve as the Value and Key of the self-attention mechanism [5], while the Query is served by a separate feature map D_{L-1}. Finally, the Value, Key and Query are input into the swin transformer block to complete the next steps. The experimental results prove that Cross Attention-Skip can better fuse features and enhance context information interaction.

3 Experiments

3.1 Dataset and Implementation Details

During the experiment, the pixels in the image need to be marked one by one, which is extremely dependent on the imaging equipment and doctors' professional knowledge. Therefore, the medical image dataset is usually small. The Synapse multi-organ segmentation dataset (Synapse) is used for algorithm verification, which includes 30 abdominal CT scans. Each image of this dataset contains eight abdominal organs. To avoid a series of situations, such as overfitting and poor model performance caused by the dataset being too small, 3D volumes are processed into 2D slices. That is, 3779 2D slices are obtained, of which 2212 slices are used for training.

TU-Net is verified on the basis of Python 3.6, PyTorch 1.5.0 and Nvidia 2080Ti GPU. During the experiment, the input image size and patch size are 224×224 and 4, so that not only more accurate segmentation results can be obtained but also the calculation amount of the algorithm can be effectively reduced. According to several experiments, the batch size is 12, the SGD optimizer with momentum 0.9 and weight $1e-4$, the parameters can be effectively optimized during the backpropagation process, and the deep and shallow features of the image can be better learned.

3.2 Experimental Results

First, the algorithm is verified using two evaluation metrics, the average Dice score (DSC) and Hausdorff Distance (HD) [25]. DSC is generally applied to compute the similarity between two samples and is sensitive to the internal filling of the segmented image. In Eqs. (1) and (2), A represents the ground truth of the segmentation image, and B represents the algorithm prediction map.

$$Dice = \frac{2|A \cap B|}{|A| + |B|} \tag{1}$$

HD (in mm) is also a measure to describe the degree of similarity between two samples of points. However, different from DSC, HD is more sensitive to the boundaries of the segmentation. Equation (2) is the two-way HD, and Eqs. (3) and (4) are the one-way HD.

$$H(A, B) = \max(h(A, B), h(B, A)) \tag{2}$$

$$h(A, B) = \max_{a \in A}\{\min_{b \in B} ||a - b||\} \tag{3}$$

$$h(B, A) = \max_{b \in B}\{\min_{a \in A} ||b - a||\} \tag{4}$$

TU-Net is verified on the Synapse dataset, the DSC and HD of eight abdominal organs are obtained after the experiment, and the performances are compared with ViT, V-Net, U-Net and Swin-Unet. The results are shown in Table 1. Due to the use of successive convolutional layers with small convolutional kernels and the improvement of Cross Attention-Skip, TU-Net achieves superior segmentation performance of DSC 77.85% and HD 24.38 mm. First, it is clear that TU-Net performs better than ViT and V-Net. Then, compared with U-Net, the same DSC of 76.85% is achieved, but HD is significantly improved by 15.32 mm. Successive convolutional layers with small convolutional kernels enhance the attention of detailed information and make the edge contour segmentation more accurate. Moreover, because of many factors, such as equipment and parameter settings, Swin-Unet only achieves DSC of 76.03% and HD of 26.46 mm under the same conditions. TU-Net has improved DSC and HD by 1.82% and 2.08 mm, respectively, achieving better performance in both interior padding and edge segmentation.

Table 1. Segmentation performance of different algorithms on the Synapse dataset

Methods	ViT	V-Net	U-Net	Swin-Unet	TU-Net
DSC	61.50	68.81	76.85	76.03	77.85
HD	39.61	–	39.70	26.46	24.38
Aorta	44.38	75.34	89.04	81.27	85.48
Gallbladder	39.59	51.87	69.72	61.44	69.20
Kidney(L)	67.46	77.10	77.77	82.17	82.77
Kidney(R)	62.94	80.75	68.60	76.17	79.59
Liver	89.21	87.84	93.43	93.42	92.64
Pancreas	43.14	40.05	53.98	52.62	54.48
Spleen	75.45	80.56	86.67	86.29	88.09
Stomach	69.87	56.98	75.58	74.89	70.85

Table 2. Ablation studies on patch embedding block and cross attention-skip

Methods	Swin-Unet		TU-Net	
Patch embedding			✓	✓
Cross attention-skip		✓		✓
DSC	76.03	77.05	76.93	77.85
HD	26.46	25.62	26.98	24.38

The segmentation results of TU-Net and Swin-Unet are mainly compared, and the visual results are shown in Fig. 4. The first column is the segmentation ground truth, the second column is the segmentation prediction map of Swin-Unet, and the third column is the prediction of TU-Net. It is clearly observed from the first and fifth rows that the

aorta   gallbladder   left kidney   right kidney   liver   pancreas   spleen   stomach

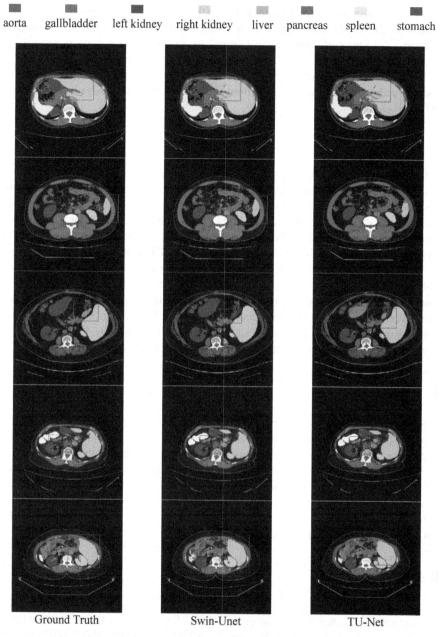

Ground Truth   Swin-Unet   TU-Net

Fig. 4. Results of different algorithms

contour edges of each organ segmentation are smoother and more refined. According to the second and third rows, Swin-Unet mispredicted segmentation of the kidney and gallbladder. Due to the improvement of Cross Attention-Skip, TU-Net integrates deep and shallow features and pays more attention to the interaction of global information, and there is no missegmentation. Accurate segmentation of the pancreas is still a challenge because it deforms particularly greatly from case to case, and the borders are not well defined. As seen from the fourth row, Swin-Unet does not predict the pancreas, while TU-Net achieves more accurate results. In conclusion, TU-Net effectively reduces false positive predictions and missed predictions and has better performance on contour edges and small organs.

3.3 Ablation Study

In this section, we mainly analyze the effect of the patch embedding block and the method of Cross Attention-Skip in the algorithm and conduct ablation studies on the Synapse dataset. The experimental results are shown in Table 2. Specifically, to explore the influence of the patch embedding block, the successive convolutional layers with small convolutional kernels in TU-Net are replaced by a patchwise convolution with the kernel size of 4×4 and the stride of 4 in Swin-Unet. The DSC of TU-Net is improved by nearly 1%, indicating that the performance of using successive convolutional layers with small convolutional kernels in TU-Net is better than large ones and reduces the running time of the algorithm during training. In addition, to explore the effectiveness of Cross Attention-Skip, it is only replaced with Attention-Skip in Fig. 3. The experimental results show that the DSC of Cross Attention-Skip is improved by 1.02%, effectively promoting the fusion of information with different resolutions in TU-Net.

4   Conclusion

A novel algorithm TU-Net is proposed for abdominal multi-organ segmentation. TU-Net is constructed by successive convolutional layers with small convolutional kernels and skip connections based on Cross Attention-Skip. It not only enhances the attention of low-level details and the interaction of global information but also reduces the computational complexity of the algorithm. In the validation based on the Synapse dataset, 8 abdominal organs with different sizes, shapes and textures are predicted effectively and accurately. Furthermore, it achieves DSC of 77.85% and HD of 24.38 mm. It is proven that TU-Net has superior segmentation performance.

References

1. Ronneberger, O., Fischer, P., Brox, T.: U-net: convolutional networks for biomedical image segmentation. In: Navab, N., Hornegger, J., Wells, W.M., Frangi, A.F. (eds.) MICCAI 2015. LNCS, vol. 9351, pp. 234–241. Springer, Cham (2015). https://doi.org/10.1007/978-3-319-24574-4_28
2. He, K., Zhang, X., Ren, S.: Deep residual learning for image recognition. In: Proceedings of the IEEE Conference on Computer Vision and Pattern Recognition, pp. 770–778 (2016)
3. Huang, G., Liu, Z., Van Der Maaten, L., Weinberger, K.: Densely connected convolutional networks. In: Proceedings of the IEEE Conference on Computer Vision and Pattern Recognition, pp. 4700–4708 (2017)

4. Szegedy, C., Liu, W., Jia, Y., Sermanet, P., Reed, S., Anguelov, D., et al.: Going deeper with convolutions. In: Proceedings of the IEEE Conference on Computer Vision and Pattern Recognition, pp. 1–9 (2015)

5. Vaswani, A., Shazeer, N., Parmar, N., et al.: Attention is all you need. In: Advances in Neural Information Processing Systems, pp. 5998–6008 (2017)

6. Yu, Q., Xie, L., Wang, Y., Zhou, Y., Fishman, E.K., Yuille, A.L.: Recurrent saliency transformation network: incorporating multi-stage visual cues for small organ segmentation. In: Proceedings of the IEEE Conference on Computer Vision and Pattern Recognition, pp. 8280–8289 (2018)

7. Zhou, Y., Xie, L., Shen, W., Wang, Y., Fishman, E.K., Yuille, A.L.: A fixed-point model for pancreas segmentation in abdominal CT scans. In: Descoteaux, M., Maier-Hein, L., Franz, A., Jannin, P., Collins, D.L., Duchesne, S. (eds.) MICCAI 2017. LNCS, vol. 10433, pp. 693–701. Springer, Cham (2017). https://doi.org/10.1007/978-3-319-66182-7_79

8. Çiçek, Ö., Abdulkadir, A., Lienkamp, S.S., Brox, T., Ronneberger, O.: 3D U-Net: learning dense volumetric segmentation from sparse annotation. In: Ourselin, S., Joskowicz, L., Sabuncu, M.R., Unal, G., Wells, W. (eds.) MICCAI 2016. LNCS, vol. 9901, pp. 424–432. Springer, Cham (2016). https://doi.org/10.1007/978-3-319-46723-8_49

9. Milletari, F., Navab, N., Ahmadi, S.: V-net: fully convolutional neural networks for volumetric medical image segmentation. In: 2016 Fourth International Conference on 3D Vision (3DV), pp. 565–571. IEEE (2016)

10. Chen, J., Lu, Y., Yu, Q., Luo, X., Adeli, E., et al.: TransUNet: transformers make strong encoders for medical image segmentation, arXiv preprint arXiv:2102.04306 (2021)

11. Schlemper, J., et al.: Attention gated networks: learning to leverage salient regions in medical images. Med. Image Anal. **53**, 197–207 (2019)

12. Zhao, H., Shi, J., Qi, X., Wang, X., Jia, J.: Pyramid scene parsing network. In: Proceedings of the IEEE Conference on Computer Vision and Pattern Recognition, pp. 2881–2890 (2017)

13. Chen, L.C., Papandreou, G., Kokkinos, I., Murphy, K., Yuille, A.: Deeplab: semantic image segmentation with deep convolutional nets, atrous convolution, and fully connected crfs. IEEE Trans. Pattern Anal. Mach. Intell. **40**(4), 834–848 (2017)

14. Carion, N., Massa, F., Synnaeve, G., Usunier, N., Kirillov, A., Zagoruyko, S.: End-to-end object detection with transformers. In: Vedaldi, A., Bischof, H., Brox, T., Frahm, J.-M. (eds.) ECCV 2020. LNCS, vol. 12346, pp. 213–229. Springer, Cham (2020). https://doi.org/10.1007/978-3-030-58452-8_13

15. Devlin, J., Chang, M.W., Lee, K., et al.: Bert: pre-training of deep bidirectional transformers for language understanding. arXiv preprint arXiv:1810.04805 (2018)

16. Zheng, S., Lu, J., Zhao, H., Zhu, X., Luo, Z., et al.: Rethinking semantic segmentation from a sequence-to-sequence perspective with transformers. In: Proceedings of the IEEE Conference on Computer Vision and Pattern Recognition, pp. 6881–6890 (2021)

17. Dosovitskiy, A., Beyer, L., Kolesnikov, A., Weissenborn, D., et al.: An image is worth 16x16 words: transformers for image recognition at scale. In: ICLR (2021)

18. Zhang, Y., Liu, H., Hu, Q.: TransFuse: fusing transformers and CNNs for medical image segmentation. In: de Bruijne, M., et al. (eds.) MICCAI 2021. LNCS, vol. 12901, pp. 14–24. Springer, Cham (2021). https://doi.org/10.1007/978-3-030-87193-2_2

19. Valanarasu, J.M.J., Oza, P., Hacihaliloglu, I., Patel, V.M.: Medical transformer: gated axial-attention for medical image segmentation. In: de Bruijne, M., et al. (eds.) MICCAI 2021. LNCS, vol. 12901, pp. 36–46. Springer, Cham (2021). https://doi.org/10.1007/978-3-030-87193-2_4

20. Chen, B., Liu, Y., Zhang, Z.: TransAttUnet: multi-level attention-guided u-net with transformer for medical image segmentation. arXiv preprint arXiv:2107.05274 (2021)

21. Cao, H., Wang, Y., Chen, J., Jiang, D., et al.: Swin-Unet: unet-like pure transformer for medical image segmentation, arXiv preprint arXiv:2105.05537 (2021)

22. Liu, Z., Lin, Y., Cao, Y., Hu, H., Wei, Y., Zhang, Z., et al.: Swin transformer: hierarchical vision transformer using shifted windows. In: Proceedings of the IEEE International Conference on Computer Vision, pp. 10012–10022 (2021)

23. Hendrycks, D., Gimpel, K.: Gaussian error linear units (gelus). arXiv:1606.08415 (2016)

24. Ba, J.L., Kiros, J.R., Hinton, G.: Layer normalization. arXiv preprint arXiv:1607.06450 (2016)

25. Fu, S., et al.: Domain adaptive relational reasoning for 3D multi-organ segmentation. In: Martel, A.L., et al. (eds.) MICCAI 2020. LNCS, vol. 12261, pp. 656–666. Springer, Cham (2020). https://doi.org/10.1007/978-3-030-59710-8_64

Reducing Video Transmission Cost of the Cloud Service Provider with QoS-Guaranteed

Pingshan Liu[1], Yemin Sun[2(✉)], Kai Huang[2], Guimin Huang[2], and Zhongshu Yu[2]

[1] Business School, Guilin University of Electronic Technology, Guilin, China
[2] Guangxi Key Laboratory of Trusted Software, Guilin University of Electronic Technology, Guilin, China
2198371783@qq.com

Abstract. With the advancement of cloud computing technology, many service providers are combining with cloud service providers to build a highly available streaming video-on-demand cloud platform and provide video services to end users. Generally, cloud service providers deploy many edge cloud CDN nodes in different geographic areas and provide video services to end users. However, when an end-user wants to watch certain videos and request video resources from surrounding edge cloud CDN nodes, the edge cloud CDN node will request missing video clips from other cloud nodes. Therefore, this will generate a large amount of additional video transmission costs and reduce the quality of service of the cloud service provider. To reduce or even minimize the video transmission cost of edge cloud CDN nodes while ensuring the quality of service (QoS). We designed a video transmission algorithm called *Netdmc* to ensure transmission quality. The algorithm can be divided into two parts. The first part is a low-latency video request algorithm based on ensuring service quality, and the second part is a video request algorithm based on minimizing video transmission costs. The simulation results demonstrate that the *Netdmc* algorithm can effectively reduce the cost of cloud service providers and ensure the quality of video services.

Keywords: Cloud service provider · Video transmission cost · QoS

1 Introduction

Due to the advancement of Internet-related technologies, the number of users is also increasing. Among them, online videos occupy a large amount of Internet traffic and have many active users [1]. Therefore, when a large number of users watch online videos, a large number of additional transmission costs are incurred. At the same time, they have a higher demand for the quality of video services.

Because cloud computing has the advantages of high availability and robust scalability, a large number of video service providers and cloud service providers have combined to build a highly available resource-sharing cloud platform [2]. Generally, cloud service providers deploy a large number of edge cloud CDN nodes in different geographic areas to provide video services to end-users. However, because edge cloud CDN nodes are

Y. Wang et al. (Eds.): ICPCSEE 2022, CCIS 1628, pp. 387–402, 2022.
https://doi.org/10.1007/978-981-19-5194-7_29

usually composed of servers with relatively poor performance and storage capacity [3], cloud service providers store some popular videos in edge cloud CDN nodes [4]. When an end-user wants to watch a video, it will request the video resource from the surrounding CDN nodes. In this way, the CDN node may have missing video clips, and it needs to request the missing video clips from other CDN nodes and then return them to the end-user. After receiving the video clip, the end-user decodes and plays it locally. At that time, the edge cloud CDN node incurred a large number of transmission costs during the video request process and increased the transmission delay, thereby reducing the smoothness of end users' video playback. Therefore, how to reduce the cost of video transmission between cloud CDN nodes while ensuring QoS is a significant problem to be solved urgently [5].

Currently, most cloud service providers lease many ISP (Internet Service Provider) lines to connect edge cloud CDN nodes in different geographic areas. Then, the ISP charges for the bandwidth cost of the traffic generated. Generally, ISPs have many other charging methods, such as bandwidth peak-based and traffic charging methods. This paper adopts a charging method based on peak bandwidth [6]. In this charging method, the ISP will collect the traffic bandwidth passing through the link every five minutes as the time interval and sort them in descending order. After the charging period ends, the maximum bandwidth traffic obtained by these adoptions is used as the charging bandwidth. The video data can be transmitted for free when the maximum historical bandwidth peak is not exceeded. Therefore, we can design routing algorithms to reduce the cost of video transmission [7]. In this paper, we develop a video transmission algorithm called *Netdmc*. The algorithm can be divided into two parts. The first part is a low-latency video request algorithm based on ensuring service quality, and the second part is a video request algorithm based on minimizing video transmission costs. Next, we prove the performance of the algorithm through simulation experiments. The simulation results demonstrate that the *Netdmc* algorithm can effectively reduce the cost of cloud service providers and ensure the quality of video services.

The rest of the paper is structured as follows. The second part discusses the related work of this paper. The third part introduces the system model and formulaically describes the structure of the model. The fourth part mainly introduces related algorithms and related pseudocodes. In the fifth part, the performance indicators and results of the simulation experiment are described.

2   Related Work

With the advancement of online video, research on streaming media systems has attracted the attention of many researchers. How to reduce the video transmission cost of cloud service providers while ensuring the service quality of end users has attracted a large number of researchers, and many different solutions have also been proposed.

Sem C. Borst designed a collaborative cache management algorithm to maximize the traffic provided by the cache and minimize bandwidth costs [8]. By dividing the large video file into multiple small video files and storing them in different edge nodes, it provides an effective mechanism to alleviate these huge bandwidth requirements. W Chu studied the problem of service assurance in content-centered networks [9] and designed a network model for content delivery. To reduce the latency associated with clients routing content to different locations, he aligned network location with content popularity to ensure that each content provider has optimized network latency when routing content to clients. MM Amble et al. designed a strategy for request routing, content placement, and content expulsion [10] to reduce user latency. They developed an algorithm with optimal node throughput to solve the problem of routing placement and eviction, and the constructed algorithm with optimal throughput produces a shorter queue length. Albert G. Greenberg and others proposed a new mechanism for jointly optimizing network and data center resources. This mechanism can improve the reliability of the system and reduce the service cost that may be increased by the geographically diverse data center network [11]. JA Chandy et al. designed a copy placement strategy based on video clips. By logically pairing resources between end-users and nodes in the system, the number of hops requested by end-users is reduced, effectively reducing transmission delay [12].

There are other algorithms related to the cost and delay of video transmission [6, 23]. Xiao introduced an online low-cost transmission scheme. He suggested that a long-term transmission request can be divided into a series of short-term requests. By reasonably predicting unpredictable traffic, it is possible to transmit video data without increasing the cost of video transmission [6]. Hu Han proposed an algorithm based on community sharing video placement using cloud CDN infrastructure [18]. The algorithm divides video viewing history and geographical location into communities, and then categorizes video viewing preferences of different end users into different communities. Zhang proposed a QoS-aware peer-to-peer coordinated control mechanism under the conditions of dynamic traffic fluctuations [5], node distance and capacity differences, and user dynamics. The author introduced a circular buffering mechanism in the paper to realize circular coordination, and realized QoS perceptual stream synchronization through two different algorithms of cycle length and cycle time to ensure video playback quality. Wang et al. proposed a propagation-based social video replication strategy based on the joint use of edge cloud and P2P clients [19]. Liu et al. considered the access/update rates of different types of user data and decided which parts of the data should be replicated [20].

Previous studies have failed to consider the problem of how to minimize the cost generated by the edge cloud CDN nodes when they request the missed video clips in time and guarantee the QoS. In this paper, we focus on the problem and propose algorithms to minimize the video transmission cost while ensuring the QoS.

3 System Model and Problem Formulation

In this section, we mainly introduce the system model and problem formulation description of the *Netdmc* algorithm. The important symbols used are shown in Table 1.

Table 1. Notations and definitions

Symbol	Description
E	A set of edges in the network
V	A set of nodes in the network
N	Number of clips in each video
ω	Traffic transmission cost per unit bandwidth
$b_{ij}(t)$	The traffic bandwidth already occupied by link $\{i, j\}$ at time t
C_{ij}	Bandwidth capacity in link $\{i, j\}$
(D_k, S_k)	The minimum cost path when requesting p^k
$f_{ij}(t)$	The maximum bandwidth that the link $\{i, j\}$ can allocate at time t
N_{GOP}	Each video clip contains N_{GOP} video frames
max $w_{ij}(t)$	The historical maximum of 5-min traffic for the link $\{i, j\}$ at time period t
$Cost_{ij}(t)$	Video transmission cost of link $\{i, j\}$ at time t
p^k	Video clip p^k
τ_k	The size of the p^k
u_k	Whether p^k will arrive on time
T_k	The time it takes for the cloud node to request the p^k
$t_{playback}$	The current playing time
ε	The size of the video clip cache window
d_k	The cutoff time of video clip k

3.1 System Model

In this paper, we use a hierarchical general network model [13]. As depicted in Fig. 1, the network model is divided into three layers. The deepest layer is the cloud data center, another layer is the edge cloud CDN nodes distributed in different geographical regions, and the furthest layer is the end-user. Cloud nodes are connected through highly available ISP links and are responsible for providing video services to end-users. A tracking server is installed in a cloud data center to track and record the historical peak bandwidth of links, video resource bitmap, cloud node transmission rate, and other related information. Each edge cloud CDN node is composed of an edge streaming media source server and an edge index server. Edge streaming source servers are responsible for caching local video clips. The edge index server is responsible for receiving and responding to client requests.

When the CDN node requests receiving the video service request of the end-user, it uses the H264 video editing method to divide the video into N different video clips [14, 17] and then requests the missing video clips from other cloud nodes. This structure reduces the pressure on the cloud data center and can distribute most of the requests to different edge cloud CDN nodes. Because this structure has good scalability and decentralization, most current streaming media-on-demand cloud platforms use this system structure [13, 16].

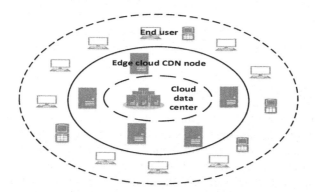

Fig. 1. Cloud system network model diagram

3.2 Formulaic Description of the Algorithm

Generally, we can describe the directed graph $G = (V, E)$ as a network system model. For each link $\{i, j\} \in E$, C_{ij} represents the maximum bandwidth capacity. (D_k, S_k) represents the minimum cost path when requesting video clip p^k. If C_{ij} is used to represent the maximum bandwidth capacity of link $\{i, j\}$, the following formula (1) exists. In formula (1), $b_{ij}(t)$ represents the traffic bandwidth already occupied by link $\{i, j\}$ at time t, and $f_{ij}(t)$ represents the bandwidth traffic that the link can allocate at time t.

$$C_{ij} = f_{ij}(t) + b_{ij}(t) \qquad (1)$$

For simplicity, we assume that the cost function $c(x)$ is a linear function $c(x) = \omega * x$, where x is the video transmission traffic bandwidth to be charged. The tracking server in the cloud data center records the value of all 5-min traffic volumes of a relay path as the charging volume and the maximum traffic calculated from 288 statistical points for the same day. Therefore, the traffic volume of the transmission link did not exceed the historical maximum traffic volume [6], and these data can be transmitted for free. In this way, the link-paid bandwidth flow can be maximized without increasing additional traffic, thereby reducing the cost of video transmission. If we use max $w_{ij}(t)$ to represent the maximum historical traffic of link $\{i, j\}\{i, j\}$ in time period t, ω represents each unit traffic cost. From the cost function $c(x)$, we can obtain formula (2). ω is the cost per GB of traffic.

$$Cost_{ij}(t) = \begin{cases} Cost_{ij}(t-1) & f_{ij}(t) \leq \max w_{ij}(t-1) \\ \omega * \max w_{ij}(t) & otherwise \end{cases} \qquad (2)$$

According to formula (2), the transmission cost of the link is determined by the maximum value of 5-min traffic. This means that link transmission is free and will not exceed the maximum traffic of the link. When the traffic at time t exceeds the previously charged capacity, the cloud service providers will charge for the overflowed bandwidth. When a CDN node requests a video clip, it is assumed that the size is τ_k. Then, there is the following formula:

$$\tau_k \leq \min(\max w_{ij}(t-1), C_{ij} - b_{ij}(t)) \tag{3}$$

$$\{i, j\} \in E, (i \in V, j \in V) \tag{3.a}$$

$$\tau_k \leq V_{ij} \tag{3.b}$$

In the set of Eq. (3), constraint (3.a) means $\{i, j\}$ is the link of the network. Constraint (3.b) represents the maximum video clip size that can be allocated by a link. In general, a CDN node can receive multiple video requests. When the CDN node is missing the video, it will divide each requested video into N video clips and then request the video clips from other cloud nodes. Each video clip is transmitted with a certain file header, such as the clip number, playback timestamp, and more identifying information. For each video clip k ($1 \leq k \leq N$), we assume that (D_k, S_k) represents the minimum cost path when requesting video clip p^k.

Generally, there are many different factors that affect the delay caused by video transmissions, such as link bandwidth capacity and excessive node load [15]. Therefore, for the purpose of reducing the delay caused by video transmission, we can ensure that as much video data as possible can be transmitted without increasing the cost. At the same time, when a CDN node receives a large number of video requests, it causes video transmission congestion. Since the urgency of different video clips is different, different transmission methods are used for different video clips. For video clips with a higher degree of urgency, we can give priority to distribution. This can ensure to the greatest extent that each video clip reaches the end-user for decoding and playback before the playback point. For video clips with different urgency levels, we designed a low-latency request algorithm for urgent video clips and a low-cost request algorithm for nonurgent videos. To ensure the video playback quality of the end-user, each requested video clip is decoded and played before the deadline for the video clip to be played. Therefore, the transmission delay of each video clip needs to be shortened as much as possible. We assume that each video clip contains N_{GOP} video frames, and v_f represents the video frame rate. Usually, for the same video, the frame rate is the same. Therefore, the video playback deadline d_k of the video clip in the sliding window can be expressed as:

$$d_k = \frac{(k-1) * N_{GOP}}{v_f} \tag{4}$$

From formula (4), we can see that the closer the video clip is to the playback deadline, the higher the urgency. Suppose we use ε to represent the window size of a certain request scheduling period, and $t_{playback}$ represents the playback point of the video clip p^k. For each video clip p^k, its urgency can be expressed as:

$$P^k_{emergency} = \frac{\varepsilon - (d_k - t_{playback})}{\varepsilon} \tag{5}$$

4 Algorithm Design and Description

To reduce the video transmission cost of cloud service providers, the corresponding algorithms are designed based on the relevant model designed in the third section, called *Netdmc*. The algorithm is based on the Dijkstra algorithm to calculate the optimal path [17]. The algorithm can be separated into two parts. The first portion is a minimum-cost algorithm for nonemergency video clips. The second portion is a low-latency algorithm for emergency video clips. In this section, we present and analyze the two parts of the algorithm.

4.1 The Low Delay Algorithm for the Emergency Video Clips

When the CDN nodes request their missing video clips from other cloud nodes, they have different policies for requesting these missing clips based on their priorities. To ensure the service quality of cloud service providers, a low delay algorithm for emergency video clips is designed. The algorithm is run when a CDN node requests an emergency video clip from other cloud nodes. Through this algorithm, the CDN nodes can request emergency video clips with low latency and then return the requested video clips to the end-user. First, the edge cloud CDN nodes visit the tracking server to obtain the video transmission rate of each link in the system and relevant information about each node. In general, the video transmission delay of a link with a low video transmission rate will increase. Therefore, in this algorithm, a link path with a high video transmission rate is found. Through this path, the missing emergency video clips can be requested. The specific algorithm description is shown in **Algorithm** 1.

In **Algorithm** 1, we initialize a queue Q to cache the nodes set with the shortest delay path. The required video clips can be requested with the shortest delay through this path. The video transmission rate of each link can be obtained through the tracking server. Finally, the algorithm returns the queue Q of the shortest path node set.

Algorithm 1 low delay transmission algorithm for emergency video clips

1	Calculate the weight w_k of the video clip p^k $\{1 \leq k \leq n\}$
2	Get the video transmission rate $v_i(i)\{i \in V\}$ of all nodes
3	Initialize the video request source node D_k
4	Initializes the queue **Q**. isEmpty
5	Initialize temporary variables max=0;
6	**Q**. put(D_k)
5	For each video $p^k \in \{p^1, p^2, ..., p^n\}$ in the sliding window **do**
6	**For** (Each node v in V) **do**
7	**If** (the video clip p^k exists for node i $\in$ V)) **do**
8	max=v_i
9	**For** (j in each **Q**) **do**
10	**If** ($v_j >$ max && formula (3)) **do**
11	max=v_j
12	If(pi>b_k)
13	b[j][v]=p^k
14	$Cost_{ij}(t) = \beta * \max w_{ij}(t)$
16	Q.put(v);
17	**Else**
18	$Cost_{iv}(t)=Cost_{iv}(t-1)$
19	Q.put(v);
20	**End else**
21	**End If**
22	**End For**
23	**End If**
24	**End For**
25	**Return Q**

4.2 The Minimum Cost Path Algorithm for the Nonemergency Video Clips

In the third section above, we briefly introduced that the cost of a cloud service provider is mainly caused by video transmission costs caused by edge cloud CDN nodes requesting missing video clips from other cloud nodes. Therefore, to reduce the cost of the cloud service providers while guaranteeing the quality of service, it is necessary to minimize the cost generated by the CDN nodes requesting the missing video clips. We have designed an algorithm that can reduce the video transmission cost of cloud nodes when requesting nonemergency video clips. The algorithm runs on the edge cloud CDN nodes. In this algorithm, the path with the lowest cost is returned. Through this least-cost path, the CDN nodes return the acquired nonurgent video clips to the end-users. The specific algorithm description is shown in **Algorithm** 2.

First, the edge cloud CDN nodes visit the tracking server in the cloud center to obtain the maximum historical value of all 5-min traffic and information about each node. In this algorithm, a queue Q is set, and the path with the least cost is saved in queue Q. D_k represents the edge cloud CDN nodes of the p^k requested. Each video clip has a playback timestamp. According to our assumption, when the value of $P_{emergency}$ is 0.7, p^k is converted to a low delay transmission delay. Video clips are requested with a timestamp and change with the time of the request. When a nonurgent video clip request reaches a certain urgency value, it will be converted to an urgent video clip. Finally, the algorithm returns to queue Q. The CDN node requests missing video clips through this path.

Algorithm 2 Minimum transmission cost for nonemergency video clips

1	Calculate the weight w_k of each video clip p^k in the sliding window
2	Initializes the queue **Q**. isEmpty
3	Initialize Minimum transmission cost min=0
4	Get the maximum historical value of all 5-minutes traffic b[i][j] of the link $\{i, j\}$ from the CDN auxiliary server
5	Gets the timestamp for each video clip $p^k \in \{p^1, p^2, \dots p^N\}$
6	**Q**.put $(\boldsymbol{D_k})$
7	**For (each p^k in $\{p^1, p^2, \dots p^N\}$) do**
8	**If ($P^k_{emergency}$>=0.7) do**
9	using **Algorithm 1**
10	**else do**
11	**For (Each node v in V) do**
12	**If** (the p^k exists for node v) **do**
13	min=b[**Q**.front][v]
14	**For** (j in each **Q**) **do**
15	**If** (b[j][v]>b&&Formula (3)) **do**
16	**If**(p^k>b_k) **do**
17	b[j][v]=p^k
18	Minimum cost min+=$\beta * \max w_{ij}$ (t)
19	$Cost_{ij}(t) = \beta * \max w_{ij}$ (t)
20	**Q**.put(v)
21	**Else**
22	Minimum cost min+= 0
23	$Cost_{iv}(t) = Cost_{iv}(t-1)$
24	**Q**.put(v)
25	**End Else**
27	**End If**
28	**End For**
29	**End If**
30	**End For**
31	**End Else**
32	**End For**

5 Evaluation

To verify the performance of the proposed *Netdmc* algorithm, simulation experiments are carried out. We mainly introduce the experimental simulation in three parts. (1) The setup of the simulation is introduced. (2) The relevant indicator description is presented in detail. (3) The effectiveness is obtained by comparing the experimental figures.

5.1 Simulation Setup

In the simulation, we use an event-driven simulator proposed by Shen [16]. The network structure of the simulation platform is a layered superposition network structure. We write a protocol that encapsulates the *Netdmc* algorithm, and implement the encapsulated protocol in the emulator. In the simulation protocol, we set up a cloud data center node, a tracking server, and 100 different edge cloud CDN nodes. At the same time, we also set up 1000 nodes to simulate end-users and request video data from edge cloud CDN nodes. The tracking server is responsible for caching the link video transmission rate, the maximum historical value of 5-min traffic, and related information of CDN nodes. The experiment set up a number of different video sizes, and the video bitrate is 2.4 Mbps. The sliding window size of the edge cloud CDN node is set to 5 min. We randomly set the capacity of the link and the maximum historical value of all 5-min traffic of the link in the initialization of the experiment. In the experiment, it is assumed that 1000 end-users request video clips from different CDN nodes and that the same CDN node may lose multiple video clips requested by different end-users. We can perform simulations with different video sizes. In the experiment, we used 2 T and 20 T video sources of two different sizes, randomly divided them into multiple video clips of different sizes, and randomly stored them in different network nodes. The relevant information from the video is shown in Table 2. To better compare the performance of the algorithm in the experiment, we chose two different video clip scheduling methods. *General simulation* represents randomly requesting the required video clips from other nodes in the network. The *Generalcdc* algorithm represents obtaining the required video clips directly from the cloud data center. In the final simulation results, we compare the results of the general simulation algorithm and the *Generalcdc* algorithm with the results of the *Netdmc* algorithm, respectively. We use the comparison result data in MATLAB to draw the relevant performance comparison result figure.

Table 2. Video test file parameters

Video resolution	HDTV(1920 * 1080)
Frame rate	30
Total video frames (QP)	51715
Quantization parameter	28-28−30

5.2 Description of Key Performance Indicators

In general, the performance of an algorithm requires several key indicators. Therefore, we have listed below several key metrics that determine the performance of the *Netdmc algorithm*.

(1) Video transmission cost per unit time

The first indicator is the video transmission cost per unit time. The video transmission cost per unit time represents the cost of video transmission per unit of time in the system. Formula (6) is the calculation formula for the unit transmission cost of the p^k requested by an edge cloud CDN node. (D_k, S_k) represents the least cost path of p^k, and T represents each of the different time periods. The tracking server globally records the video transmission cost generated by the system in unit time and records this value regularly. Finally, the cost per unit time of video transmission of the system is compared under different algorithm simulations. T represents the time interval.

$$Cost_{average}(t) = \sum_{\{i,j\}}^{E} \frac{Cost_{ij}(t)}{T} \tag{6}$$

(2) Video transmission cost

The second indicator is the cost of video transmission. When a CDN node requests video clips from other cloud nodes, a certain video transmission cost will be incurred. The tracking server records the video transmission cost generated by each link of the system every other time period. Formula (7) is the cost calculation formula for video transmission in the system. Finally, through this indicator, the video transmission cost of the system under the simulation of the *Netdmc algorithm* and the other two algorithms can be verified.

$$Cost(t) = \sum_{\{i,j\}}^{E} Cost_{ij}(t) \tag{7}$$

(3) Continuity index

The third indicator is the continuity index. It can indicate the end users' video playback fluency index. Generally, the larger the continuity index is, the superior the algorithm effect. The end-users request N video clips each time, and the size of the video clips is variable. Equation (8) represents the on-time arrival rate of all clips of the requested video. u_k indicates whether the video clip arrives on time and γ_k indicates the size of the p^k. u_k is 1 means that the video clips arrived on time, and u_k is 0 means that it did not arrive on time. When the CDN node requests a video clip, if the video request delay is higher than the difference between the video playback point and the deadline, the video clip will arrive on time; otherwise, the video clip will not arrive on time.

$$\beta = \frac{\sum_{k=1}^{N} u_k \tau_k}{\sum_{k=1}^{N} \tau_k} \tag{8}$$

(4) Average transmission rate

The fourth indicator is the average downlink transmission rate of video clips. The tracking server records the downlink rate of the transmitted video clips each period and then calculates its average value. The specific formula calculation is shown in Formula (9). In formula (9), $b_i(t)$ represents the average video transmission rate of node $\{i\}$ in t statistical intermediate periods, and $v_{average}$ represents the average transmission rate. The secondary server in the cloud center will record the video transmission rate b_i of the node at intervals of a one-time unit. Usually, the larger the value is, the superior the algorithm performance will be.

$$v_{average} = \frac{1}{V} \sum_{i}^{V} b_i \tag{9}$$

5.3 Simulation Results

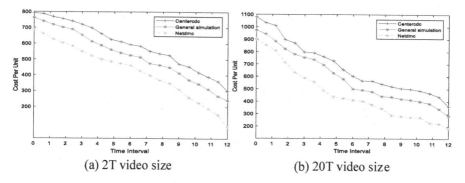

(a) 2T video size   (b) 20T video size

Fig. 2. Video transmission cost per unit time

In the experiment, we set two different video sizes 2 T and 20 T. The simulation experiments are carried out under the three algorithms of *General* simulation, *Netdmc*, and *Centercdc*. Formula (6) is the definition of video transmission cost per unit time. Figure 2(a) demonstrates the experimental results of 2 T, and Fig. 2(b) demonstrates the experimental results of 20 T. From Fig. 2, we can see that the video transmission cost per unit time under the *General simulation* algorithm and the *Centercdc* algorithm is higher than that of the *Netdmc algorithm*. Therefore, it can be seen that the *Netdmc algorithm* can effectively reduce the video transmission cost per unit time of the system, and the effect is better than the other two algorithms.

Formula (6) is the formula definition of the video transmission cost in the system. Figure 3 shows the video transmission cost results under different algorithm simulations. As depicted in Fig. 3(a), the video transmission cost under the *Netdmc* algorithms is 14.2% and 28.7% lower than that of the *General Simulation* algorithm and the *Centercdc* algorithm, respectively. In Fig. 3(b), compared with the *General Simulation* algorithm and the *Centercdc algorithm*, the video transmission cost under the simulation of the *Netdmc algorithm* is reduced by 21.5% and 35.05%, respectively.

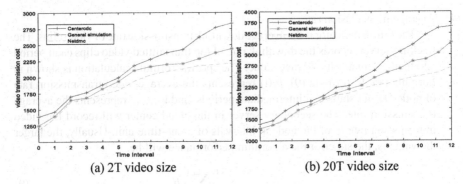

(a) 2T video size   (b) 20T video size

Fig. 3. Video transmission cost

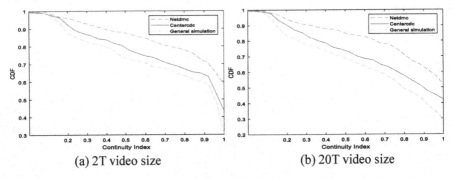

(a) 2T video size   (b) 20T video size

Fig. 4. Continuity index

In the experiment, we set two video sizes of 2 T and 20 T and simulated the experiment under different algorithms. In the experiment, 1000 end-users are set, and each end-user requests N video clips to the CDN nodes. β of the continuity index is obtained by formula (8). The continuity index under different algorithms is shown in Fig. 4. Obviously, *Netdmc* can achieve the best performance among the three algorithms. As depicted in Fig. 4(a), the continuity index of the *Netdmc algorithm* is 0.9 with 0.7 users. The continuity index of the *general simulation* algorithm reached 0.9 with 0.58 end users, and the continuity index of the *Centercdc* algorithm reached 0.9 with 0.62 end users. In Fig. 4(b), the *Netdmc algorithm* still has an end-user continuity index of 0.8 reaching 0.73, while the continuity index of the other two algorithms is between 0.42–0.58.

In the experiment, we set two video sizes of 2T and 20T and simulated the experiment under different algorithms. Equation 8 is the formula definition of the average downlink rate. Figure 5 demonstrates the experimental results under simulation of the three algorithms. As depicted in Fig. 5(a), the average downlink rate of a video clip requested by CDN nodes is higher than that of the other two algorithms during the simulation of the *Netdmc algorithm*. In Fig. 5(b), the video transmission rate per unit time of the simulation of the *Netdmc algorithm* is relatively higher than that of the other two algorithms. The video transmission rate per unit time of the 20T video simulation is lower than that of the 2T video simulation.

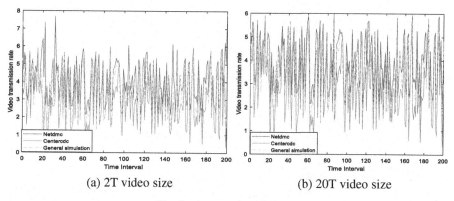

(a) 2T video size   (b) 20T video size

Fig. 5. Average downlink rate

6 Conclusions

With the growth of online videos, video service providers and cloud service providers have merged to build a highly available video service platform. Cloud service providers deploy multiple edge cloud CDN nodes in different geographic regions to provide video service to end-users. When the CDN nodes receive a video clip request from the end-user, it will request the missing video clip from other cloud nodes. This not only incurs a high cost but also reduces the smoothness of video playback for end users. To reduce the cost of low cloud service providers while ensuring the quality of service, the cost of missing video clips from other cloud nodes must be reduced or even minimized. Therefore, we propose a set of video transmission algorithms for minimizing video transmission cost generated by edge cloud CDN nodes with QoS-guaranteed, called *Netdmc*. The algorithm can be separated into two parts. The first portion is a minimum-cost algorithm for nonemergency video clips. The second portion is a low-latency algorithm for emergency video clips.

Finally, the *Netdmc algorithm* is encapsulated in the protocol and run in the simulation experiment. Compared with the simulation results of general simulation algorithms, it can be seen that the *Netdmc* algorithm not only improves the fluency of end-user video playback but also reduces the bandwidth cost of service providers.

Acknowledgment. This research was supported by the national key research and development program of China (No. 2020YFF0305300), National Natural Science Foundation (No. 61762029, No. 61662012, No. U1811264), Guangxi Key Laboratory of Trusted Software (No. kx201726).

References

1. Zhu, J., Zheng, Z., Zhou, Y., Lyu, M.R.: Scaling service-oriented applications into geo-distributed clouds. In: 2013 IEEE Seventh International Symposium on Service-Oriented System Engineering, pp. 335–340 (2013)
2. Cherkasova, L.: Optimizing the reliable distribution of large files within CDNs. In: 10th IEEE Symposium on Computers and Communications (ISCC 2005), pp. 692–697 (2005)

3. Mahimkar, A., et al.: Bandwidth on demand for inter-data center communication. In: Proceedings of the 10th ACM Workshop on Hot Topics in Networks, pp. 1–6 (2011)

4. Jiao, L., Li, J., Xu, T., Du, W., Fu, X.: Optimizing cost for online social networks on geo-distributed clouds. IEEE/ACM Trans. Networking **24**, 99–112 (2014)

5. Zhang, J., Zhang, Y.: QoS-awareness peer coordination control for topology-converging P2P live streaming. Multimedia Tools Appl. **76**(22), 23835–23858 (2016). https://doi.org/10.1007/s11042-016-4092-9

6. Dong, X., Laiping, Z., Zhou, X., Li, K., Qiu, T., Guo, D.: An online cost-efficient transmission scheme for information-agnostic traffic in inter-datacenter networks. IEEE Trans. Cloud Comput. **10**(1), 202–215 (2019)

7. Golubchik, L., Khuller, S., Mukherjee, K., Yao, Y.: To send or not to send: reducing the cost of data transmission. In: 2013 Proceedings IEEE INFOCOM, pp. 2472–2478 (2013)

8. Borst, S., Gupta, V., Walid, A.: Distributed caching algorithms for content distribution networks. In: 2010 Proceedings IEEE INFOCOM, pp. 1–9 (2010)

9. Chu, W., Wang, L., Xie, H., Zhang, Z.-L., Jiang, Z.: Network delay guarantee for differentiated services in content-centric networking. Comput. Commun. **76**, 54–66 (2016)

10. Amble, M.M., Parag, P., Shakkottai, S., Ying, L.: Content-aware caching and traffic management in content distribution networks. In: IEEE (2011)

11. Greenberg, A., Hamilton, J., Maltz, D.A., Patel, P.: The cost of a cloud: research problems in data center networks, vol. 39, pp. 68–73. ACM New York, NY, USA (2008)

12. Chandy, J.A.: A generalized replica placement strategy to optimize latency in a wide area distributed storage system. In: Proceedings of the 2008 International Workshop on Data-Aware Distributed Computing, pp. 49–54 (2008)

13. Li, B., Zhang, X., Liu, J., Yum, T.-S.P.: Coolstreaming: a data-driven overlay network for efficient live media streaming. In: Proceedings of IEEE INFOCOM 2005 (2005)

14. Wiegand, T., Sullivan, G.J., Bjontegaard, G., Luthra, A.: Overview of the H. 264/AVC video coding standard. IEEE Trans. Circuits Syst. Video Technol. **13**, 560–576 (2003)

15. Shorfuzzaman, M., Graham, P., Eskicioglu, R.: Distributed placement of replicas in hierarchical data grids with user and system qos constraints. In: 2011 International Conference on P2P, Parallel, Grid, Cloud and Internet Computing, pp. 177–186 (2011)

16. Zhang, Z., Zhang, M., Greenberg, A.G., Hu, Y.C., Mahajan, R., Christian, B.: Optimizing cost and performance in online service provider networks. In: NSDI, pp. 33–48 (2010)

17. Shen, Y., Hsu, C.-H., Hefeeda, M.: Efficient algorithms for multi-sender data transmission in swarm-based peer-to-peer streaming systems. IEEE Trans. Multimedia **13**, 762–775 (2011)

18. Hu, H., et al.: Community based effective social video contents placement in cloud centric CDN network. In: 2014 IEEE International Conference on Multimedia and Expo (ICME), pp. 1–6 (2014)

19. Wang, Z., et al.: Propagation-based social-aware replication for social video contents. In: Proceedings of the 20th ACM International Conference on Multimedia, pp. 29–38 (2012)

20. Liu, G., Shen, H., Chandler, H.: Selective data replication for online social networks with distributed datacenters. IEEE Trans. Parallel Distrib. Syst. **27**, 2377–2393 (2015)

21. de Almeida, D.F., Yen, J., Aibin, M.: Content delivery networks-q-learning approach for optimization of the network cost and the cache hit ratio. In: 2020 IEEE Canadian Conference on Electrical and Computer Engineering (CCECE), pp. 1–5 (2020)

22. Ding, C., Zhou, A., Huang, J., Liu, Y., Wang, S.: ECDU: an edge content delivery and update framework in mobile edge computing. EURASIP J. Wirel. Commun. Netw. **2019**(1), 1–9 (2019). https://doi.org/10.1186/s13638-019-1590-2

23. Triukose, S., Rabinovich, M.: Client-centric content delivery network. In: 2016 Fourth IEEE Workshop on Hot Topics in Web Systems and Technologies (HotWeb), pp. 1–6 (2016)

Association Strategy Graph Convolutional Neural Network for Human Skeletal Behavior Recognition

Tinglong Liu[✉]

Center for Information Technology, Dalian Polytechnic University, Dalian 116034, Liaoning, China
liutl@dlpu.edu.com

Abstract. Aiming at the problem that the joint point partition strategy expresses the important information of the human body in the human body behavior recognition of bones cannot fully express the behavior, an RCTR-GCN human bone behavior recognition model of the correlation strategy is proposed. First, by adding an association strategy of a refined graph convolutional network model (CTR-GCN) of the smart channel topology, it can dynamically learn different topological structures and efficiently amplify the characteristics of the connection points in different channels while improving the key joint points of associated characteristics. Then, the network model redefines each channel by learning a shared topology and uses a specific channel relationship to unify the model through theoretical analysis; finally, redefining the model structure effectively reflects the associated information of local nodes within the channel. Action recognition has stronger aggregation capabilities. The results show that the recognition accuracy in the commonly used NTU RGB + D and NW-UCLA datasets reaches 93.6% (X-View), 97.6% (X-Sub), and 97.2%, respectively. The experimental results show that the accuracy rate is improved.

Keywords: Vertices · Channelwise topology refinement graph convolution net · Association strategy · Skeleton action recognition · Feature extraction

1 Introduction

Bone-based behavior recognition plays a significant role as a research topic. Its research and development are due to the development of neural networks, supervised learning and related sensors.

At the beginning of the study, convolutional neural networks or recurrent neural networks were used to predict sequences and images composed of human joint features. However, this method has great limitations, ignoring the connection and internal relationship between the nodes in the human body. These connections are important information about human behavior. Continuity and dynamics play an important role in behavior identification. Kamel et al. [1] proposed a human behavior recognition method based on a CNN depth map of posture data. The input data are divided into two forms.

© The Author(s), under exclusive license to Springer Nature Singapore Pte Ltd. 2022
Y. Wang et al. (Eds.): ICPCSEE 2022, CCIS 1628, pp. 403–412, 2022.
https://doi.org/10.1007/978-981-19-5194-7_30

To maximize feature extraction, the channels are divided into three. This method can effectively improve the identification accuracy. Pham et al. [2] proposed a deep residual neural network based on a CNN and designed a new network structure that can effectively improve the rate of action recognition. Later, [3–10] improved convolutional neural networks in adaptive, size, model structure and other aspects, but there are still shortcomings in accuracy and performance. Bruna et al. [11] a graph convolutional neural network (GCN) was proposed to model the relationship between human body nodes. Custom topologies make it difficult to obtain relational models between unnatural join points, which limits the representational power of GCN. Methods [12–15] enhance the data through the three-dimensional changes of the time series and learn the topological structure during training to enhance the expressive ability of the model. In the full range of the model, the same expression structure is forced to cluster, which leads to the problem of poor robustness. In the case of different behavioral characteristics, the correlation of joint motion will also be different, which results in the optimal solution of the model topology not being found if the same topology is used. In response to this problem, the optimization method of Chen [16] and others is to avoid the same parameter topology and use a separate method for training, but there are problems that the number of parameters is too large, the optimization is difficult, the training time is too long, and the model cannot be effectively derived. In addition, the parameterized topology remains the same for all samples, which makes sample correlation impossible to model.

This paper proposes a detailed graph convolutional network model for intelligent channel topology based on an association strategy. The model does not simply learn the model on different channels but implements the topology of channels under an intelligent association strategy. By expanding the domain value of the node, the connection between the associated node and the final node is strengthened, thus strengthening the connection of the whole body part. Each sample can dynamically infer the correlation and obtain the direct subtle relationship of each channel node. At the same time, each channel is modeled independently, and the minimum parameters are used to reduce the difficulty of modeling. In this way, the local information and the whole information perception ability of the model are stronger. The main contributions of this paper are as follows:

First, compared with the CTR-GCN method, the proposed association strategy enhances the internal correlation between nodes based on the intelligent topology-based refined convolutional network and greatly improves the spatial recognition accuracy of joint information. Second, a large number of experimental results show that the crT-GCN association strategy is superior to the existing methods in skeleton-based motion recognition.

2   Related Work

Convolutional neural networks are remarkable achievements in image processing. However, the performance of nonimage data of skeletal behavior is not ideal. Thus, a graph convolutional network is proposed [11]. Graph convolutional neural networks are divided into spectral methods and spatial methods. The spectral method is suitable for graphs with the same structure. Spatial methods can directly perform convolution operations on graphs, but they also face the challenge of dealing with different size thresholds.

In various GCN models, the commonly adopted feature updating rules are as follows: convert features into high-level representations according to the topology aggregation characteristics of the graph. In bone behavior recognition, GCN can be classified as follows according to different topologies:

(1) It is divided into shared and nonshared methods by determining whether different channels share topology.
(2) According to whether the topological structure is dynamically adjusted in the process of toppling, it can be divided into static methods and dynamic methods.

Reference [12] proposed an ST-GCN in the static strategy. With unchanged topological structure. The main feature of this model is that it remains unchanged during training and testing while being able to define the model structure in advance. This strategy strengthens the relationship between the relative positions of the body and is more conducive to improving the temporal and spatial association of skeleton node information. Lei et al. [23] proposed a 2S-AGCN bone behavior recognition model. Its model graph topology is unified or learned independently by the BP algorithm. At the same time, the end-to-end method is adopted to increase the universality of the model and make the model have first- and second-order characteristics at the same time to improve the accuracy of the model. Li et al. [17] used the innovative method of the A-links inference module in the dynamic method to obtain the coherence of human actions, which enhanced the expressive ability of the model, and only the related joints were identified as having the greatest correlation. Their common feature is to strengthen the relationship of local features to enhance the overall correlation. Reference [14] is also a dynamic model structure, but it adds the correlation between all joints to enhance the generalization ability. To overcome the limitations of shared methods in this regard, non shard topology methods are proposed, which employ different channels or groups to distinguish different structures. A representative example is the DCGCN proposed in [18], which uses individually set parameters to parameterize different channel groups. However, DCGCN has many parameters and is difficult to optimize. Cheng et al. [24] proposed shift-GCN to solve the problem of complex calculation of the GCN graph method, which can provide a flexible acceptance domain for graph operation by using transition graph operation and lightweight point-by-point operation. Chen et al. [25] proposed the CTR-GCN model, which can dynamically learn different topologies and efficiently amplify the features of connection points in different channels. At the same time, it has a stronger representation ability and effectively improves the recognition accuracy.

At present, there are single labels, distance partitions and space configuration partitions for joint connections of the human body. These three partitions consider the connection between adjacent nodes but cannot fully take into account the important role of the connection between the relative positions of human body parts in behavior recognition. The model wants to obtain important feature information of bone joints. A CTR-GCN model, RCTR-GCN, was proposed based on the original partition strategy to improve the recognition rate of the whole model.

3   Algorithms and Construction

First, the relevant symbols are defined, the RCTR-GCN model of the associated policy is given, and the structure of the model is analyzed.

3.1   Equations

The human skeleton diagram is a diagram of node joints, and edges are bones. The graph is denoted by. is a set of N nodes. ε is an edge set. Adjacency matrix. Elements represent relationships. is, χ is N nodes feature set∘Represented by, is. The topology convolution uses w for feature transformation. Features are updated by aggregating features, and the formula is as follows:

$$z_i = \sum_{v_j \in N(v_i)} a_{ij} x_j W \tag{1}$$

Custom or as a training parameter in static methods. In the dynamic method, it is generated by the input sample model.

3.2   RCTR-GCN Model Definition of the New Association Policy

Based on the joint information of CTR-GCN, a new nearest neighbor association strategy is used for repartition. The new nearest neighbor strategy adopts a maximum of two adjacent points and fully considers the correlation characteristics of different actions. On the basis of the original model, it can dynamically learn different topologies and efficiently amplify the features of connection points in different channels. Subdivision using temporal and spatial domains in a single channel. Divide the domain set according to the distance between the node and the root node. In this section, we set $D = 2$ to divide the domain set into three subsets: (1) the root node; (2) a subset of neighbor nodes with a distance D of 1 from the root; and (3) a distance D from the root of 2's subset of neighbor nodes. This partition is based on the fact that the joints in the human body's behavior activities are mainly manifested in the form of local activities, and the recent joint participation is higher. The model is more sensitive to behavior recognition perception and can improve the recognition accuracy of the model by reinforcing relevant information with the latest subset set. The topology of the different partitioning policies is shown in Fig. 1.

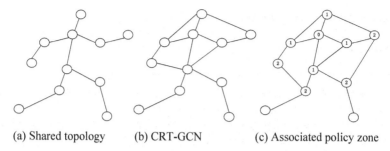

(a) Shared topology   (b) CRT-GCN   (c) Associated policy zone

Fig. 1. Different partition strategy topologies

The following is represented by graph convolution [19]. The single channel is represented by the identity matrix of E in the node of the root. The adjacency matrix is represented as X. In a single channel, the proposed association strategy formula is expressed as follows:

$$z_i = \sum_{v_j \in N(v_i)} a_{ij} \Lambda_j^{-\frac{1}{2}} x_j \Lambda_j^{-\frac{1}{2}} W \tag{2}$$

In the grouping adjacency matrix, there are the following relations:

$$X + E = \sum_j X_j \tag{3}$$

Here, to unify the evaluation criteria, different graphs need to be convolved into a unified form. For dynamic shared topology neural networks, dynamic topology has better generalization ability. The dynamic topological distance relationship depends on the input sample.

3.3 The Relational Architecture of the Model

We construct an association policy model architecture based on the RCTR-GNC. The neighborhood of each joint is set up as a diagram of the whole human skeleton diagram. The validity of the study has been detailed in the previous introduction. The model consists of 10 basic blocks to form the overall architecture. The TCN and GCN modules use residual error and adaptive technology, and the initial input channel value is 64. The TCN module uses multiscale temporary convolution, including a layer of two-dimensional convolution, a layer of standardization, a layer of ReLU and a layer of temporary convolution. The average pooling layer and the softmax classifier layer are added to complete the classification task. In the spatial model module, we use 3 RCTR-GCs to obtain the inner connections of joint points and summarize their results as output. Add normalization and dropout operations to the model to prevent model training from overfitting. To model different actions in continuous time, this paper adds a temporal multiscale modeling module.

4 Experiment and Analysis

4.1 Datasets

NTU RGB + D. NTU RGB + D [20] is a large human action recognition dataset. This dataset contains 56880 human skeleton behavior sequences. The samples were collected from 40 volunteers and divided into 60 categories. Each sample contains one action and is guaranteed to have up to 2 subjects, which are captured by three Microsoft Kinect v2 cameras from different views simultaneously.

Northwestern-UCLA. The Northwestern-UCLA dataset [21] is simultaneously acquired from multiple angles by 3 Kinect v2 cameras. It contains 1494 video clips covering 10 action categories. Each action is performed by 10 different subjects. Our dataset is divided into the following: the training set is from two cameras, and the test set is from the other camera.

4.2 Comparison of RCTR-GCN Experiments with Different Association Strategies

All experiments are performed on the Paddle deep learning framework on a Tesla V100 GPU. The training model uses SGD with a power of 0.9 and a weight decay of 0.0004. The number of trainings is set to 65, and the learning rate is set to 0.1.

This paper presents a new grouping-based associative partitioning strategy. The nodes are divided into three subsets: the set of root joint points, the set of adjacent points with a distance of 1 from the root node, and the set of adjacent points with a distance of 2 from the root node. The following compares the unified partition, distance partition and sparse partition with the grouping-associated partition proposed in this paper and compares the experimental results.

(1) NTU-RGB + D (X-Sub)

Table 1 shows the action recognition performance of the skeleton-based NTU-RGC + D dataset. The accuracy of several partitioning strategies is compared by using the x-sub method. It can be seen that under the evaluation of NTU-RGB + D, the final training result has 0.75% point improvement in accuracy compared with top-5, the best training result among the three existing partitioning strategies.

(2) NTU-RGB + D (X-View)

Table 2 shows the human action recognition indicators on this dataset. Experimental methods from different perspectives are used to compare the accuracy of different algorithms. From the analysis of the data in the table, it can be seen that the recognition rate obtained by the partition strategy proposed in this paper is higher than that of all previous recognition effects. The accuracy of top-1 and top-5 reached 96.42% and 99.63%, respectively.

Table 1. NTU-RGB + D (X-Sub) experimental results.

Epoch	Uniform	Distance	Spatial	Ccorrelative
10	90.80	88.73	90.25	90.57
20	95.84	95.64	78.37	95.83
30	95.76	96.54	96.30	96.56
40	96.63	96.70	96.42	97.25
50	96.93	96.48	96.73	96.54
60	97.53	97.60	97.67	97.68
70	97.34	97.54	97.53	97.52
80	97.86	97.35	97.63	97.63

Table 2. NTU−RGB + D (X−View) experimental results

Epoch	Uniform		Distance		Spatial		Correlative	
	Top1	Top5	Top1	Top5	Top1	Top5	Top1	Top5
10	66.58	93.86	64.86	93.52	66.69	94.69	60.35	90.53
20	80.67	98.65	80.76	97.87	81.96	98.52	84.30	98.62
30	82.49	98.79	83.36	98.54	83.58	98.86	85.37	98.96
40	84.65	98.84	84.66	98.96	86.52	99.21	96.42	99.63

4.3 Comparison with Other Popular Algorithms

Many of the most advanced approaches employ multistream fusion frameworks. We fused the results of four modes, namely, joint, bone, joint motion and bone motion. Compare our algorithm with other advanced popular algorithms. The performance is shown in Tables 3 and 4. Several current advanced methods are compared on the NTU-RGB + D dataset. The comparison results of the ST-GCN, 2S-AGCN, Shift-GCN, D-GCN, CTR-GCN, CTR-GCN, and RCTR-GCN algorithms show that, compared with the best method, the new correlation partition RCTR-GCN improves the recognition accuracy of the x-View and X-Sub experiments by 1.2 and 0.8% points.

Table 3. Comparison of several recognition technologies on the NTU-RGB + D dataset

方法	X-View/%	X-Sub/%
ST-GCN	81.5	88.5
2s-AGCN	88.5	95.1
Shift-GCN	90.7	96.5
D-GCN	91.1	97.1
CTR-GCN	92.4	96.8
RCTR-GCN	93.6	97.6

Comparing several methods on the Northwestern-UCLA dataset, compared with the more commonly used Shift-GCN model, the Top1 recognition accuracy is improved by 2.6% points, increased by 97.2%, and increased by 0.7% points.

Table 4. Comparison of existing recognition technologies on the Northwestern-UCLA dataset

Method,	Top1/%
ST-GCN	89.2
2S-AGCN	93.3
Shift-GCN	94.6
D-GCN	95.3
CTR-GCN	96.5
RCTR-GCN	97.2

The experimental results on the following datasets show that the algorithm in this paper has the highest accuracy top 1, which shows that the method in this paper has the characteristics of high efficiency in human behavior recognition based on bones.

5 Conclusion

We put forward a new kind of channel based on the strategy of grouping associated topological refinement RCTR convolution networks - GCN, through changing the structure of the model, increase the correlation properties of adjacent nodes, maximize action feature extracting, which based on the improved model of gesture recognition accuracy and the experimental results show that the model based on skeleton of action recognition performance is good. Mathematical analysis and experimental results show that the new associative grouping strategy RCTR-GCN has better representation ability than other existing graph convolutions. Under the guidance of the new zoning strategy, how to obtain the direct connection between different nodes under specific actions in the next step has become a subject that can be further studied. Next, the focus of this work is to improve the end-side performance of the model, reduce the size of the model and improve the efficiency.

References

1. Kamel, A., Sheng, B., Yang, P., et al.: Deep convolutional neural networks for human action recognition using depth maps and postures. IEEE Trans. Syst. Man Cybern. Syst. **49**(9), 1806–1819 (2018)
2. Pham, H.H., Khoudour, L., Crouzil, A., et al.: Exploiting deep residual networks for human action recognition from skeletal data. Comput. Vis. Image Underst. **03**(170), 51–66 (2018)
3. Gao, Z., Xuan, H.Z., Zhang, H., et al.: Adaptive fusion and category-level dictionary learning model for multiview human action recognition. IEEE Internet Things J. **6**(6), 9280–9293 (2019)
4. Defferrard, M., Bresson, X., Vandergheynst, P.: Convolutional neural networks on graphs with fast localized spectral filtering. In: Advances in Neural Information Processing Systems, pp. 3844–3852 (2016)
5. Paulose, N., Muthukumar, M., Swathi, S., et al.: Recurrent neural network for human action recognition using star skeletonization. Int. Res. J. Eng. Technol. **6**(3), 123–130 (2019)
6. Fernando, B., Gavves, E., Oramas, J.M., et al.: Modeling video evolution for action recognition. In: Proceedings of the IEEE Conference on Computer Vision and Pattern Recognition, pp. 5378–5387 (2015)
7. Ji, X., Cheng, J., Feng, W., et al.: Skeleton embedded motion body partition for human action recognition using depth sequences. Signal Process. **143**(C), 56–68 (2018)
8. Zhao, Y., Xiong, Y., Wang, L., et al.: Temporal action detection with structured segment networks. In: Proceedings of the IEEE International Conference on Computer Vision, pp. 2914–2923 (2017)
9. Akula, A., Shah, A.K., Ghosh, R.: Deep learning approach for human action recognition in infrared images. Cogn. Syst. Res. **50**(1), 146–154 (2018)
10. Liu, J., Shahroudy, A., Perez, M., Wang, G., Duan, L.Y., Kot, A.C.: Ntu rgb+d 120: a large-scale benchmark for 3d human activity understanding. IEEE Trans. Pattern Anal. Mach. Intell. **42**(10), 2684–2701 (2020)
11. Bruna, J., Zaremba, W., Szlam, A., LeCun, Y.: Spectral networks and locally connected networks on graphs. In: ICLR, vol. 21, no. 5, p. 14 (2014)
12. Yan, S., Xiong, Y., Lin, D.: Spatial temporal graph convolutional networks for skeleton-based action recognition. In: Proceedings of the Thirty-Second AAAI Conference on Artificial Intelligence (2018)
13. Cheng, K., Zhang, Y., He, X., Chen, W., Cheng, J., Lu, H.: Skeleton-based action recognition with shift graph convolutional network. In: Proceedings of the IEEE/CVF Conference on Computer Vision and Pattern Recognition, vol. 01, no. 06, pp. 188–193 (2020)
14. Shi, L., Zhang, Y., Cheng, J., Lu, H.: Skeleton-based action recognition with directed graph neural networks. In: Proceedings of the IEEE Conference on Computer Vision and Pattern Recognition, vol. 01, no. 06, pp. 7912–7921 (2019)
15. Huang, Z., Shen, X., Tian, X., Li, H., Huang, J., Hua, X.S.: Spatiotemporal inception graph convolutional networks for skeleton-based action recognition. In: Proceedings of the 28th ACM International Conference on Multimedia, vol. 19, no. 8, pp. 2122–2130 (2020)
16. Chen, Y., Dai, X., Liu, M., Chen, D., Yuan, L., Liu, Z.: Dynamic convolution: attention over convolution kernels. In: Proceedings of the IEEE/CVF Conference on Computer Vision and Pattern Recognition, vol. 1, no. 1, pp. 11030–11039 (2020)

17. Li, M., Chen, S., Chen, X., Zhang, Y., Wang, Y., Tian, Q.: Actional-structural graph convolutional networks for skeleton-based action recognition. In: Proceedings of the IEEE/CVF Conference on Computer Vision and Pattern Recognition, vol. 1, no.1, pp. 3595–3603 (2019)
18. Chen, Y., Dai, X., Liu, M., Chen, D., Yuan, L., Liu, Z.: Dynamic convolution: attention over convolution kernels. In: Proceedings of the IEEE/CVF Conference on Computer Vision and Pattern Recognition, vol. 6, no. 1, pp. 11030–11039 (2020)
19. Kipf, T.N., Welling, M.: Semisupervised classification with graph convolutional networks. In: ICLR, vol. 22, no. 1, pp. 124–135 (2017)
20. Shahroudy, A., Liu, J., Ng, T.T., Wang, G.: Ntu rgb+ d: a large scale dataset for 3d human activity analysis. In: Proceedings of the IEEE conference on computer vision and pattern recognition, vol. 6, no. 2, pp. 1010–1019 (2016)
21. Li, C., Zhong, Q., Xie, D., et al.: Skeleton-based action recognition with convolutional neural networks. In: 2017 IEEE International Conference on Multimedia & Expo Workshops, vol. 3, no. 1, pp. 597–600 (2017)
22. Zhang, P., Lan, C., Zeng, W., Xing, J., Xue, J., Zheng, N.: Semantics-guided neural networks for efficient skeleton-based human action recognition. In: Proceedings of the IEEE/CVF Conference on Computer Vision and Pattern Recognition, vol. 17, no. 3, pp. 1112–1121 (2020)
23. Shi, L., Zhang, Y., Cheng, J., Lu, H.: Two-stream adaptive graph convolutional networks for skeleton-based action recognition. In: The IEEE Conference on Computer Vision and Pattern Recognition (CVPR), vol. 9, no. 7, pp. 12026–12035 (2019)
24. Cheng, K., Zhang, Y., He, X., et al.: Skeleton-based action recognition with shift graph convolutional network. In: 2020 IEEE/CVF Conference on Computer Vision and Pattern Recognition (CVPR), vol. 6, no. 3, pp. 180–189 (2020)

Unsupervised Style Control for Image Captioning

Junyu Tian[1], Zhikun Yang[1], and Shumin Shi[1,2(✉)]

[1] School of Computer Science and Technology, Beijing Institute of Technology, Beijing, China
{tjy,yzk,bjssm}@bit.edu.cn
[2] Beijing Engineering Research Center of High Volume Language Information Processing and Cloud Computing Applications, Beijing, China

Abstract. We propose a novel unsupervised image captioning method. Image captioning involves two fields of deep learning, natural language processing and computer vision. The excessive pursuit of model evaluation results makes the caption style generated by the model too monotonous, which is difficult to meet people's demands for vivid and stylized image captions. Therefore, we propose an image captioning model that combines text style transfer and image emotion recognition methods, with which the model can better understand images and generate controllable stylized captions. The proposed method can automatically judge the emotion contained in the image through the image emotion recognition module, better understand the image content, and control the description through the text style transfer method, thereby generating captions that meet people's expectations. To our knowledge, this is the first work to use both image emotion recognition and text style control.

Keywords: Image caption · Image sentiment recognization · Text style transfer

1 Introduction

In recent years, image semantic understanding and generating natural language descriptions have been rapidly developing topics in the fields of natural language understanding and computer vision, and there is a huge challenge at their intersection: automatic generation of relevant descriptions from a given image, namely, image captioning.

Although image captioning [1,9,17,25] has achieved exciting results, there is still much room for improvement. The current models generally overpursue excellent evaluation indicators, which makes the expression style of the generated text description too boring, which is difficult to satisfy people's expectations for demand for image captions with diverse styles. There are a thousand Hamlets in the eyes of a thousand people. Different people have different opinions on the same picture, and each user's expression habits are different, so the generated description will also have a different style, as shown in Fig. 1. Therefore, it is clearly not

Y. Wang et al. (Eds.): ICPCSEE 2022, CCIS 1628, pp. 413–424, 2022.
https://doi.org/10.1007/978-981-19-5194-7_31

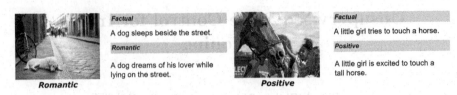

Fig. 1. Example of stylized image caption.

enough for image captioning models to generate monotonic descriptions. An image should be able to be described in different styles, such as positive or negative emotions, in an informal or formal way. Because language style is an important indicator of expressing emotions and conveying information, it is very important to people. In practical applications, stylized image captioning is also of great value. For example, in the process of online recommendation, pictures can be automatically matched with stylized texts. These interesting texts can arouse people's interest, thereby increasing the number of clicks on websites and articles.

Of course, stylized image captioning also faces many practical difficulties. The first is image semantic understanding, and generating natural language descriptions is an old and fresh challenge for natural language understanding and computer vision. There is an unbridgeable semantic gap between text and images. How to solve this semantic gap has always been a problem that puzzles scholars. Second, training the image captioning model requires considerable paired training data. Now, the more common image captioning datasets styles are neural and objective, similar to MSCOCO [10], without obvious language style and emotion. Existing image captioning datasets containing text styles are difficult to meet people's needs in both scale and quantity.

To overcome the existing difficulties mentioned, we propose an unsupervised and style-controllable method for image caption generation. The language model is controlled by the language style control module to generate image captions that meet people's desired style.

We innovatively integrate the image captioning model and the language style control model and use the precollected emotions, topics and other information as language style knowledge to control the model and produce corresponding image captions. To make full use of image and text semantic information, we use the powerful CLIP pretraining model to obtain image features with richer semantics. In addition, we have also improved the style control module so that it can effectively control the generated results of the language model. Specifically, the PPLM [4] model is added to the text generation module of the image caption model. The hidden layer gradient of the language generation module is controlled by the PPLM so that it propagates in the direction we need to generate natural sentences.

In summary, our contributions are mainly focused on the following points:

1) We present a novel model in this paper that combines image captioning with language style control and can generate a variety of styles of captions.

2) We use the image sentiment recognition method to judge the emotion contained in the image, and can automatically generate a stylized description suitable for the image.
3) A large number of experiments on multiple datasets show that this method achieves good results.

2 Related Work

2.1 CLIP Pretrained Model

Some scholars have proposed the CLIP multimodal pretraining model [17] to learn the concept of vision under language supervision. CLIP consists of a text encoder and a visual encoder, which is trained on 400M noisy image-text pairs scraped from the Internet. Train on more diverse and large-scale data sources, not constrained by a fixed set of labels, and have the ability to generalize to unseen objects and concepts. CLIP shows strong Zeroshot performance on benchmark datasets such as ImageNet classification. On this basis, to effectively use the powerful effect of the CLIP pretraining model, Sheng Shen et al. [21] integrate CLIP with the existing image captioning model and directly uses CLIP as the visual encoder in a multimodal visual language model and fine-tuning for image captioning tasks. A few months ago, Ron Mokady et al. [15] mapped the visual features extracted by CLIP to the language space as a prefix for the GPT-2 [18] language model, with good results.

2.2 Stylized Image Captioning

From now on, much meaningful work has emerged in stylized image captioning. There are two main types of work: models [3,26] using supervised stylized corpora and models [12] using semisupervised stylized image caption data. The goal of SentiCap [13] is to generate both positive and negative styles, mainly by simulating word inflections using two parallel LSTMs and word-level supervision. StyleNet [14] proposes a caption database characterized by the ability to generate narrative sentences. The author constructs a mapping from traditional sentences to story sentences and uses two-stage training, from images to keywords, and then from keywords to story sentences, reducing the cost of annotation. An adversarial learning network was proposed by MSCAP [7] for multistyle image subtitle tasks on multiple corpora. They propose a style recognizer to distinguish whether the input sentence is real or not. Recognizers and generators are trained in adversarial ways to achieve more natural and human-like captions.

2.3 Text Style Control

Wang et al. [24] proposed an unsupervised text style transformation framework and proposed the FGIM algorithm to modify the decoder's latent representation by changing the attributes of the text attribute classifier. Fu et al. [5] used

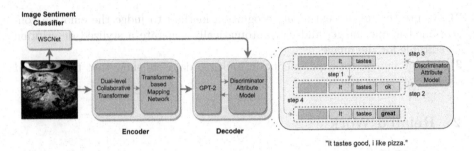

Fig. 2. There is the overall architecture of our model. After the target image is extracted and fused with CLIP features, the obtained image features are mapped to the GPT-2 space, and the discriminator is used. The Attribute Model judges the generated words and retrograde and controls the GPT-2 through the gradient to generate the required captions.

an adversarial network to let the model learn text representations and style representations independently. In addition, because of the shortage in effective evaluation indicators, this paper proposes two new indicators for evaluating style transfer: conversion strength and content preservation. A novel generative model proposed by Zhou et al. [29] can dynamically perform style transfer according to the style correlation of generated words. A style transfer decoder is designed that takes word-level style correlation and semantic information as input and performs finetune training through multiple loss items to realize the style transfer function.

3   Approach

3.1   Overview

Facing the reality of the lack of paired stylized datasets, we propose an unsupervised method to generate image captions with stylistic and emotional tendencies. The overall framework of the proposed unsupervised stylized image caption generation model is shown in Fig. 2. It contains three basic parts, namely, image encoder E, image style classifier C, and stylized caption decoder D, wherein caption generator D is composed of a language generation model G and a style control module P. Faced with an image x, we first classify it using an image sentiment analysis model and use the obtained style attribute s to guide the next step of stylized caption generation. We use the powerful CLIP pretraining model to extract visual features from images and extract global features and grid features, denoted $CLIP_{global}$ and $CLIP_{grid}$, respectively.

$$CLIP_{global}, CLIP_{grid} = CLIP(x) \tag{1}$$

We fuse $CLIP_{global}$ and $CLIP_{grid}$ using a dual-level collaborative transformer and obtain fused features.

$$CLIP_{fuse} = DLCT(CLIP_{global}, CLIP_{grid}) \tag{2}$$

Then, the fused visual information is mapped to the latent space of GPT-2 through a light mapping network denoted as MAP.

$$GPT_{prefix} = MAP(CLIP_{fuse}) \tag{3}$$

While the GPT-2 [18] language model uses GPT_{prefix} to generate captions, we use the PPLM language style control model to perturb the hidden layer of GPT-2 and make the generated caption y meet our requirements by controlling the direction of the gradient.

$$y = GPT(GPT_{prefix}), PPLM \tag{4}$$

3.2 Image Encoder

The semantic gap has always been a huge difficulty faced by visual language models. How to effectively integrate the semantics of images and texts has troubled the majority of artificial intelligence scholars. Faced with this problem, we followed the work proposed by Radford et al. [17] and Shen et al. [21] and combined the CLIP multimodal pretraining technique with the image captioning model. The CLIP model contains a large amount of image and text semantic information, which can be used as a bridge between image semantics and text semantics. It can make the model encoder and decoder interact better and obtain the relevant semantic information more effectively to improve the model performance effectively. Because our work is unsupervised, the information that the CLIP learns on large-scale data is also valuable to us. To mine the information contained in the image as much as possible, inspired by Luo et al. [11], we extract global and grid features and use the dual-level collaborative transformer (DLCT) for fusion to obtain information-dense visual features. The visual features of the image are then mapped to the potential space of GPT-2 using a lightweight transformer-based mapping network.

3.3 Stylized Caption Decoder

Our model is based on the ClipCap image captioning model [15]. ClipCap uses a prefix that uses visual encodings for image captioning by a transformer-based mapping network and then generates image captions by fine-tuning the language model. When generating image captions, the pretrained language model starts with the CLIP prefix and generates captions one by one. This effectively reduces the semantic distance between the visual and textual worlds. To generate stylized text, we plug PPLM [4] into the pretrained language model in ClipCap, enabling control over image captioning. It is worth noting that the KL divergence coefficient is relatively small compared with PPLM, because GPT-2 language model guided by Prompt has already possessed some prior knowledge. If the constraint is too strong, it will be difficult to generate stylized image caption effectively. PPLM contains a style classifier to judge whether the words generated by the language model meet the requirements and uses the disturbance mechanism to modify the latent state of the language model through the gradient return to affect the output of the language model and generate the text that conforms to the emotional style of the image while ensuring the smoothness of the generated text. Since the classifier in PPLM does not have the ability to judge romantic and humorous styles, we retrain a classifier with romantic and humorous sentences from FlickrStyle 10K [6].

3.4 Image Sentiment Recognition

We found that although the direct use of the PPLM can generate rich and diverse captions to a large extent, many images themselves convey very distinct characteristics; in other words, not all images fit every style. Therefore, we use the results of image sentiment classification [27,28] to judge the style of the image and automatically select the style suitable for the image. We use the pretrained WSCNet model [20] as our image sentiment classifier to judge whether the image is positive or negative. This image sentiment classifier is only used for scenarios where positive or negative captions are automatically generated. When generating romantic or humorous captions, select the desired attributes and do not make style judgments on the images.

3.5 Training Strategy

We follow the classic practice of image captioning, using word-level cross-entropy loss to train the model. Our cross-entropy loss is defined as:

$$L_x = - \sum_{i=1}^{N} \sum_{j=1}^{L} log P_\theta(C_j^i | CLIP_{prefix}, C_1^i, ..., C_{j-1}^i) \qquad (5)$$

where θ is a learnable parameter and C is the generated token. The style control module is added after the completion of the training phase to ensure that the model can accurately understand the image content and can also generate styled captions without supervision.

4 Experiment

4.1 Datasets

We train our model on a classic MSCOCO 2014 dataset and then test it on FlickrStyle10K and SentiCap.

MSCOCO 2014 is a large-scale visual task dataset. The dataset has 164,062 images, and each image corresponds to 5 captions. Among them, the training set, verification set and test set contain 82,783, 40,504 and 40,775 images. To compare with the results obtained in other works, we adopted the dataset partitioning method of Karpathy et al. [8]. Their method merges the original training and validation sets in the aggregate 123,287 images. They used 113,287 images as the training set, 5000 images as the validation set and 5000 images as the test set.

The SentiCap dataset [13] is an image caption dataset based on the MSCOCO dataset, which contains images marked by positive and negative emotion labels. The training set for the POS subset contained 998 images and 2,873 positive captions, and the test set contained over 673 images and 2,019 captions. The training set for the NEG subset contained 997 images with 2,468 negative captions, in addition, 1,509 captions and more than 503 images for testing. Each test image has three positive and three negative titles. For SentiCap, we selected 500 positive and 500 negative images for testing.

The FlickrStyle10K dataset contains 10,000 images from Flickr. It is important to note that only the 7,000 training set is publicly available. In particular, for 7,000 images, each image is annotated with 5 factual captions, 1 humorous caption, and 1 romantic caption. We randomly select 1,000 of them as the test set.

4.2 Evaluation Metrics

We evaluate our stylized image captioning model from two aspects: image relevance and language style.

We use the widely used automated evaluation metrics METEOR [2], CIDEr [23] and BLEU-1, BLEU-3 [16] to measure the relevance of captions and images. These indicators are primarily based on N-gram overlap, but they are not very suitable indicators for evaluating stylized image captions because stylized image captions generally have more flexibility in choosing words to describe images.

We evaluated language style in terms of style accuracy and fluency.

Fluency. We evaluate the fluency of the generated captions according to the target style. We use the SRILM [22] to test the fluency of the sentences generated by the model. SRILM uses a triplet language model trained on the corresponding dataset to calculate the confusion of the generated sentences. The smaller the SRILM score represents, the lower the confusion of the title, meaning the smoother the whole sentence and the better the performance of the model.

Style accuracy. We measure how often generated captions have the correct target style according to a pretrained text style discriminator. To do this, we choose TextCNN [19] as the style classifier.

4.3 Implementation Details

For image feature extraction, we choose ViT-B/32 for the visual model. Feature fusion of CLIP global and grid image features We follow the work of Luo et al. [11] using the same parameters. For the mapping network, we set the CLIP embedding as prefix length $L = 40$ constant tokens and freeze the parameters of

GPT-2. For GPT-2, we use Wolf's implementation and choose the version of the GPT-2 medium. We trained with 50 batches of 10 epochs. We use AdamW with weight decay fix introduced by Loshchilov et al., a learning rate of 2e-5 and 5000 warmup steps for optimization. For the PPLM text style control module, when generating positive and negative style captions, we chose 30 as the maximum length of the generated sentence, set 0.03 to the step size of each modification of the hidden state, set 10 to the number of iterations of each modification, and set the other parameters constant.

4.4   Results

We compare our model with several state-of-the-art methods for stylized image captioning, including StyleNet, MSCap, and MemCap. Our model achieves good results even in the unsupervised situation and even outperforms StyleNet in both positive and negative caption generation, but in certain the performance of these indicators is slightly inferior to MSCap and MemCap. In particular, some evaluation metrics measured by n-gram degree are not satisfactory, largely because our method is unsupervised, the model is not trained on SentiCap and FlickrStyle 10K results, and the humor and the romantic style is too variable, resulting in a large deviation of the generated captions in comparison with the ground truth. Even so, our model still achieves good results in fluency and classification accuracy. Compared to humorous and romantic styles, our model achieves relatively good results in both positive and negative classification accuracy and performance. We think that because the positive and negative classifiers are relatively more mature, the amount of training data is relatively large, and the difficulty of distinguishing is less than distinguishing romantic and humorous styles, the style control module has a better control effect on captions. For romantic and humorous captions, due to the relatively small amount of relevant manual annotation data, the knowledge learned by the model in the previous training is not sufficient, which is also one of the research directions of our future work (Table 1).

4.5   Ablation Studies

We conduct an ablation study to validate the contribution of each component. The results of the ablation study are presented in Table 2. UP-DOWN stands for commonly used BUTD visual features, CLIP stands for only CLIP global features, fused CLIP stands for fused CLIP grid and global semantic features. We analyzed the results as follows:

In terms of visual features, it is shown that using the CLIP feature significantly outperforms the common BUTD feature [1]. We think this is because CLIP has been sufficiently pretrained on a large-scale dataset to learn more information. We found that the use of feature fusion also improves the model performance to a certain extent, and we believe that this fusion mechanism greatly increases the utilization of image information, thereby improving the effect.

Table 1. Experimental results on SentiCap and FlickrStyle10K.

Method	Style	B-1	B-3	M	C	ppl	cls
StyleNet	Positive	45.3	12.1	12.1	36.3	24.8	45.2
	Negative	43.7	10.6	10.9	36.6	24.9	56.6
	Romantic	13.3	1.5	4.5	7.2	52.9	37.8
	Humorous	13.4	0.9	4.3	11.3	48.1	41.9
MSCap	Positive	46.9	16.2	16.8	55.3	19.6	92.5
	Negative	45.5	15.4	16.2	51.6	19.2	93.4
	Romantic	17.1	2.0	5.4	10.1	20.4	88.7
	Humorous	16.3	1.9	5.3	15.2	22.7	91.3
Memcap	Positive	50.8	17.1	16.6	54.4	13	99.8
	Negative	48.7	19.6	15.8	60.6	14.6	93.1
	Romantic	19.7	4.0	7.7	19.7	19.7	91.7
	Humorous	19.8	3.9	7.2	18.5	17.0	97.1
Ours	Positive	45.2	13.1	11.8	38.3	24.6	47.0
	Negative	44.5	12.4	12.2	38.4	24.2	59.4
	Romantic	12.0	1.3	4.2	8.6	55.3	40.7
	Humorous	11.3	1.1	4.0	10.2	49.5	43.3

Table 2. Ablation evaluation

Method	Style	B-1	B-3	M	C	ppl	cls
BUTD	Positive	42.7	11.9	10.3	32.6	33.9	38.2
	Romantic	9.8	0.9	3.5	7.3	63.1	31.8
CLIP	Positive	43.8	12.6	11.3	33.1	32.7	41.6
	Romantic	10.6	1.1	3.8	7.9	61.8	34.4
Fusion CLIP	Positive	45.2	13.1	11.8	38.3	24.6	47.0
	Romantic	12.0	1.3	4.2	8.6	55.3	40.7

4.6 Human Evaluation

Automatic evaluation metrics often do not reflect the quality of stylized captions in the minds of users. Therefore, we manually evaluate the generated captions in terms of text fluency, image text relevance, and text style appropriateness. For the results produced by each style in the test set, we randomly select 50 images for manual evaluation. We recruited 5 volunteers to rate the captions, each for all images. They were asked to rate the generated captions on the basis of image text relevance and text style appropriateness. Fluency of text content refers to whether the generated captions are smooth and fluent, in line with people's reading habits, On a scale of 0(poor) to 5(good). Image-text relevance refers to whether the caption accurately conveys the information of the image,

On a scale of 0(poor) to 5(good). Style appropriateness refers to whether the title appropriately possesses the desired style, On a scale of 0(poor) to 5(good) (Table 3).

Table 3. Human evaluation of the relevance, fluency, and stylistic accuracy of generated captions.

	Pos	Neg	Roman	Humor	Avg
Fluency	4.2	4.1	3.8	3.8	3.975
Relevancy	4.1	3.9	3.8	4.0	3.95
Accuracy	3.9	3.7	2.9	2.70	3.3

5 Conclusion

We propose an unsupervised stylized image captioning model with low training cost and high adaptability compared to other image captioning models using supervised the data. In addition to the four styles mentioned above, it can also be widely used to generate other styles of text generation, depending on what the text style classifier has learned. Since the cost of training a text style classifier is much lower than training a large-scale image caption generation model, our model is also more advantageous in training time.

Acknowledgment. This work is supported by the National Key Research & Development Program (Grant No. 2018YFC0831700) and National Natural Science Foundation of China (Grant No. 61671064, No. 61732005).

References

1. Anderson, P., et al.: Bottom-up and top-down attention for image captioning and visual question answering. In: Proceedings of the IEEE Conference on Computer Vision and Pattern Recognition, pp. 6077–6086 (2018)
2. Banerjee, S., Lavie, A.: METEOR: an automatic metric for MT evaluation with improved correlation with human judgments. In: Proceedings of the ACL Workshop on Intrinsic and Extrinsic Evaluation Measures for Machine Translation and/or Summarization, pp. 65–72 (2005)
3. Chen, T., et al.: "factual"or "emotional": stylized image captioning with adaptive learning and attention. In: Proceedings of the European Conference on Computer Vision (ECCV), pp. 519–535 (2018)
4. Dathathri, S., et al.: Plug and play language models: a simple approach to controlled text generation (2019)
5. Fu, Z., Tan, X., Peng, N., Zhao, D., Yan, R.: Style transfer in text: exploration and evaluation. In: Proceedings of the AAAI Conference on Artificial Intelligence, vol. 32 (2018)

6. Gan, C., Gan, Z., He, X., Gao, J., Deng, L.: Stylenet: generating attractive visual captions with styles. In: Proceedings of the IEEE Conference on Computer Vision and Pattern Recognition, pp. 3137–3146 (2017)

7. Guo, L., Liu, J., Yao, P., Li, J., Lu, H.: MSCap: multi-style image captioning with unpaired stylized text. In: Proceedings of the IEEE/CVF Conference on Computer Vision and Pattern Recognition, pp. 4204–4213 (2019)

8. Karpathy, A., Fei-Fei, L.: Deep visual-semantic alignments for generating image descriptions. In: Proceedings of the IEEE Conference on Computer Vision and Pattern Recognition, pp. 3128–3137 (2015)

9. Li, X., et al.: OSCAR: object-semantics aligned pre-training for vision-language tasks. In: Vedaldi, A., Bischof, H., Brox, T., Frahm, J.-M. (eds.) ECCV 2020. LNCS, vol. 12375, pp. 121–137. Springer, Cham (2020). https://doi.org/10.1007/978-3-030-58577-8_8

10. Lin, T.-Y., et al.: Microsoft COCO: common objects in context. In: Fleet, D., Pajdla, T., Schiele, B., Tuytelaars, T. (eds.) ECCV 2014. LNCS, vol. 8693, pp. 740–755. Springer, Cham (2014). https://doi.org/10.1007/978-3-319-10602-1_48

11. Luo, Y., et al.: Dual-level collaborative transformer for image captioning. arXiv preprint arXiv:2101.06462 (2021)

12. Mathews, A., Xie, L., He, X.: Semstyle: learning to generate stylised image captions using unaligned text. In: 2018 IEEE/CVF Conference on Computer Vision and Pattern Recognition (2018)

13. Mathews, A., Xie, L., He, X.: Senticap: generating image descriptions with sentiments. In: Proceedings of the AAAI Conference on Artificial Intelligence, vol. 30 (2016)

14. Mathews, A., Xie, L., He, X.: Semstyle: learning to generate stylised image captions using unaligned text. In: Proceedings of the IEEE Conference on Computer Vision and Pattern Recognition, pp. 8591–8600 (2018)

15. Mokady, R., Hertz, A., Bermano, A.H.: ClipCap: clip prefix for image captioning (2021)

16. Papineni, K., Roukos, S., Ward, T., Zhu, W.-J.: Bleu: a method for automatic evaluation of machine translation. In: Proceedings of the 40th Annual Meeting of the Association for Computational Linguistics, pp. 311–318 (2002)

17. Radford, A., et al.: Learning transferable visual models from natural language supervision. In: International Conference on Machine Learning, pp. 8748–8763. PMLR (2021)

18. Radford, A., Jeffrey, W., Child, R., Luan, D., Amodei, D., Sutskever, I., et al.: Language models are unsupervised multitask learners. OpenAI blog **1**(8), 9 (2019)

19. Rakhlin, A.: Convolutional neural networks for sentence classification. GitHub (2016)

20. She, D., Yang, J., Cheng, M.-M., Lai, Y.-K., Rosin, P.L., Wang, L.: WSCNet: weakly supervised coupled networks for visual sentiment classification and detection. IEEE Trans. Multimedia **22**(5), 1358–1371 (2019)

21. Shen, S., et al.: How much can clip benefit vision-and-language tasks? arXiv preprint arXiv:2107.06383 (2021)

22. Stolcke, A.: SRILM-an extensible language modeling toolkit. In: Seventh International Conference on Spoken Language Processing (2002)

23. Vedantam, R., Lawrence Zitnick, C., Parikh, D.: Cider: consensus-based image description evaluation. In: Proceedings of the IEEE Conference on Computer Vision and Pattern Recognition, pp. 4566–4575 (2015)

24. Wang, K., Hua, H., Wan, X.: Controllable unsupervised text attribute transfer via editing entangled latent representation. In: Advances in Neural Information Processing Systems, vol. 32 (2019)
25. Yao, T., Pan, Y., Li, Y., Mei, T.: Exploring visual relationship for image captioning. In: Proceedings of the European Conference on Computer Vision (ECCV), pp. 684–699 (2018)
26. You, Q., Jin, H., Luo, J.: Image captioning at will: a versatile scheme for effectively injecting sentiments into image descriptions. arXiv preprint arXiv:1801.10121 (2018)
27. Zhao, S., Ding, G., Huang, Q., Chua, T.-S., Schuller, B.W., Keutzer, K.: Affective image content analysis: a comprehensive survey. In: IJCAI, pp. 5534–5541 (2018)
28. Zhao, S., et al.: Affective image content analysis: two decades review and new perspectives. IEEE Trans. Pattern Anal. Mach. Intell. (2021)
29. Zhou, C., et al.: Exploring contextual word-level style relevance for unsupervised style transfer. arXiv preprint arXiv:2005.02049 (2020)

AM-PSPNet: Pyramid Scene Parsing Network Based on Attentional Mechanism for Image Semantic Segmentation

Dikang Wu, Jiamei Zhao, and Zhifang Wang[✉]

Department of Electronic Engineering, Heilongjiang University, Harbin 150080, China
wangzhifang@hlju.edu.cn

Abstract. In this paper, AM-PSPNet is proposed for image semantic segmentation. AM-PSPNet embeds the efficient channel attention (ECA) module in the feature extraction stage of the convolutional network and makes the network pay more attention to the channels with obvious classification characteristics through end-to-end learning. To recognize the edges of objects and small objects more effectively, AM-PSPNet proposes a deep guidance fusion (DGF) module to generate global contextual attention maps to guide the expression of shallow information. The average crossover ratio of the proposed algorithm on the Pascal VOC 2012 dataset and Cityscapes dataset reaches 78.8% and 69.1%, respectively. Compared with the other four network models, the accuracy and average crossover ratio of AM-PSPNet are improved.

Keywords: Semantic segmentation · Efficient channel attention · Deep guide fusion

1 Introduction

In the field of computer vision, the application of neural networks mainly includes image recognition, target detection, and semantic segmentation. Semantic segmentation is the classification of each pixel in the image to determine the category of each point (such as background, person or car). Compared with image recognition and target location and detection, semantic segmentation not only provides the classification information of objects but also extracts the location information, which lays the foundation for other computer vision tasks [1, 2]. For example, in the field of autonomous driving, the system can automatically and quickly classify images to avoid obstacles. In medical image analysis, the semantic segmentation system automatically generates a simple disease report to help doctors diagnose. In precision agriculture, the machine performs semantic segmentation of crops and weeds in the image, realizes the weeding behavior of the machine, and accurately reduces the number of herbicides sprayed, greatly improving agricultural efficiency [3–5].

Traditional semantic segmentation methods generally divide images by extracting the grayscale information, texture shape, color and other shallow features of the image.

It divides the information of the same semantic category as the same region by fixing a range that has the same semantic category. The traditional methods are threshold segmentation, edge detection and region segmentation [6]. When the background is complex and contains multiple objects, the segmentation effect of traditional methods is not obvious, and the segmentation result is rough.

With the improvement of computer performance and the rapid development of deep learning in the field of computer vision, many image semantic segmentation methods based on deep learning have been proposed [7, 8]. The output of the full convolutional network (FCN) [9] is changed from a two-dimensional vector of fixed length to a two-dimensional space feature graph. The FCN adds an upsampling structure and then predicts each pixel. However, there are problems such as pixel loss in the upsampling process, resulting in rough segmentation results. To solve this problem, SegNet [10] was proposed and restored image details by retaining the maximum index during decoding. U-Net [11], the low-resolution features obtained by the encoder and the high-resolution features extracted by the decoder end are fused at the upsampling stage of the feature map to restore the refined features of the object and refine the edge information of the object. However, the above networks do not pay more attention to spatial context information, resulting in complex semantic information in the image, which easily confuses the target. To solve this problem, DeepLabv3 [12] uses atrous convolution of different expansion rates, and PSPNet [13] uses multiscale pooling features to aggregate multiscale contextual information. However, atrous convolution may lose some pixel position information, and PSPNet may cause information loss, making segmentation inaccurate. Therefore, the proposed DeepLabV3 + [12] in 2008 improved DeepLabV3 by introducing low-level features in the decoding stage. Nevertheless, the use of shallow information is very important in the process of feature fusion, and the segmentation effect is not obvious due to the insufficient use of shallow information. Excessive use of shallow features may lead to information redundancy. Semantic segmentation algorithms that treat all pixels equally are obviously different from human visual mechanisms. To enhance the influence on the region of interest in images and reduce information redundancy, researchers use an attention mechanism as the main method to solve such problems [14–17].

In this paper, a network model based on the attention mechanism AM-PSPNet is proposed. In this model, PSPNet is the backbone network, and the ECA attention module is added in the encoding stage, which can effectively learn the channel attention of each convolution block, reduce the noise and weight the feature channels, and improve the feature extraction performance of the network. In the decoding stage, the DGF module uses deep features to guide the expression of shallow features, strengthen the learning of important features, and restore the shallow features of image edge and texture information to achieve better pixel location and finer details.

2   AM-PSPNet

Based on PSPNet, AM-PSPNet is proposed in this paper. The ECA module and DGF module are added into the model in the encoding stage and decoding stage, respectively, which improves the feature extraction ability of the network and refines the classification results. The structure of the entire network is shown in Fig. 1.

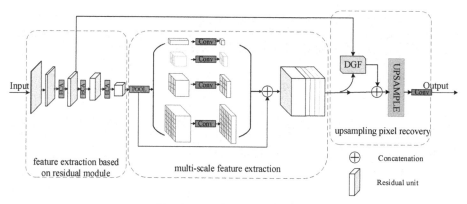

Fig. 1. AM-PSPNet framework.

AM-PSPNet is composed of three subnetworks: feature extraction based on a residual module, multiscale feature extraction and upsampling pixel recovery. The feature extraction subnetwork uses ResNet50 as the basic feature extraction. The network has five convolution modules of different structures. To avoid damage to the original ResNet structure, this paper adds ECA attention after the third, fourth and fifth convolution modules of ResNet so that the network can extract the discriminant features of images in the channel dimension. The multiscale feature extraction subnetwork uses a pyramid pooling module (PPM) to aggregate the context information of the multiscale to obtain the global context information. The DGF module is used in the upsampling recovered pixel subnetwork to guide the classification of shallow features more accurately through global contextual information.

2.1 Efficient Channel Attention Module

Adding the attention module to the existing convolutional neural network can bring performance improvement [18]. Most existing methods focus on more complex attention modules for better performance but result in increased computational burden on the network. To balance the relationship between network performance and complexity, the ECA module is introduced in this paper. The ECA module [19] adds little algorithm complexity while increasing network performance.

The ECA module has an improvement on the squeeze-and-excitation (SE) module [20], and the SE module can learn the channel attention of each convolutional block, which brings significant performance improvement to the deep convolutional neural network architecture. The SE module is used to control the complexity of the network, but dimension reduction can have a negative effect on predicting channel attention, and it is not necessary to obtain dependencies between all channels [21]. As a result, the ECA module efficiently captures local cross-channel interactions. As shown in Fig. 2, the ECA module implements global average pooling between channels without dimension reduction. It captures local cross-channel interactions through fast one-dimensional convolution of kernel size k. k is the coverage of the cross-channel interaction. Then, the sigmoid function is used to generate the weight proportion of each channel. The channel attention feature is obtained by multiplying the given input by the channel weight. The size k can be determined by the adaptive function according to the size of the input channel C, and its calculation equation can be expressed as

$$k = \Phi(C) = \left| \frac{\log_2 c}{\gamma} + \frac{b}{\gamma} \right|_{odd} \tag{1}$$

$$C = \varphi(k) = 2^{(\gamma*k-b)} \tag{2}$$

In the equation, $|t|_{odd}$ is the odd number closest to t, the constant r is set to 2, and the constant b is set to 1.

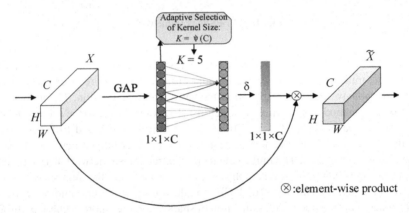

Fig. 2. ECA module.

2.2 Deep Guidance Fusion Module

Usually, multiscale context information is extracted by PPM. However, multistage spatial pooling will lose many fine information. Therefore, features in the deep layer of the network have strong semantic expression ability but poor pixel accuracy, while shallow features contain more pixel information. The direct superposition of deep features and shallow features easily produces considerable noise, while the segmentation accuracy of the model is reduced.

This paper proposes the deep guidance fusion module. As shown in Fig. 3, the DGF is embedded behind the PPM, and it performs global average pooling on deep features to produce attention maps. The shallow features are convolved with 3×3 to reduce the feature mapping channels from the CNN. Then, the shallow features are multiplied by the global attention force to screen out effective information. Finally, the output is added to the deep feature elements and upsampling to produce the final prediction results. To reconcile the contradiction between improving performance and reducing complexity, the output of the third stage is selected as a shallow feature in the feature extraction stage after many experiments.

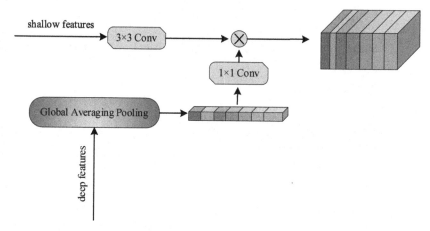

Fig. 3. DGF module.

3 Experiments and Analysis

3.1 Experimental Design

The PASCAL VOC 2012 dataset and the Cityscapes dataset are used to evaluate the performance of AM-PSPNet. First, ablation studies are carried out on the PASCAL VOC 2012 dataset, and then experiments are carried out on two datasets to compare the performance of the network.

To better reflect the performance of the model. Pixel accuracy (PA) is the ratio of correctly segmented pixel points to total pixel points, and mean intersection over union (mIoU) is the ratio of intersection and union of ground truth and prediction graph. The calculation equations are as follows:

$$PA = \frac{\sum_{i=o}^{N} n_{ii}}{\sum_{i=o}^{N} \sum_{j=0}^{N} n_{ij}} \tag{3}$$

$$mIoU = \frac{1}{N}(\sum_{i=1}^{N} \frac{n_{ii}}{\sum_{j=1}^{N} n_{ij} + \sum_{j=1}^{N} (n_{ji} - n_{ii})}) \tag{4}$$

where N is the number of category labels and n_{ji} is the total number of pixels of true category i but predicted category j. n_{ii} and n_{ij} are similar to n_{ji}.

3.2 Ablation Study

To test the importance and performance of each part of the model, an ablation study is designed. To simplify ablation studies, all methods are performed on the PAS CAL VOC 2012 dataset using the same experimental environment to compare performance in different configurations. Resnet-50 is used as the feature extraction network in this paper. The clipping size of the input data is set to 380 × 380, and the batch size is set

Table 1. Ablation study on the PASCAL VOC 2012 dataset

Model	Stage2	Stage3	Stage4	mIoU(%)
ResNet-50				77.4
ResNet-50 + DGF				77.9
ResNet-50 + DGF	✓			78.1
ResNet-50 + DGF		✓		78.4
ResNet-50 + DGF			✓	78.2
ResNet-50 + DGF		✓	✓	78.5
ResNet-50 + DGF	✓	✓	✓	78.8

to 8. The performance of each module is compared in a fair way, and the corresponding results are shown in Table 1.

In this paper, feature extraction is divided into five stages. Stages 2, 3 and 4 in Table 1 indicate whether to add the ECA attention module in the second, third and fourth stages of network feature extraction, respectively, and " + DGF" indicates that the DGF module is added in the network decoding stage. Table 1 shows that the addition of the DGF module is beneficial to the improvement of network performance. When the ECA attention module is added to the second, third and fourth stages of network feature extraction, the network performance is the best.

3.3 Performance Evaluation on PASCAL VOC 2012

The validity of AM-PSPNet is verified using PASCAL VOC 2012, which is a public standard dataset commonly used in the field of semantic segmentation. It contains 1464, 1456 and 1449 images used for training, testing and verification, respectively. There are four categories of human, animal, vehicle and indoor objects, and there are 20 categories in total. There are 21 semantic categories, including one background category.

To accurately measure the performance of the model, AM-PSPNet, FCN-8S, U-Net, PSPNet and DeepLabV3 are experimentally verified on the PASCAL VOC 2012 dataset. The prediction results are shown in Table 2 and Table 3.

Table 2. Semantic segmentation results on the PASCAL VOC 2012 dataset

Model	PA (%)	mIoU (%)
FCN-8s	90.5	64.6
U-Net	91.8	70.4
DeepLabV3	94.3	77.8
PSPNet	94.2	77.4
AM-PSPNet	94.6	78.8

Table 3. Each category results on the PASCAL VOC 2012 testing set.

Model	FCN-8s	U-Net	DeepLabV3	PSPNet	AM-PSPNet
Background	90.3	90.7	93.4	93.3	93.8
Aeroplane	79.4	81.2	88.5	89.8	92.9
Bicycle	35.2	37.8	44.2	42.3	43.3
Bird	74.2	83.9	90.5	91.6	89
Boat	61.2	62.2	70.2	73.2	74.6
Bottle	61.4	68.2	80.6	77.9	79.6
Bus	79	91	91.8	90.1	92.3
Car	77.2	80.2	88.3	86.6	90.8
Cat	79.6	83.6	93.2	91.6	94
Chair	27.5	32.8	43.8	42.6	38.2
Cow	65.9	79.4	88.1	87.2	89.4
Diningtable	47	56.5	55.6	56.6	56.9
Dog	71.6	80.3	89.2	86.8	89.8
Horse	63.7	74.9	88.1	84.1	87.9
Motorbike	74.2	79.5	85.7	85.6	87.4
Person	79.2	81.7	86.3	86	86.9
Pottedplant	48.3	56.7	65.2	66.8	66.9
Sheep	69.3	78.6	85.2	85.3	88.4
Sofa	38.5	37.9	50.6	51.4	51.1
Train	72.2	85.4	82.8	85.9	85.1
Tvmonitor	62.6	58.3	73.9	72.9	76.7
mIoU	64.6	70.5	77.8	77.4	78.8

As seen from Tables 2 and 3, PSPNet achieves good prediction results compared with other semantic segmentation models, but AM-PSPNet achieves better prediction results; the PA is 94.6%, and the mIoU is 78.8%, which are 0.4% and 1.4% higher than the prediction results of PSPNet. AM-PSPNet obtains the highest accuracy for 15 of all categories of segmentation results. Compared with PSPNet, the segmentation results of 19 categories are improved, among which the segmentation effect of object categories with indistinguishable boundaries is significantly improved. For example, the segmentation results of the network for horse and sheep categories improved by 3.8% and 3.1%, respectively, compared with PSPNet. The use of the ECA module enhances the feature class resolution of the network, and the DGF module is helpful in restoring the image edge detail features. The experiment verifies the effectiveness of these two modules in AM-PSPNet.

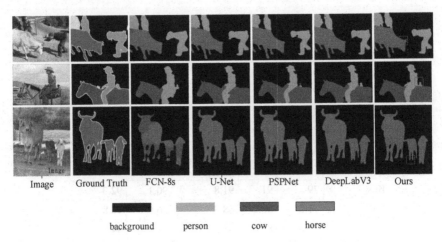

Image   Ground Truth   FCN-8s   U-Net   PSPNet   DeepLabV3   Ours

background   person   cow   horse

Fig. 4. Comparison of prediction results.

To display the segmentation effect of the model more intuitively, the comparison of the visualization results of each model is shown in Fig. 4. By observing the first line, it can be seen that PSPNet segmentation of cow horns is rough, while AM-PSPNet better retains segmentation details and makes the prediction more accurate and clearer. Compared with the picture in the second line, PSPNet failed to predict the distant figure completely, and several network models missed segmentation with serious loss of details. AM-PSPNet accurately expresses the details of the image. Compared with the picture in the third line, it can be seen that AM-PSPNet has a more delicate prediction of cattle legs. Compared with FCN-8s, U-Net, DeepLabV3 and PSPNet, the overall predicted contour of AM-PSPNet is smooth and delicate, and the predicted result is closer to the ground truth.

3.4   Performance Evaluation on Cityscapes

This paper also evaluated AM-PSPNet on the Cityscapes Dataset, which has 5000 images of driving scenes in urban environments, with 19 categories, recording street scenes in 50 different cities.

Resnet-50 is used as the backbone feature extraction network for network training. Affected by the GPU memory capacity, 380 × 380 is selected as the cutting size of the input in this paper, and the batch size is set to 6. The prediction results are shown in Table 4. The Am-pspnet proposed in this paper is superior to other networks, with mIoU reaching 69.1% and PA reaching 95.2%, improving by 1.6% and 1.1%, respectively, compared with the original network.

Table 4. Semantic segmentation results on the Cityscapes dataset

Model	PA (%)	mIoU (%)
FCN-8s	90.6	55.6
U-Net	92.4	61.7
DeepLabV3	94.7	68.7
PSPNet	94.1	67.5
AM-PSPNet	95.2	69.1

4 Conclusions

This paper uses AM-PSPNet as the backbone network, and the DGF module is proposed to guide shallow feature expression through deep features and achieve better pixel positioning. The ECA module is added in the feature extraction stage to improve the performance of the convolutional neural network architecture by learning the channel attention of each convolutional block. The experiment is carried out on the PASCAL VOC 2012 dataset and Cityscapes dataset. That, AM-PSPNet has good performance compared with FCN-8s, U-Net, PSPNet and DeepLabV3.

References

1. Wang, J., Liu, B., Xu, K.: Semantic segmentation of high-resolution images. Sci. China Inf. Sci. **60**(12), 1–6 (2017). https://doi.org/10.1007/s11432-017-9252-5
2. Yan, B., Niu, X., Bare, B., Tan, W.: Semantic segmentation guided pixel fusion for image retargeting. IEEE Trans. Multimedia **22**, 676–687 (2020)
3. Zhao, Y., Qi, M., Li, X., Meng, Y., Yu, Y., Dong, Y.: P-LPN: toward real time pedestrian location perception in complex driving scenes. IEEE Access **8**, 54730–54740 (2020)
4. Cheng, Z., Qu, A., He, X.: Contour-aware semantic segmentation network with spatial attention mechanism for medical image. Vis. Comput. **38**(3), 749–762 (2021). https://doi.org/10.1007/s00371-021-02075-9
5. Zhang, R., Chen, J., Feng, L., Li, S., Yang, W., Guo, D.: A refined pyramid scene parsing network for polarimetric SAR image semantic segmentation in agricultural areas. IEEE Geosci. Remote Sens. Lett. **19**, 1–5 (2022)
6. Bai, S., Wang, C.: Information aggregation and fusion in deep neural networks for object interaction exploration for semantic segmentation. Knowl. Based Syst. **218**, 106843 (2021)
7. Hao, S., Zhou, Y., Zhang, Y., Guo, Y.: Contextual attention refinement network for real-time semantic segmentation. IEEE Access **8**, 55230–55240 (2020)
8. Ji, J., Lu, X., Luo, M., Yin, M., Miao, Q., Liu, X.: Parallel fully convolutional network for semantic segmentation. IEEE Access **9**, 673–682 (2021)
9. Shelhamer, E., Long, J., Darrell, T.: Fully Convolutional networks for semantic segmentation. IEEE Trans. Pattern Anal. Mach. Intell. **39**, 640–651 (2015)
10. Badrinarayanan, V., Kendall, A., Cipolla, R.: SegNet: a deep convolutional encoder-decoder architecture for image segmentation. IEEE Trans. Pattern Anal. Mach. Intell. **39**, 2481–2495 (2017)

11. Ronneberger, O., Fischer, P., Brox, T.: U-Net: convolutional networks for biomedical image segmentation. In: Navab, N., Hornegger, J., Wells, W.M., Frangi, A.F. (eds.) MICCAI 2015. LNCS, vol. 9351, pp. 234–241. Springer, Cham (2015). https://doi.org/10.1007/978-3-319-24574-4_28

12. Chen, L.-C., Zhu, Y., Papandreou, G., Schroff, F., Adam, H.: Encoder-decoder with atrous separable convolution for semantic image segmentation. In: Ferrari, V., Hebert, M., Sminchisescu, C., Weiss, Y. (eds.) ECCV 2018. LNCS, vol. 11211, pp. 833–851. Springer, Cham (2018). https://doi.org/10.1007/978-3-030-01234-2_49

13. Zhao, H., Shi, J., Qi, X., Wang, X., Jia, J.: Pyramid scene parsing network. In: IEEE Computer Society (2016)

14. Lin, Z.K., Sun, W., Tang, B., Li, J.D., Yao, X.Y., Li, Y.: Semantic segmentation network with multipath structure, attention reweighting and multiscale encoding. Vis. Comput. 1–12 (2022). https://doi.org/10.1007/s00371-021-02360-7

15. Li, H., Qiu, K., Chen, L., et al.: SCAttNet: semantic segmentation network with spatial and channel attention mechanism for high-resolution remote sensing images. IEEE Geosci. Remote. Sens. Lett 18(5), 905–909 (2021)

16. Xia, Z., Kim, J.: Mixed spatial pyramid pooling for semantic segmentation. Appl. Soft Comput. 91, 106209 (2020)

17. Wang, Z., Wang, J., Yang, K., Wang, L., Su, F., Chen, X.: Semantic segmentation of high-resolution remote sensing images based on a class feature attention mechanism fused with Deeplabv3+. Comput. Geosci. 158, 104969 (2022)

18. Yin, J., Xia, P., He, J.: Online hard region mining for semantic segmentation. Neural Process. Lett. 50(3), 2665–2679 (2019). https://doi.org/10.1007/s11063-019-10047-3

19. Wang, Q., Wu, B., Zhu, P., Li, P., Zuo, W., Hu, Q.: ECA-Net: efficient channel attention for deep convolutional neural networks. In: 2020 IEEE/CVF Conference on Computer Vision and Pattern Recognition (CVPR), pp. 11531–11539 (2020)

20. Jie, H., Li, S., Gang, S., Albanie, S.: Squeeze-and-excitation networks. In: IEEE Transactions on Pattern Analysis and Machine Intelligence (2017)

21. Wang, Y.-N., Tian, X., Zhong, G.: FFNet: feature fusion network for few-shot semantic segmentation. Cogn. Comput. 14(2), 1–12 (2022). https://doi.org/10.1007/s12559-021-09990-y

Author Index